AF779154

Agricultural Innovation and Sustainable Development

Agricultural Innovation and Sustainable Development

Editor

Michael Blakeney

MDPI • Basel • Beijing • Wuhan • Barcelona • Belgrade • Manchester • Tokyo • Cluj • Tianjin

Editor
Michael Blakeney
School of Law, University of
Western Australia
Australia

Editorial Office
MDPI
St. Alban-Anlage 66
4052 Basel, Switzerland

This is a reprint of articles from the Special Issue published online in the open access journal *Sustainability* (ISSN 2071-1050) (available at: http://www.mdpi.com).

For citation purposes, cite each article independently as indicated on the article page online and as indicated below:

LastName, A.A.; LastName, B.B.; LastName, C.C. Article Title. *Journal Name* **Year**, *Volume Number*, Page Range.

ISBN 978-3-0365-3447-3 (Hbk)
ISBN 978-3-0365-3448-0 (PDF)

© 2022 by the authors. Articles in this book are Open Access and distributed under the Creative Commons Attribution (CC BY) license, which allows users to download, copy and build upon published articles, as long as the author and publisher are properly credited, which ensures maximum dissemination and a wider impact of our publications.
The book as a whole is distributed by MDPI under the terms and conditions of the Creative Commons license CC BY-NC-ND.

Contents

About the Editor

Michael Blakeney is Professor of Law at the University of Western Australia and holds visiting professorships in intellectual property (IP) at The University of Ferrara, Queen Mary University of London and Zhongnan University of Economics and Law, Wuhan. He formerly worked in the Asia Pacific Bureau of the World Intellectual Property Organization (WIPO). He is an IP arbitrator with the International Court of Arbitration. He has directed a number of EC projects concerned with the establishment of an IP infrastructure in a number of new EU Member States and EU Applicant States. He has also directed and been a short-term expert (STE) in a number of projects to assist the development of countries in their accession to the World Trade Organization. He has been a STE in a number of EC projects for ASEAN and in a number of ASEAN member states. Professor Blakeney has advised the Asian Development Bank, the Consulting Group for International Agricultural Research, European Commission (EC), European Patent Office, Food and Agricultural Organization, World Intellectual Property Organization, and a number of university and public research institutes on intellectual property management.

sustainability MDPI

Editorial

Agricultural Innovation and Sustainable Development

Michael Blakeney [1,2]

1 School of Law, The University of Western Australia, Crawley, WA 6009, Australia; michael.blakeney@uwa.edu.au
2 Institute of Agriculture, The University of Western Australia, Crawley, WA 6009, Australia

Global agriculture is confronted by a number of substantial challenges, with some of them being existential. The principal challenge is climate change, which threatens desertification, with attendant increases in salinity, insect and animal infestations, and floods (as well as droughts). These impacts will remove some areas from cultivation at a time when the global population is projected to increase from 7.2 billion persons today to 9.7 billion persons in 2050. The agricultural sector in developing countries and less-developed countries is a source of employment and livelihood for the majority of the population, a major foreign exchange earner and supplier of raw materials to local industry, and has important potential for the economic development of those countries. However, the sector has been weak. It has not succeeded in ensuring secured income to small farmers and food security at national and household levels, or in supporting other sectors and serving as a basis to boost socio-economic development. The reasons behind this include the use of agricultural technologies and practices that have been transmitted from generation to generation with no or little improvement; storage and transportation difficulties; and structural problems with the marketing of agricultural products.

Significant sustainable agricultural innovation is required to deal with these challenges. Intellectual property rights (IPRs) may be of crucial importance for modern agriculture. They serve to make R&D in agriculture attractive by encouraging investment in new technologies and generating tradeable assets. The principal IPRs relevant to agricultural innovation are: (i) patents, which protect inventions; (ii) plant variety rights, which protect the breeding of new and distinct plant varieties; and (iii) trademarks and geographical indications, which facilitate the marketing of products by providing protection for the symbols of their manufacturing or geographic origin. Also relevant, but of lesser significance, are: (iv) layout designs of integrated circuits, which are relevant to smart agricultural technologies; (v) confidential information law, which protects know-how or trade secrets; and (vi) copyright, which protects works of cultural creativity, such as books, articles, scientific papers, and arrangements of data.

Through patenting, genetic sequences which enable plants to withstand agricultural stresses have been identified and protected. Plant variety rights protection encourages the breeding of new varieties which can also meet contemporary agricultural challenges. Trademarks and geographical indications can be used to identify the sources of agricultural products, which can reassure consumers by their traceability. The protection of layout designs of integrated circuits can encourage the investment in digital farming technologies, which allow for the monitoring of inputs such as water, fertilizers and herbicides in order to harmonize them with changing climatic conditions. Confidential know how is particularly important in applying proprietary technologies to agriculture. Finally, copyright protects a wide variety of works which can have an agricultural application. These range from scientific books, articles, and research papers through to questionnaires, surveys, training manuals, and even to computer programs encompassing algorithms which are used to monitor crops in the field.

The significance of IPRs in encouraging sustainable agricultural innovation is slowly being recognized. The World Intellectual Property Organization (WIPO), which is the

Citation: Blakeney, M. Agricultural Innovation and Sustainable Development. *Sustainability* **2022**, *14*, 2698. https://doi.org/10.3390/su14052698

Received: 10 February 2022
Accepted: 23 February 2022
Published: 25 February 2022

Publisher's Note: MDPI stays neutral with regard to jurisdictional claims in published maps and institutional affiliations.

Copyright: © 2022 by the author. Licensee MDPI, Basel, Switzerland. This article is an open access article distributed under the terms and conditions of the Creative Commons Attribution (CC BY) license (https://creativecommons.org/licenses/by/4.0/).

United Nations' specialized agency dealing with IPRs, has adopted a 'Development Agenda' which takes into account the Sustainable Development Goals (SDG) of the United Nations.

This Special Issue of *Sustainability* on agricultural innovation and sustainable development contains a number of case studies on the impacts of climate change upon agriculture and on the adaptive agricultural practices which have been adopted to deal with agricultural stresses in a number of key countries.

The vulnerability of energy and food source crops to climate change impacts has led a search for sustainable new and orphan crops. Rahiel Hagos, Abdulwahab Saliu Shaibu, Lei Zhang, Xu Cai, Jianli Liang, Jian Wu, Runmao Lin and Xiaowu Wang look at the available literature about Ethiopian mustard as an alternative energy source and its sustainable economic importance as a new promising Brassicacea crop. The literature suggests that Ethiopian mustard has many advantages and can be adopted to replace crops that are susceptible to adverse environmental conditions (although the focus has been largely on it as an energy source). The authors suggest a shift of focus to Ethiopian mustard edible oil varieties, in which case further agronomic, quality, and genomic studies on oilseed nutritional traits for efficient breeding and utilization are required (Contribution 1).

China has carried out remarkable work in feeding 22% of the world's population with only 9% of the world's arable land, but it has paid a huge environmental price with problems such as water eutrophication, soil acidification, air pollution, and the reduction of biodiversity. China has realized the need to coordinate the relationship between agricultural development and environmental protection and promote the green transformation of agriculture. Different methods have been adopted in different provinces in the process of agricultural green development with varying degrees of effectiveness. Hongpeng Guo, Shuang Xu and Chulin Pan, focus on the spatial complexity of agricultural green development (AGD) in different regions. They construct an evaluation index system for the level of AGD from four dimensions: social development, economic benefits, resource input, and ecological environment. The article uses an improved entropy weight method to evaluate the level of AGD with panel data of 31 provinces in China from 2007 to 2018. Their study provides a reference for understanding the status of China's agricultural green development level and policy recommendations on how to improve the level of agricultural green development. The results imply that some effective policy measures, such as prompting the integrated development of the three major industries and optimizing the industrial structure, should be taken to coordinate "green" with "development" from national and regional perspectives (Contribution 2).

Phenological variables are closely correlated with rice (*Oryza sativa* L.) yields, as they play important roles in influencing and controlling the carbon allocations between plant organs. However, their impacts on rice yields and their relative importance compared with climatic variables have not yet been investigated thoroughly. Yahui Guo, Wenxiang Wu, Yumei Liu, Zhaofei Wu, Xiaojun Geng, Yaru Zhang, Christopher Robin Bryant, and Yongshuo Fu assess the impacts and the relative importance of climatic and phenological variables on the yields of early mature rice using the trial data from 75 agricultural stations across China, spanning from 1981 to 2010. They found that phenological variables were the dominating influencing factors on rice yields at 63% of the sites, suggesting that the relative importance of phenology to rice yields may be even higher than that of climate. The climatic variables were closely correlated with rice yields. However, their results highlight that phenology should be precisely evaluated in crop models to improve the accuracy of simulating their response to climate change, and that due to the limited understanding of phenological processes, manipulative experiments are urgently needed to improve our understanding of rice phenology and rice yield response to ongoing climate change (Contribution 3).

The per capita food output in sub-Saharan Africa (SSA) has declined considerably, exacerbating food and income insecurity. With climate change, there has come an increase in pests and diseases. In the SSA region, the propagation system is characterized by both formal and informal plant material supplies. For bananas (*Musa* spp.), the majority of smallholder farmers depend on the informal supply (including home-saved material from previous season harvests) and propagating material sourced from friends, neighbors, relatives to either expand or establish new banana plantations. The high prevalence of pests and disease in this material has necessitated research to increase the use of high-quality, formal supplies of plant material. One such effort is the development of tissue culture (TC) planting materials, which are always free of pests and diseases. Lucy Mulugo, Florence Birungi Kyazze, Paul Kibwika, Bonaventure Aman Omondi, and Enoch Mutebi Kikulwe utilize the Double-Hurdle model on cross-sectional data of 174 banana farmers in Central Uganda to analyze the drivers for uptake of TC banana plant materials. Their study demonstrates that seed security factors, along with farmer competencies, social influence, and socioeconomic factors, influence farmer decisions on uptake of TC technology for banana production. Their findings emphasize the need for more involvement of extension services and research institutions in the education and promotion of TC plants in farming communities with attention given to banana varieties that are acceptable and adaptable to the environmental conditions confronting farmers (Contribution 4).

The United Nations Intergovernmental Panel on Climate Change has suggested that the cultivation of traditional crops might have something to offer in encouraging adaptation to climate change. Rice cultivation is significant for a substantial proportion of the world's population. However, traditional rice cultivars and cultivation are in decline in most rice-growing areas (mainly as a result of their low productivity). Packed with nutritionally, environmentally, and locally superior qualities, traditional cultivars hold the key for sustainability in rice cultivation. Jayasree Krishnankutty, Michael Blakeney, Rajesh K. Raju, and Kadambot H. M. Siddique explored the dynamics of traditional rice cultivation in Kerala, India. They examined the economic, institutional and socio demographic factors involved in the production and marketing of traditional rice. They found that holding size and institutional support were the main factors governing the marketing behavior of farmers. The study also found that traditional farmers are ageing, have a lower education, and use limited marketing channels. However, the majority of them were satisfied with their farm enterprise. By streamlining market support mechanisms and processing facilities, traditional rice would most likely gain momentum in key areas (Contribution 5).

Although traditional agriculture has a contribution to make to climate adaptation, it is characterized by the low adoption of new agricultural technology. This may be due to the expense and the unwillingness of farmers to try it due to the risk it entails. The Tanzanian government's strategy to bring about greater efficiency in the agricultural sector includes strengthening various agricultural development strategies and technical cooperation with local and international development partners to increase awareness and make technology easier to access. George Mgendi, Shiping Mao, and Fangbin Qiao analyzes the effect of training programs on the yield of smallholder rice farmers in the Mvomero district of Tanzania. The results indicate that the yield outcome among trained and non-trained farmers with water access for irrigation was significantly more than double. However, the yield difference between trained and non-trained farmers was insignificant in non-irrigated plots. These findings have policy implications for agricultural development in developing countries where training programs alone may not be a panacea for smallholder farmers' productivity improvement. Therefore, respective governments, policymakers, and other agricultural stakeholders should consider both farm and non-farm factors altogether, which may increase agricultural training effectiveness to address the challenges of low yields (Contribution 6).

As a kind of infrastructure that enables production of both food and energy, dams are a key component of climate change adaptation. They are meant to improve agricultural productivity and reduce vulnerability to droughts. In semi-arid regions experiencing desertification, water management infrastructure for food and energy production is necessary to sustain a growing human population and promote economic development in a changing climate. However, dams face many criticisms, ranging from cost overruns and population displacement to ecosystem disruption and increased transmission of infectious disease.

Andrea J. Lund, David Lopez-Carr, Susanne H. Sokolow, Jason R. Rohr and Giulio A. De Leo propose four agricultural innovations for promoting equity, health, sustainable development, and climate resilience in dammed watersheds: (1) restoring migratory aquatic species; (2) removing submerged vegetation and transforming it into an agricultural resource; (3) restoring environmental flows; and (4) integrating agriculture and aquaculture. They conclude that as investment in dams accelerates in low- and middle-income countries, appropriately addressing their livelihood and health impacts can improve the sustainability of agriculture and economic development in a changing climate (Contribution 7).

The Asian rice–wheat cropping system feeds billions across the globe. However, the productivity and long-term sustainability of this system are threatened by stagnant crop yields and greenhouse gas emissions from flooded rice production. The negative environmental consequences of excessive nitrogen fertilizer use are further exacerbating the situation, along with the high labor and water requirements of transplanted rice. Residue burning in rice has also severe environmental concerns. Under these circumstances, many farmers in South Asia have shifted from transplanted rice to direct-seeded rice and reported water and labor savings and reduced methane emissions. Aman Ullah, Ahmad Nawaz, Muhammad Farooq, and Kadambot H. M. Siddique argue that there is a need for adopting the precision agriculture techniques for the sustainable management of nutrients and that allelopathic crops could be useful in the rotation for weed management, the major yield-reducing factor in direct-seeded rice and that legume incorporation might be a viable option for improving soil health. This is particularly the case as governments in South Asia have imposed a strict ban on the burning of rice residues. They argue that the soil/climatic conditions and farmer socio-economic conditions must be considered while promoting these technologies in the rice-wheat system in South Asia (Contribution 8).

The challenge of sustainable agriculture development in light of population growth, resource shortage, ecological deterioration, and climate change has led many governments to support agricultural science, technology, and innovation (ASTI). However, innovation performance is dependent not only on the available innovation resources but also and maybe most importantly on their efficient and productive use. Innovation efficiency, which is the ability to translate inputs into innovation outputs, has become very important and attractive to scholars and governments. The limited attention to innovation efficiency at the national level could be a potentially significant omission from a policy-oriented perspective, since measuring the ASTI efficiency helps to both identify the best innovation practitioners for benchmarking and propose ways to improve efficiency by pinpointing areas of weakness. The agriculture of the G20 countries account for 60% of global arable land and 80% of global agricultural trade, thus having a significant effect on global agriculture development. An efficiency-oriented innovation analysis will enhance the understanding of the operational quality related to the transformation process of limited innovation investments for improving innovation outputs. Xiangyu Guo, Canhui Deng, Dan Wang, Xu Du, Jiali Li, and Bowen Wan sought to measure the static-dynamic efficiency of agricultural science, technology, and innovation (ASTI) and identify the efficiency determinants across the G20 countries. The empirical results indicated that: (1) the efficiency range of the G20 developing countries was relatively larger than the G20 developed countries; (2) the total factor productivity change (TFPC) of ASTI showed an alternating trend of "decline–growth–continuous decline–growth recovery", where the G20 developed countries experienced "growth–decline–growth" and the G20 developing countries underwent a fluctuating upward trend; and (3) The G20 developed countries usually had advantages in capacity, while the G20 developing countries performed better in efficiency (Contribution 9).

Digital technologies offer a potential solution to improve the economic, social, and environmental sustainability of agri-food systems around the globe. While developed countries have led the innovation and adoption of digital agriculture, the potential impact in developing countries, including in the Middle East and North Africa (MENA) regions is very substantial. Rachel A. Bahn, Abed Al Kareem Yehya, and Rami Zurayk synthesize existing evidence to review the potential and current contribution of digital technologies to the agri-food sectors in MENA. They conclude that digital agriculture shows promise in addressing the key challenges facing the agri-food sector across MENA countries, particularly facilitating improvements in primary production, supply chain and logistics performance, and optimized use of scarce natural resources. They note that the available evidence shows that adoption of digital agriculture is at early stages, generally led by high-value agricultural production targeting domestic markets in Gulf countries and export markets in Mashreq countries. They suggest that public policies should not only foster the adoption of digital technologies in MENA but also ensure equity of access, transparency of use, data protection, and labor protection and that policymakers should move beyond traditional, production-centric views to privilege social and environmental sustainability (Contribution 10).

Farmers' adoption of agricultural practices and technologies that contribute to achieving sustainable intensification, sustainable development, and food security require some degree of risk taking and risk management by farmers. The risks confronted by farmers include those associated with climate change as well as market input and output price fluctuations and inadequate access to insurance. Despite smallholder farmers usually being thought to have homogeneous risk averse attitudes, Omotuyole Isiaka Ambali, Francisco Jose Areal, and Nikolaos Georgantzis point out that there is evidence that this is not the case. Thus, they stress that identifying and understanding the heterogeneity in farmers' risk preferences is crucial to guide policy formulation and implementation on risk management and investment decisions concerning the adoption of technology, the adoption of new crop varieties, or the adoption of sustainable agricultural practices. While existing studies have identified the socio-economic factors driving farmers' risk attitudes, spatial variables that may correlate with decisions have often been ignored in the risk models due to the difficulties involved in their measurement. The authors studied unobserved spatial heterogeneity in farmer's risk preferences by incorporating spatial dependency into a farmer's risk preference model using data from a survey conducted with Nigerian farmers between March and May 2016. They found that unobserved spatial heterogeneity (e.g., soil, topographic farmers emulating each other) was present in farmer's risk preferences (along with socio-demographic variables such as age, gender, marital status, and religion and farm characteristics such as farm size and road quality), and that these results are relevant for policy decision-making processes (Contribution 11).

Serena Mariani seeks to investigate the role of EU legislation in shaping innovation in cereal varieties. The research focuses on intellectual property and agricultural law. Her paper looks at the role played by European Community plant variety protection and EU legislation on the marketing of seed and plant propagating material in shaping innovation and stimulating plant breeding of new cereal varieties. She adopts a focus on cereal varieties due to the substantial socio-economic impact of innovation in this field, as well as strategic scientific and environmental implications. Her study concludes that it is necessary to simplify the existing legal framework by coordinating intellectual property and agricultural law, providing for legislative review and better coherence in order to effectively shape innovation and meet the changing demands of society and the sustainability challenges (Contribution 12).

The agricultural challenges that were outlined at the beginning of this editorial is trenchantly illustrated by the food security of Bangladesh which is largely depends on rice production (92% of the total food grain production). Although the country's agro-climatic conditions are perfect for cultivating rice all year, the national average rice yield (2.60 t/ha) is much lower than the potential national yield (5.40 t/ha). The population of Bangladesh is currently 162.7 million and projected to be 189.85 million by the year 2030, requiring an increase of the current rice yield to 3.74 t/ha to keep the production of rice in line with the growing population of the country. Inefficient and often imbalanced fertilizer use impedes farmers from achieving expected yields. Thahamina Bagum, Md. Kamal Uddin, Salim Hassan, Nitty Hirawaty Kamarulzaman, Md. Zulfikar Rahman, and Ahmad Numery Ashfaqul Haque explore the contribution of selected factors that influence farmers' work performance and determine the highest contributing factors on farmers' work performance towards fertilizer application in rice. They used a multistage simple random sampling method to select 355 farmers from twenty-one rice production areas of Bangladesh. Data, collected using a structured questionnaire, were subjected to multiple linear regression analysis to explore the contribution of selected factors and identify the highest contributing factors towards farmers' work performance. The results of their study revealed that the motivation of farmers was found to be the highest contributing factor, followed by knowledge influencing their work performance and concluded that farmers need to be equipped with essential knowledge and motivation crucial to strengthening their work performance as this will increase rice production (Contribution 13).

Contribution List:

1. Hagos, R.; Shaibu, A.S.; Zhang, L.; Cai, X.; Liang, J.; Wu, J.; Lin, R.; Wang, X. Ethiopian Mustard (Brassica carinata A. Braun) as an Alternative Energy Source and Sustainable Crop. *Sustainability* **2020**, *12*, 7492. https://doi.org/10.3390/su12187492.
2. Guo, H.; Xu, S.; Pan, C. Measurement of the Spatial Complexity and Its Influencing Factors of Agricultural Green Development in China. *Sustainability* **2020**, *12*, 9259. https://doi.org/10.3390/su12219259.
3. Guo, Y.; Wu, W.; Liu, Y.; Wu, Z.; Geng, X.; Zhang, Y.; Bryant, C.R.; Fu, Y. Impacts of Climate and Phenology on the Yields of Early Mature Rice in China. *Sustainability* **2020**, *12*, 10133. https://doi.org/10.3390/su122310133.
4. Mulugo, L.; Kyazze, F.B.; Kibwika, P.; Omondi, B.A.; Kikulwe, E.M. Seed Security Factors Driving Farmer Decisions on Uptake of Tissue Culture Banana Seed in Central Uganda. *Sustainability* **2020**, *12*, 10223. https://doi.org/10.3390/su122310223.
5. Krishnankutty, J.; Blakeney, M.; Raju, R.K.; Siddique, K.H.M. Sustainability of Traditional Rice Cultivation in Kerala, India—A Socio-Economic Analysis. *Sustainability* **2021**, *13*, 980. https://doi.org/10.3390/su13020980.
6. Mgendi, G.; Mao, S.; Qiao, F. Is a Training Program Sufficient to Improve the Smallholder Farmers' Productivity in Africa? Empirical Evidence from a Chinese Agricultural Technology Demonstration Center in Tanzania. *Sustainability* **2021**, *13*, 1527. https://doi.org/10.3390/su13031527.
7. Lund, A.J.; Lopez-Carr, D.; Sokolow, S.H.; Rohr, J.R.; De Leo, G.A. Agricultural Innovations to Reduce the Health Impacts of Dams. *Sustainability* **2021**, *13*, 1869. https://doi.org/10.3390/su13041869.
8. Ullah, A.; Nawaz, A.; Farooq, M.; Siddique, K.H.M. Agricultural Innovation and Sustainable Development: A Case Study of Rice–Wheat Cropping Systems in South Asia. *Sustainability* **2021**, *13*, 1965. https://doi.org/10.3390/su13041965.
9. Guo, X.; Deng, C.; Wang, D.; Du, X.; Li, J.; Wan, B. International Comparison of the Efficiency of Agricultural Science, Technology, and Innovation: A Case Study of G20 Countries. *Sustainability* **2021**, *13*, 2769. https://doi.org/10.3390/su13052769.
10. Bahn, R.A.; Yehya, A.A.K.; Zurayk, R. Digitalization for Sustainable Agri-Food Systems: Potential, Status, and Risks for the MENA Region. *Sustainability* **2021**, *13*, 3223. https://doi.org/10.3390/su13063223.
11. Ambali, O.I.; Areal, F.J.; Georgantzis, N. On Spatially Dependent Risk Preferences: The Case of Nigerian Farmers. *Sustainability* **2021**, *13*, 5943. https://doi.org/10.3390/su13115943.
12. Mariani, S. Law-Driven Innovation in Cereal Varieties: The Role of Plant Variety Protection and Seed Marketing Legislation in the European Union. *Sustainability* **2021**, *13*, 8049. https://doi.org/10.3390/su13148049.
13. Bagum, T.; Uddin, M.K.; Hassan, S.; Kamarulzaman, N.H.; Rahman, M.Z.; Haque, A.N.A. Contribution of Selected Factors on Farmers' Work Performance towards Fertilizer Application in Rice of Bangladesh. *Sustainability* **2021**, *13*, 10795. https://doi.org/10.3390/su131910795.

Funding: This research received no external funding.

Institutional Review Board Statement: Not applicable.

Informed Consent Statement: Not applicable.

Data Availability Statement: Not applicable.

Conflicts of Interest: The authors declare no conflict of interest.

sustainability

Review

Ethiopian Mustard (*Brassica carinata* A. Braun) as an Alternative Energy Source and Sustainable Crop

Rahiel Hagos [1,2], Abdulwahab Saliu Shaibu [3,4], Lei Zhang [1], Xu Cai [1], Jianli Liang [1], Jian Wu [1], Runmao Lin [1] and Xiaowu Wang [1,*]

1. Institute of Vegetables and Flowers, Chinese Academy of Agricultural Sciences, Key Laboratory of Biology and Genetic Improvement of Horticultural Crops of the Ministry of Agriculture, Sino-Dutch Joint Laboratory of Horticultural Genomics, Beijing 100081, China; rahag21@gmail.com (R.H.); zhangleo1116@163.com (L.Z.); caixu0518@163.com (X.C.); liangjianli@caas.cn (J.L.); wujian@caas.cn (J.W.); linrunmao@caas.cn (R.L.)
2. Department of Dryland Crop and Horticultural Sciences, Mekelle University, College of Dryland Agriculture and Natural Resources, Endayesus Main Campus, P.O. Box 231, Mekelle, Tigray, Ethiopia
3. The National Engineering Laboratory for Crop Molecular Breeding, MOA Key Laboratory of Soybean Biology (Beijing), Institute of Crop Sciences, Chinese Academy of Agricultural Sciences, 12 Zhongguancun South Street, Beijing 100081, China; asshuaibu.agr@buk.edu.ng
4. Department of Agronomy, Bayero University, P.O. Box 700001, Kano, Nigeria

* Correspondence: wangxiaowu@caas.cn; Tel.: +86-10-82105971; Fax: +86-10-62174123

Received: 13 August 2020; Accepted: 6 September 2020; Published: 11 September 2020

Abstract: Energy and food source crop demand claims to be vulnerable to climate change impacts. The new and orphan crops, which in the past have received only limited research attention but are sustainable to environmental systems, are needed. In this review, we summarize the available literature about Ethiopian mustard as an alternative energy source and its sustainable economic importance as a new promising Brassicacea crop for new opportunities in the face of producing sustainable environment and energy development. Ethiopian mustard has many advantages and can be adopted to replace crops that are susceptible to adverse environmental conditions. Ethiopian mustard is becoming a new promising Brassicaceae crop with the current global energy demand increases. However, researchers have only focused on energy source production which has resulted in developing high erucic acid varieties. This results partly in limited studies on developing Ethiopian mustard edible oil varieties. The adoption and scaling-up of this promising crop as an oilseed crop in developing countries and Mediterranean conditions can sustain the impact of climate change with the demand for food and energy debate concepts. Indeed, further agronomic, quality and genomic studies on oilseed nutritional traits for efficient breeding and utilization are needed.

Keywords: alternative energy source; Ethiopian mustard; sustainability

1. Introduction

Ethiopian mustard (*Brassica carinata* A. Braun) is mainly originated in the highlands of Ethiopia [1]. It is locally known as "Gomenzer" ("Yehabesha Gomen") and "Hamli Adri" in Amharic and Tigrigna languages, respectively. It can well be adopted in the Mediterranean climate areas [2] and it is a drought and heat tolerant oilseed crop [3,4]. It is used as a food and oilseed crop [5], especially in the arid and semi-arid areas [6].

Nowadays, Ethiopian mustard has disseminated to most semi-arid climate countries, such as European and South Asian countries and is used as an alternative energy crop in marginal lands. It is used for biofuel, bio-industrial processes and soil remediations. This is because different Ethiopian mustard breeding programs are elucidated to develop and adopt a candidate variety for better adaptability and oil yield production [7,8] in different arid and semi-arid areas.

Biochemically, Ethiopian mustard has high erucic and linoleic acids [2] with less saturated fatty acids. Hence, having these characteristics makes it a desirable oilseed crop that can be processed into biofuel [8–10]. Additionally, the oil extracted from the Ethiopian mustard is considered as a nonfood oil because of the high concentration of erucic acid (35–51%) [11,12]. Therefore, the crop has the potential to supplement part of the renewable energy demand without displacing feed and food crops as recently used by some countries, such as the USA, Canada, Italy and Spain. For example, it is used as a bio-industrial crop and biofuel for jets (air-crafts) [6], and hence different researchers stated that the crop can be used for smart agriculture "from the seed to sky" or "from the field to flight" platforms [9,13].

Therefore, this review aims to elucidate the status of Ethiopian mustard, as one of the Brassicacea species used as alternative energy sources and sustainable crops, and the benefits associated with its adoption as a new and promising crop for energy sources in new vulnerable agricultural systems. The review also tends to motivate researchers working on other underutilized edible oilseed crops so as to develop environmental sustainability to face the impact of energy source imbalances at the globe.

2. Ethiopian Mustard as a Preference Crop

Nowadays, many growers and researches are actively searching and prefer a more resilient oilseed Brassica species [14], such as the Ethiopian mustard which can withstand in highly vulnerable environmental conditions even no other crops growing areas (i.e., marginal lands) [8,15]. Therefore, it is one of the most interesting Brassicaceae crops, which can be used currently for energy purposes in the Mediterranean regions [2,16] and can also be considered as suitable for marginal and contaminated areas, as it is a strong candidate for phytoremediation properties [6,17]. It is also found that the crop has better agro-ecological adaptability and productivity than canola (*Brassica napus*) and Indian mustard (*Brassica juncea*) under unfavorable environmental conditions and even in low cropping systems [2,18]. This is preferred because producers can cultivate the crop without difficulty with the production cost and can indirectly develop an environmentally sustainable business.

Ethiopian mustard is now a new promising energy crop for most Mediterranean, arid and semi-arid climate countries [6,16]. Indeed, the presence of high yield levels requires fewer inputs, and the ability to adapt and resist abiotic and biotic stresses [19] makes it appreciated in terms of agronomy and energy balances. For instance, its adaptation has considerably expanded and increased its production in some drier areas of the USA (California) [8,20,21], Canada [6,19], Italy [18,22], Spain [23] and South Asian countries [24], due to the increasing demand for bioenergy and oilseed productions in these countries [2,12,19,25] and the incidence of global climate change effects on sustainable agricultural production and environmental systems.

3. Cultural Practices of Ethiopian Mustard Effects on Environmental Sustainability

Ethiopian mustard is suitable for crop rotation and intercropping with food crops, such as wheat [26], chickpea [27], barley and sorghum [28]. Mulvaney et al. [8] reported that neither row spacing nor seeding rate affected the Ethiopian mustard oil concentration, seed and yield traits. It also has a role in the low farming systems as a potential rotational crop for cereals and pulses [29]. It is noteworthy that the crop is widely used as biofumigant to break soil-borne diseases and pests; this is why farmers in Australia and New Zealand used it as a rotating crop during crop husbandry systems [9,30]. Ngwene et al. [31] also reported that health-related phytochemicals of Ethiopian mustard can be modified by intercropping with nightshade, an African indigenous vegetable crop. The benefits of growing Ethiopian mustard as a winter crop has two advantages over other crops: first, it can increase the income generation of farmers [8,32] and second, it helps in environmental sustainability [26]. For instance, in honeybee production, it decreases the use of plenty of herbicides and maintains soil biotypes. Therefore, growing the Ethiopian mustard followed by summer crops and pastures has an advantage on the sustainable development for many producers, such as a livestock feed [8,30], soil amendments and biofumigation, and it is also likely to be used as a source of food for insect pollinators. In addition, Ethiopian mustard when cultivated as a winter crop, for instance, in Canada helps as a cover crop (green manure) to reduce soil erosion, herbicide use and to balance nutrient losses [9,33,34]; it then later increases soil organic matter, maintains soil biotypes and crop diversification.

The traditional practices of mixed cropping systems involving Ethiopian mustard have gained admiration in recent years in the form of intercropping with suitable modification in planting patterns. The intercropping of Ethiopian mustard with chickpea in the semi-arid tropics of India, has improved the consumptive use, moisture use efficiency (MUE) and moisture extraction pattern, and when combined with moisture conservation practices (MCP) and proper phosphorus (P) and sulfur (S) nutrients, better significant results have been obtained [27].

On the contrary, reducing the mechanical operation of Ethiopian mustard with minimum tillage and low production inputs has no significant impact on yield production and seed oil content [18,35]. Indeed, Debiase et al. [36] found that the replacement of conventional tillage and mineral fertilizer with minimum tillage and sewage sludge can be sustainable for the cultivation of Ethiopian mustard without significant changes in heavy metal concentrations in the Mediterranean environments. It can be emphasized that Ethiopian mustard can be grown under limited environmental conditions with minimum cultural practices and a lower cost of production. Furthermore, it has been shown that seeding rate does not affect the yield and nutritional properties of Ethiopian mustard [8].

3.1. Responses of Ethiopian Mustard to Heavy Metal Contaminated Soils

The major environmental factors that currently reduce plant productivity are drought, salinity and nutrient imbalances [37–39]. Saline soil types are one of the abiotic stresses that influence food and energy imbalances [40,41], which causes a formidable global challenge to existing crop production. Therefore, identifying plant species and cultivars that can contend with such environmental challenges is a key remedy for establishing a sustainable food and energy balance with climate change [42].

The genus Brassica species have significant importance to resist contaminated soils and can be used for phytoremediation purposes [26,43–45]. For instance, Akhtar et al. [46] studied the effect of salinity in four Brassica species: mainly, Ethiopian mustard, Indian mustard, Chinese cabbage and rapeseed. As a result, rapeseed and Ethiopian mustard were more tolerant and yielded better than Indian mustard and Chinese cabbage for seed yield, stem diameter, plant height, number of siliques and 1000 seed weight per plants.

The Ethiopian mustard has shown a high tolerance index and adaptability in multicontaminated soils with zinc (Zn), nickel (Ni), lead (Pb), manganese (Mn) and cadmium (Cd) [17,43]. Gezahegn et al. [47] also reported the highest accumulation of heavy metals in leafy vegetables of Ethiopian mustard than cabbage, cauliflower, lettuce and Swiss chard. On the other hand, they found no accumulation of Pb and arsenic (As) metals as compared to other leafy vegetables. Likewise, a pot culture experiment was also used to test the phytoextraction of Zn, copper (Cu), Ni and Pb from a contaminated soil among five *Brassica* species (Indian mustard, rapeseed, Ethiopian mustard, oilseed rape (canola) and black mustard) [17]. Accordingly, they found that among these *Brassica* species, the Ethiopian mustard has shown the highest concentration (mg kg^{-1}) as well as uptake (μg pot^{-1}) of Ni and Pb at maturity, and it is a promising phytoextractor for Zn, Ni and Pb when compared to Indian mustard.

3.2. The Response of Ethiopian Mustard to Plant Pests and Diseases

Ethiopian mustard is reported to be resistant to many diseases affecting crucifers in Canada, such as blackleg (*Leptosphaeria maculans*; [48]), white rust (*Albugo candida*; [49]), alternaria (*Alternaria brassicae*; [50]), sclerotinia stem rot (*Sclerotinia sclerotiorum* (Lib.) Massee) and aster yellows (*Candidatus phytoplasma asteris*) [51]. However, it is less susceptible to insect herbivores than oilseed rape (canola). However, it is also susceptible to clubroot (*Plasmodiophora brassicae*), a soil-borne fungus-like pathogen [52], but it has been widely reported for its ability to reduce soil-borne plant pathogens better than other Brassicaceae species. In addition, the biological control methods through environmentally sound and effective means of reducing or mitigating pests using natural enemies for the Ethiopian mustard volunteers have not yet been developed [53].

In the USA, it has been evaluated further for weed risk assessment; nevertheless, it has low or no weed risk assessment. However, it can germinate and grow as a contaminant from discarded birdseed and bird-seed screenings due to the presence of uncertainty in seed dormancy. Taking these into account, Ethiopian mustard needs further research and breeding to protect from diseases such as sclerotinia that happens when it rotates with other energy crops [28] to obtain a better yield.

4. Economic Sustainability of Ethiopian Mustard

4.1. As Sources of Secondary Metabolites (Glucosinolates)

Different *Brassica* species have shown their allelopathic potential against noxious weeds of agroecosystems and are considered as a sustainable tool for integrated weed management [54]. The leaves of Ethiopian mustard are rich in glucosinolates, that is, a secondary metabolite that has been identified in Brassicaceae and is a volatile sulfur-containing compound that has antimicrobial and allelopathic effects [55,56] (Table 1).

Glucosinolates can also prevent microorganisms and insect infestation and nematode invasion in the roots through a process known as biofumigation. The presence of glucosinolate in seed cake is a limiting factor in Ethiopian mustard, although it is used as high protein feed for animals [57]. However, glucosinolates are also repellants to most insect pests due to their presence in the volatile oils released from Brassicaceae [30], of which Ethiopian mustard is used as a trap crop (an alternative pest management strategy) for massive butterfly infestation in Indian mustard by utilizing the pests' preference for these plants [58]. Additionally, Odongo et al. [59] and Martínez-Valdivieso [60] reported that the leaves and seeds of Ethiopian mustard are also rich in nutrients and health-promoting secondary plant metabolites, i.e., glucosinolates, especially 2-propenyl glucosinolate (sinigrin) as well as phenolic compounds. However, the concentration was different based on the plant parts used, stage of maturity and growth conditions.

Table 1. Secondary metabolites (allelochemicals) reported in Ethiopian mustard.

S.N.	Type of Allelochemicals	Activities	References
1	2-propenyl-glucosinolate (Sinigrin)	Health promoting effect, antimicrobial effects	[12,59,61]
2	Flavonoids	Antioxidant activity	[59]
3	Carotenoids	Antioxidant activity	[59]
4	Allyl Isothiocyanate	Insecticidal activity; weed suppresses (reduces the use of synthetic chemicals)	[59]

Ethiopian mustard has a larger sinigrin content (5.83 mg·g^{-1}) [61] than cabbage (0.26 mg·g^{-1}), Chinese cabbage (0.77 mg·g^{-1}), Broccoli (0.33 mg·g^{-1}) and Korean leaf mustard (5.56 mg·g^{-1}) [62], which is the predominant glucosinolate content in the leaves of the crop. On the other hand, the presence of high erucic acid [12,63], makes Ethiopian mustard have an antinutritional effect such as tannins and sinigrin. Increasing the processability of orphan-crop entails focusing on the aspects of chemical composition that affect their products. Therefore, improved processability means decreasing the presence of antinutritional compounds that are present in the edible portions of crop plants [64]. However, through different breeding programs, there are few Ethiopian mustard varieties developed with zero erucic acids [11,13,65] and a high content of glucosinolates (100–200 µmoles g^{-1})—almost exclusively sinigrin [66]. Therefore, farmers, private companies and researchers can use these improved varieties based on their agronomic performance and the need for sustainable research preferences.

4.2. As an Alternative Source of Energy Crop

Among the botanical species yielding high quantities of erucic acid, the Brassicaceae species is the most important genus in the production of this oil per hectare, particularly rapeseed and Ethiopian mustard species [7,67]. In the 1990s, European countries such as Italy searched and collected Ethiopian mustard lines from India, Germany and the Netherlands to screen for better agronomic performance and high potential oil productivity in winter conditions and marginal soils. Meanwhile, the Ethiopian mustard cultivar "Sincron" has been found as a candidate biodiesel production oilseed crop in Italy [7]. Likewise, the North Florida Research and Education Center (NFREC) has also developed a new commercial variety of Ethiopian mustard called AAC A120, which has a high yielding performance (i.e., both in seed and oil production), is disease resistant, early maturing and well adapted in the southeast USA. Hence, this variety may increase diversification, generate revenue and improve ecosystem sustainability. However, the target was pick up to develop region-specific agronomic traits and targeted for double crop production in the southeastern USA [9]. There are also other improved new promising Brassicaceae species adopted as supplementary oilseed crops (Table 2).

Increasing the demand for high erucate oils for industrial feedstocks, enforce researchers to develop high erucic acid (HEA) Brassicaceae cultivars with increased proportions of erucic acid and very-long-chain fatty acids (VLCFAs) to fulfill the industrial oil market needs [6,19,63]. Although more research and development for food and industrial use has been conducted using canola and Indian mustard has been practiced for food and industrial use, other Brassicaceae relatives such as camelina, Ethiopian mustard and *Sinapis alba* have also shown promise as potential Hydrotreated Renewable Jet fuel (HRJ) feedstock [2,20]. Therefore, the advocating and development of Ethiopian mustard as an alternative biorefinery and bio-industrial oil platform using traditional and molecular breeding techniques and tools is essential. Currently, Ethiopian mustard is also considered as a new molecular farming platform for delivering bio-industrial oil feedstock [2,19] that is improving through genetic modifications to produce more very long-chain fatty acids and oil contents in seeds of the crop.

In East Africa, particularly in Ethiopia, the Ethiopian mustard is cultivated as a leafy vegetable crop with limited commercial oil production. However, it is more commercially adapted as an alternative energy source and biofuel product in developed countries. This is why the interest of the crop is changing nowadays from more food (leafy vegetable) to more sustainable biodiesel crop production [20] for bio-industrial processes, as bio-energy sources increase from time to time in developed countries [25,28], which started in the mid-1980s [53].

Most governmental and nongovernmental organizations such as European countries, the University of Florida, North Florida Research and Education Center (NFREC) in Quincy, in South Eeastern USA and Agrisoma Biosciences Inc. company (i.e., the world's largest Ethiopian mustard breeding program company located in the University of Saskatchewan, Canada) have developed varieties for energy sources [9]. Meanwhile, researchers in the USA, such as California and Florida, Canada and European countries, have emphasized Ethiopian mustard as an alternative energy source (jet fuel) crop and bio-industrial feedstock, i.e., for more biodiesel crop production.

On the other hand, the oil extracted from Ethiopian mustard can be utilized as vegetable oil, later converted into jet fuel and biodiesel such as other oilseed crops via the existing technologies [20]. For instance, in 2018, Australian Qantas Airways, together with Canadian agritech company Agrisoma Biosciences, flew the first dedicated biofuel-powered flight between Australia and the USA, using a jet biofuel produced from the Ethiopian mustard oilseed [9]. Therefore, the Ethiopian mustard is an excellent raw material for the sustainable production of biofuels. It is becoming an interesting new complementary source of biofuel and industrial feedstock to mitigate the impact of global climate change effects on sustainable agricultural development by hindering the pollution that happened due to the use of excess petroleum gasoline in developed countries.

Table 2. The most promising oilseed Brassica species with a briefing of some agronomical characteristics and their industrial uses.

Cultivated *Brassica* Species	Native Areas	Positive Characteristics	Negative Characteristics	Stage of Commercial Development	Industrial Uses	References
Brassica carinata	Ethiopia	Drought resistant, low pod shattering, low bird predation, an alternative and interesting winter oilseed crop, a meal rich in glucosinolates and resistant to most pathologies, as a trap crop	Limited availability of commercial varieties, low seed yield stability, low harvesting index and low cold tolerance	Seeds of few improved varieties are available on the market, such as Dodalla, C90-14, AAC A100, AAC A110, AAC A120, AVANZA 641, 080814 EM	Biofuels, food processing, lubricant, plasticizers, antislipping agents, cosmetics, nylon erucamide (polyethylene, paints, etc.), jet fuel	[9,14,19,20,58]
Brassica napus	South-east Asia	Low erucic acid and glucosinolate, adapt to cool environment	susceptible to high heat and temperature	Seeds of many improved varieties are available on the market, with different environmental adaptability (e.g., DK3042 RR, Gem, Invigor L130 and SC28)	Consumable oil production, used in stews, soups and as a flavor enhancer	[20,25,30]
Brassica juncea	India	Drought resistant, domesticated crop, low-fiber meal, good adaptability to hot weather condition	Poor seed yield stability and high N requirements	Seeds of many improved varieties are available on the market, with different environmental and adaptability (e.g., Oasis, Pacific Gold)	Biofuels, food processing, lubricant, plasticizers, antislipping agents, cosmetics, nylon erucamide (polyethylene, paints, etc.), jet fuel, antidiabetic activity	[20,25,30]

4.3. As a New Promising Crop Offer for New Opportunities

Ethiopian mustard as a unique energy feedstock crop is preferable added to crop rotation by growers because they are unable to grown other crops due to the vulnerability of agricultural production to vulnerable environmental conditions. Thus, Ethiopian mustard attracts the interest of researchers in many countries, such as Canada, Spain and Italy, because of its good agronomic qualities. Impressively, this has been introducing Ethiopian mustard as a sustainable, scalable and economic oilseed and feedstock crop in other unfavorable environmental conditions. It is noteworthy that a crop with excellent agronomic traits will perform well in stressful conditions and has a ready market demand [13].

It has been proven that the crop residue of the Ethiopian mustard (straw) has a potential option for energy production (i.e., biofuel or for electricity and heat cogeneration), especially in the rural areas of developing countries that could increase the income of farmers. Besides this, the Ethiopian mustard pellet made from crop residues has been found to be modestly suited for energy use in terms of pellet quality compared to a wood pellet; it is then possible to sustain the valuable outputs of local agro-energy chains [16]. Additionally, the meal remaining after oil extraction is also protein rich (30–45%) and used as sources of protein in livestock feed [8], poultry, swine and aquaculture industries (if its glucosinolate level is reduced) or as an organic fertilizer [5,11,30]. These concepts can make the crop a new promising crop and have further benefits to adapt in unfavorable environmental conditions to support growers, consumers and researchers' perspectives of the chance to use the good agronomic characteristics of this crop for sustainable development.

As aforementioned, Ethiopian mustard has many advantages and dual purposes. Therefore, the cultivation of Ethiopian mustard can overcome problems in intensive nonrotational farming practices which have an impact on environmental sustainability. Indeed, not only this, but it has also been used for bioremediation effects as a traditional medicine to treat ailments and diseases such as analgesic activity [30] and anticancer activity [59]. Besides this, Ethiopian mustard has more production potential than rapeseed both under low-input and adverse environmental conditions [2,22], and it is usually cultivated as an alternative energy source and is well-indicated for biofuel production, such as in Italy [16].

4.4. Future Prospects of Ethiopian Mustard

Conventionally, Ethiopian mustard is propagated by seeds and can hybridize successfully with other Brassicaceae species [68,69], which makes it an alternative oilseed crop for most Mediterranean, arid and semi-arid climate countries. For instance, the Ethiopian mustard genotypes demonstrate a potential as a winter biofuel crop in South East USA. There are essential traits to enable the adoption of Ethiopian mustard as a sustainable winter crop without interrupting the normal cycle of cultivation. Therefore, the inclusion of Ethiopian mustard as a winter crop will help bring underutilized fallow land under cultivation thereby potentially adding additional revenue streams for the producer while providing ecosystem services [70].

Ethiopian mustard as compared to other Brassica species has useful genes for the resistance to abiotic and biotic stresses and, therefore, has been used to intercross with other counterparts as a recurrent parent for the introgression of genes to improve and widen the gene pool of *Brassica* species germplasms for further breeding programs [71,72]. On the other hand, there is a limited genetic improvement due to the narrow genetic variation [73] of this crop, which has not been studied at a genomic level to date and lacks sufficient literature [74]. Indeed, the presence of a narrow genetic base shapes the population structure and linkage disequilibrium in Ethiopian mustard [75].

Based on the information reviewed, around 1308 Ethiopian mustard germplasms have been found in different gene banks of the world. As we have contacted different institutes, the majority of the germplasms are found in the Ethiopia Biodiversity Institute (http://www.ebi.gov.et); World Vegetable Center (https://avrdc.org); European Search Catalogue for Plant Genetic Resources (https://eurisco.ipk-gatersleben.de/apex/f?p=103:1:::); Plant Gene Resources of Canada (http://www.agr.gc.ca/pgrc-rpc); and the US National Plant Germplasm System, Germplasm Collection Gene Banks, (http://www.ars-grin.gov/npgs), with the total number of 653, 386, 139, 92 and 38 accessions, respectively, and are available in gene banks (personal communication, 2018). With these genetic resources, it is possible to conduct further research on molecular breeding and variety improvement. Khedikar et al. [75] reported that the assessment of genetic diversity, population structure and linkage disequilibrium in the worldwide Ethiopian mustard collection will enhance future crop improvement programs.

The six *Brassica* species and their closest relative species of Radish except Ethiopian mustard (*Brassica carinata* A. Braun) have been well sequenced by different researchers [76–80]. These species obviously promoted their genetic dissection of agronomic traits and were used for variety development. Bearing this in mind, to avoid Ethiopian mustard being neglected, it is important to start research on genomic sequencing and genome-wide association study (GWAS) as well as phylogenetic analysis to improve this sustainable oilseed *Brassica* species and also improve the nutritional quality for better economic and sustainability use.

5. Conclusions

Ethiopian mustard is native to the central highlands of Ethiopia and adjoining east African countries. In Ethiopia, it has been consumed as a food (leafy vegetable) and oilseed crop since ancient times. It can adapt in the arid and semi-arid conditions, and it is used as an alternative energy (biofuel and industrial oil) source crop with the current vulnerable climate change in crop husbandry systems. Indeed, there is a huge demand for a suitable crop that can be a clean energy source without competing for land use with food crops. Ethiopian mustard has emerged as one of such low-cost candidate crops with multiple desirable agronomic traits. Nowadays, it is becoming a new promising Brassicaceae vegetable, and it can be used for multiple purposes, such as increasing farmers' income and environmentally sustainable business. The leaves are used as a source of phenolic compounds and glucosinolates, and the seeds are used as an alternative energy source. However, researchers have mainly focused on energy production in terms of improving high erucic acid varieties rather than developing edible oil varieties. As a result, the production of edible oil from the Ethiopian mustard is neglected from crossbreeding improvement techniques. With this framework, Ethiopian mustard has the potential for many applications in food, energy and environmental sustainability. This creates interest in studying the agro-economic performance of cereal-based rotations and understanding the genetic architecture controlling these traits. It is also noteworthy that Ethiopian mustard can be a resilient crop for sustainable ecological services on oilseed production as it can tolerate biotic and abiotic stresses effectively more than other Brassicaceae crops. However, further study on leaf and seed nutritional compositions using genomic approaches is needed in order to enhance its use as an alternative oilseed crop for sustainable edible oil production with diversified crop rotations.

Cross-crop knowledge application can be further supported by the unification of advanced breeding-method designs, novel gene-editing approaches to generate new genetic variation and innovative participatory domestication models that facilitate stakeholder input. The adoption and scaling-up of this promising crop as a sustainable and alternative energy source crop in developing countries and Mediterranean conditions can create a resilience to the impact of environmentally unsustainable crop production systems with the demand for food and energy (fuel) debate concepts.

Author Contributions: All authors contributed equally to reviewing different journal articles, proceeding papers, reports and the relevant and latest published papers to gather information and design the platform and concepts of this review paper. All authors contributed to the drafting of this review article, revising it critically for intellectual content, and finally, they approved the final version of the review for publication. All authors have read and agreed to the published version of the manuscript

Funding: This study was funded by the Organization for Women in Science for the Developing World (OWSD).

Acknowledgments: We would like to express our sincere gratitude to the fellowship, Organization for Women in Science for the Developing World (OWSD) for financial support and encouraging young female scientists from Science and Technology Lagging Countries (STLCs), and China Academy of Agricultural Sciences and the Institute of Vegetables and Flowers for hosting and supporting teaching materials.

Conflicts of Interest: The authors declare that they have no conflict of interest.

References

1. Getinet, A.; Rakow, G.; Raney, J.P.; Downey, R.K. Development of zero erucic acid Ethiopian mustard through an interspecific cross with zero erucic acid Oriental mustard. *Can. J. Plant Sci.* **1994**, *74*, 793–795. [CrossRef]
2. Cardone, M.; Mazzoncini, M.; Menini, S.; Rocco, V.; Senatore, A.; Seggiani, M.; Vitolo, S. *Brassica carinata* as an alternative oil crop for the production of biodiesel in Italy: Agronomic evaluation, fuel production by transesterification and characterization. *Biomass Bioenergy* **2003**, *25*, 623–636. [CrossRef]
3. Ferreres, E.; Fernandez, M.; Minguez, I.; Dominguez, J. Productivity of *B. juncea* and *B. carinata* in relation to rapeseed. Agronomic studies. In Proceedings of the 6th International Rape-Seed Congress, Paris, France, 17–19 May 1983; pp. 293–299.
4. Enjalbert, J.N.; Zheng, S.; Johnson, J.J.; Mullen, J.L.; Byrne, P.F.; McKay, J.K. Brassicaceae germplasm diversity for agronomic and seed quality traits under drought stress. *Ind. Crops Prod.* **2013**, *47*, 176–185. [CrossRef]
5. Rakow, G.; Getinet, A. *Brassica carinata* an Oilseed Crop for Canada. *Acta Hortic.* **1998**, *459*, 419–428. [CrossRef]
6. Marillia, E.F.; Francis, T.; Falk, K.C.; Smith, M.; Taylor, D.C. Palliser's promise: *Brassica carinata*, an emerging western Canadian crop for delivery of new bio-industrial oil feedstocks. *Biocatal. Agric. Biotechnol.* **2014**, *3*, 65–74. [CrossRef]
7. Bozzini, A.; Calcagno, F.; Soare, T. "Sincron", a new Brassica carinata cultivar for biodiesel production. *Helia* **2007**, *30*, 207–214.
8. Mulvaney, M.J.; Leon, R.G.; Seepaul, R.; Wright, D.L.; Hoffman, T.L. *Brassica carinata* Seeding Rate and Row Spacing Effects on Morphology, Yield, and Oil. *Agron. J.* **2019**, *111*, 528–535. [CrossRef]
9. Seepaul, R.; Bliss, C.M.; Wright, D.L.; Marois, J.J.; Leon, R.; Dufault, N.; Olson, S.M. Carinata, the Jet Fuel Cover Crop: Production Recommendations for the Southeastern United States. 2016. Available online: http://edis.ifas.ufl.edu (accessed on 20 January 2019).
10. OFI (Oils and Fats International). Qantas Flies First Dedicated Trans-Pacific Biofuel Flight Using Carinata Oilseed. 2018. Available online: http://www.oilsandfatsinternational.com (accessed on 3 September 2020).
11. Alemayehu, N.; Becker, H. Variation and inheritance of erucic acid content in *Brassica carinata* germplasm collections from Ethiopia. *Plant Breed.* **2001**, *120*, 331–335. [CrossRef]
12. Getinet, A.; Rakow, G.; Raney, J.P.; Downey, R.K. The inheritance of erucic acid content in Ethiopian mustard. *Can. J. Plant Sci.* **1997**, *77*, 33–41. [CrossRef]
13. AAFC (Agriculture and Agri-Food Canada). Carinata: A Unique Energy Feedstock Crop. A Carinata Production Manual. 2013. Available online: http://www.pgfbiofuels.com/CarinataProductionManual2013.pdf (accessed on 21 March 2019).
14. Alberti, P.K. Development of Best Management Practices for Production of Ethiopian Mustard (*Brassica carinata*) in South Dakota. Master's Thesis, South Dakota State University, Brookings, South Dakota, 2017.
15. Fiorentino, G.; Ripa, M.; Mellino, S.; Fahd, S.; Ulgiati, S. Life cycle assessment of *Brassica carinata* biomass conversion to bioenergy and platform chemicals. *J. Clean. Prod.* **2014**, *66*, 174–187. [CrossRef]
16. Licata, M.; La Bella, S.; Leto, C.; Bonsangue, G.; Gennaro, M.C.; Tuttolomondo, T.; Mario, L. Agronomic evaluation of ethiopian mustard (*Brassica carinata* A. Braun) germplasm and physical-energy characterization of crop residues in a semi-arid area of sicily (Italy). *Chem. Eng. Trans.* **2017**, *58*, 535–540.
17. Purakayastha, T.J.; Viswanath, T.; Bhadraray, S.; Chhonkar, P.K.; Adhikari, P.P.; Suribabu, K. Phytoextraction of zinc, copper, nickel and lead from a contaminated soil by different species of Brassica. *Int. J. Phytoremediat.* **2008**, *10*, 61–72. [CrossRef]
18. Montemurro, F.; Diacono, M.; Scarcella, M.; Andrea, L.D.; Boari, F.; Santino, A.; Mastrorilli, M. Agronomic performance for biodiesel production potential of *Brassica carinata* A. Braun in Mediterranean marginal areas. *Ital. J. Agron.* **2016**, *11*, 57–64. [CrossRef]
19. Taylor, D.; Kevin, C.; Falk, C.; Don, P.; Joe, H.; Vivijan, B.; Elzbieta, M.; Ashok, J.; Elizabeth-France, M.; Tammy, F.; et al. *Brassica carinata*—A new molecular farming platform for delivering bio-industrial oil feedstocks: Case studies of genetic modifi cations to improve very long-chain fatty acid and oil content in seeds David. *Biofuels Bioprod. Biorefining* **2010**, *4*, 538–561. [CrossRef]

20. Gesch, R.W.; Isbell, T.A.; Oblath, E.A.; Allen, B.L.; Archer, D.W.; Brown, J.; Hatfield, J.L.; Jabro, J.D.; Kiniry, J.R.; Long, D.S.; et al. Comparison of several Brassica species in the north central U.S. for potential jet fuel feedstock. *Ind. Crops Prod.* **2015**, *75*, 2–7. [CrossRef]
21. Seepaul, R.; Small, I.M.; Mulvaney, M.J.; George, S.; Leon, R.G.; Paula-Moraes, S.V.; Geller, D.; Marois, J.J. *Carinata, the Sustainable Crop for a Bio-Based Economy: 2018–2019 Production Recommendations for the Southeastern United States*; University of Florida Institute of Food and Agricultural Sciences: Gainesville, FL, USA, 2019.
22. Stamigna, C.; Chiaretti, D.; Chiaretti, E.; Prosini, P.P.; Centro, E.; Casaccia, R.; Maria, S. Oil and furfural recovery from *Brassica carinata*. *Biomass Bioenergy* **2012**, *39*, 478–483. [CrossRef]
23. Bouaid, A.; Martinez, M.; Aracil, J. Production of biodiesel from bioethanol and *Brassica carinata* oil: Oxidation stability study. *Bioresour. Technol.* **2009**, *100*, 2234–2239. [CrossRef] [PubMed]
24. Katiyar, R.K.; Saran, G.; Giri, G. Evaluation of *Brassica carinata* as a New Oilseed Crop in India. *Exp. Agric.* **1986**, *22*, 67–70. [CrossRef]
25. Zanetti, F.; Monti, A.; Berti, M.T. Challenges and opportunities for new industrial oilseed crops in EU-27: A review. *Ind. Crops Prod.* **2013**, *50*, 580–595. [CrossRef]
26. Basili, M.; Rossi, M.A. *Brassica carinata*-derived biodiesel production: Economics, sustainability and policies. The Italian case. *J. Clean. Prod.* **2018**, *191*, 40–47. [CrossRef]
27. Lal, B.; Rana, K.S.S.; Rana, D.S.S.; Shivay, Y.S.S.; Sharma, D.K.K.; Meena, B.P.P.; Gautam, P. Biomass, yield, quality and moisture use of *Brassica carinata* as influenced by intercropping with chickpea under semiarid tropics. *J. Saudi Soc. Agric. Sci.* **2019**, *18*, 61–71. [CrossRef]
28. Zegada-lizarazu, W.; Monti, A. Energy crops in rotation. A review. *Biomass Bioenergy.* **2011**, *35*, 12–25. [CrossRef]
29. Mekonnen, T.W. Phenotypic Variability of Ethiopian Mustard (*Brassica carinata* A. Braun) Genotypes in South Gondar. *Adv. Res. J. Plant Anim. Sci.* **2014**, *2*, 42–48.
30. Rahman, M.; Khatun, A.; Liu, L.; Barkla, B.J. Brassicaceae mustards: Traditional and agronomic uses in Australia and New Zealand. *Molecules* **2018**, *23*, 231. [CrossRef] [PubMed]
31. Ngwene, B.; Neugart, S.; Baldermann, S.; Ravi, B.; Schreiner, M. Intercropping Induces Changes in Specific Secondary Metabolite Concentration in Ethiopian Kale (*Brassica carinata*) and African Nightshade (*Solanum scabrum*) under Controlled Conditions. *Front. Plant Sci.* **2017**, *8*, 1–12. [CrossRef]
32. Chris, B.; Kan, J.K. *Carinata Management Handbook*; Agrisoma: Tifton, GA, USA, 2017.
33. Guerrero-Diaz, M.M.; Lacasa-Martinez, C.M.; Hernandez-Piñera, A.; Martinez-Alarcon, V.; Lacasa Plasencia, A. Evaluation of repeated biodisinfestation using Brassica carinata pellets to control *Meloidogyne incognita* in protected pepper crops. *Span. J. Agric. Res.* **2013**, *11*, 485–493. [CrossRef]
34. Pane, C.; Villecco, D.; Roscigno, G.; De Falco, E.; Zaccardelli, M. Screening of plant-derived antifungal substances useful for the control of seedborne pathogens. *Arch. Phytopathol. Plant Prot.* **2013**, *46*, 1533–1539. [CrossRef]
35. Bhattarai, D. *Brassica carinata* Growth and Yield Response to Nitrogen and Sulfur Fertilizers and Impacts on Selected Soil Parameters and GHG Fluxes. Master's Thesis, South Dakota State University, Brookings, SD, USA, 2019.
36. Debiase, G.; Traversa, A.; Montemurro, F.; Mastrangelo, M.; Fiore, A.; Ventrella, G.; Brunetti, G. Minimum tillage and organic fertilization for the sustainable management of *Brassica carinata* A. (Braun) in the Mediterranean environment. *Environ. Sci. Pollut. Res. Int.* **2018**, *25*, 33556–33565. [CrossRef]
37. Dellal, I.; McCarl, B.A. The economic impacts of drought on agriculture: The case of Turkey. *Options Mediterr.* **2010**, *95*, 169–174.
38. Fahad, S.; Bajwa, A.A.; Nazir, U.; Anjum, S.A.; Farooq, A.; Zohaib, A.; Huang, J. Crop production under drought and heat stress: Plant responses and management options. *Front. Plant Sci.* **2017**, *8*, 1–16. [CrossRef]
39. Machado, R.M.A.; Serralheiro, R.P. Soil salinity: Effect on vegetable crop growth. Management practices to prevent and mitigate soil salinization. *Horticulturae* **2017**, *3*, 30. [CrossRef]
40. Hamdia, M.A.; Shaddad, M.A.K. Salt Tolerance of Crop Plants. *J. Stress Physiol. Biochem.* **2010**, *6*, 64–90.
41. Raziuddin, F.; Hassan, G.; Akmal, M.; Shah, S.S.; Mohammad, F.; Zhou, W.J. Effects of Cadmium and Salinity on Growth and Photosynthesis Parameters of Brassica Species. *Pak. J. Bot.* **2011**, *43*, 333–340.
42. Raza, A.; Razzaq, A.; Mehmood, S.S.; Zou, X.; Zhang, X.; Lv, Y.; Xu, J. Impact of climate change on crops adaptation and strategies to tackle its outcome: A review. *Plants* **2019**, *8*, 34. [CrossRef]

43. Carmina, G.; Rafael, C.; Juan Navarro-Avin, C.B.; Alfonso, G.; Ramon, S.; David, J.; Walker, M.P.B. Tolerance and accumulation of heavy metals by Brassicaceae species grown in contaminated soils from Mediterranean regions of Spain. *Environ. Exp. Bot.* **2006**, *56*, 19–27.
44. Marchiol, L.; Sacco, P.; Assolari, S. Reclamation of Polluted Soil: Phytoremediation Potential of Crop-Related Brassica Species. *Water Air Soil Pollut.* **2004**, *158*, 345–356. [CrossRef]
45. Mourato, M.P.; Moreira, I.N.; Leitão, I.; Pinto, F.R.; Sales, J.R.; Martins, L.L. Effect of heavy metals in plants of the genus Brassica. *Int. J. Mol. Sci.* **2015**, *16*, 17975–17998. [CrossRef]
46. Akhtar, J.; Saqib, M.; Mahmood, K.; Agri, S.; Centre, R. Effect of Salinity on Yield, Growth and Oil Contents of Four Brassica Species. *Pak. J. Agric. Sci.* **2002**, *39*, 76–79.
47. Gezahegn, W.W.; Srinivasulu, A.; Aruna, B.; Banerjee, S.; Sudarshan, M.; Lakshmi, P.V.; Narayana, A.D.P.R. Study of Heavy Metals Accumulation in Leafy Vegetables of Ethiopia. *J. Environ. Sci. Toxicol. Food Technol.* **2017**, *11*, 57–68. [CrossRef]
48. Gugel, R.K.; Séguin-Swartz, G.; Pétrie, G.A. Pathogenicity of three isolates of leptosphaeria maculons on brassica species and other crucifers. *Can. J. Plant Pathol.* **1990**, *12*, 75–82. [CrossRef]
49. Naresh, M. Epidemiology and forecasting for the management of rapeseed-mustard diseases. *J. Mycol. Plant Pathol.* **2014**, *44*, 131–147.
50. Chavan, V.; Kamble, A. Induction of total phenolics and defence-related enzymes during beta-aminobutyric acid-induced resistance in *Brassica carinata* against *Alternaria* blight. *Arch. Phytopathol. Plant Prot.* **2014**, *47*, 2200–2212. [CrossRef]
51. Saskatchewan Mustard Development Commission. Carinata Production: A Guide to Best Management Practices. 2013. Available online: http://www.saskmustard.ca/grower/growing/pdfs/Carinata_Production_Manual_080213.pdf (accessed on 20 November 2019).
52. USDA (United State Department of Agriculture). *Weed Risk Assessment for Brassica carinata*; Animal and Plant Health Inspection Service: Miami, FL, USA, 2014.
53. CFIA (Canadian Food Inspection Agency). The Biology of *Brassica carinata* (A.) Braun (*Abyssinian cabbage*). 2017. Available online: http://www.inspection.gc.ca/plants/plants-with-novel-traits/applicants/directive-94-08/biolog (accessed on 27 October 2018).
54. Rehman, S.; Shahzad, B.; Bajwa, A.A.; Hussain, S.; Rehman, A.; Cheema, S.A.; Li, P. Utilizing the Allelopathic Potential of Brassica Species for Sustainable Crop Production: A Review. *J. Plant Growth Regul.* **2019**, *38*, 343–356. [CrossRef]
55. Barba, F.J.; Nikmaram, N.; Roohinejad, S.; Khelfa, A.; Zhu, Z.; Koubaa, M. Bioavailability of Glucosinolates and Their Breakdown Products: Impact of Processing. *Front. Nutr.* **2016**, *3*, 3–24. [CrossRef]
56. Sanlier, N.; Guler Saban, M. The Benefits of Brassica Vegetables on Human Health. *J. Hum. Health Res.* **2018**, *1*, 104.
57. Semahegn, Y. Genetic variability, Correlation and path Analysis Studies in Ethiopian Mustard (*Brassica carinata* A. Brun) Genotypes. *Int. J. Plant Breed. Genet.* **2011**, *5*, 328–338.
58. Kumar, S. Potential of Ethiopian mustard, *Brassica carinata* as a trap crop for large white butterfly, *Pieris brassicae* infesting Indian mustard, *Brassica juncea*. *J. Pest Sci.* **2017**, *90*, 129–137. [CrossRef]
59. Odongo, G.A.; Schlotz, N.; Herz, C.; Hanschen, F.S.; Baldermann, S.; Neugart, S.; Lamy, E. The role of plant processing for the cancer preventive potential of Ethiopian kale (*Brassica carinata*). *Food Nutr. Res.* **2017**, *61*, 1–11. [CrossRef]
60. Teklehaymanot, T.; Wang, H.; Liang, J.; Wu, J.; Lin, R.; Zhou, Z.; Wang, X. Variation in Plant Morphology and Sinigrin Content in Ethiopian Mustard (*Brassica carinata* L.). *Hortic. Plant J.* **2019**, *5*, 205–212. [CrossRef]
61. Kim, H.W.; Ko, H.C.; Baek, H.J.; Cho, S.M.; Jang, H.H.; Lee, Y.M.; Kim, J.B. Identification and quantification of glucosinolates in Korean leaf mustard germplasm (*Brassica juncea* var. integrifolia) by liquid chromatography–electrospray ionization/tandem mass spectrometry. *Eur. Food Res. Tech.* **2016**, *242*, 1479–1484. [CrossRef]
62. Jadhav, A.; Katavic, V.; Marillia, E.F.; Giblin, E.M.; Barton, D.L.; Kumar, A.; Taylor, D.C. Increased levels of erucic acid in Brassica carinata by co-suppression and antisense repression of the endogenous FAD2 gene. *Metab. Eng.* **2005**, *7*, 215–220. [CrossRef] [PubMed]
63. Jamnadass, R.; Mumm, R.H.; Hale, I.; Hendre, P.; Muchugi, A.; Dawson, I.K.; Van Deynze, A. Enhancing African orphan crops with genomics. *Nat. Genet.* **2020**, *52*, 356–360. [CrossRef] [PubMed]

64. Martínez-Valdivieso, D.; Font, R.; Río-Celestino, D. Prediction of Agro-Morphological and Nutritional Traits in Ethiopian Mustard Leaves (*Brassica Carinata* A. Braun) by Visible-Near-Infrared Spectroscopy. *J. Foods* **2019**, *8*, 6. [CrossRef] [PubMed]
65. Sheikh, F.A.; Banga, S.; Banga, S.S.; Najeeb, S. Development of Ethiopian mustard (*Brassica carinata*) with broad genetic base through interspecific hybridization with elite lines of *Brassica napus* and *Brassica juncea*. *Sustain. Dev.* **2011**, *3*, 77.
66. PROTA Foundation. *Plant Resources of Tropical Africa, Vegetable Oils*; van der Vossen, H.A., Mkamilo, G.S., Eds.; PROTA: Wageningen, Netherlands, 2007. Available online: https://books.google.so/books/about/Plant_Resources_of_Tropical_Africa.html?id=_1mkQwAACAAJ&pgis=1 (accessed on 15 April 2019).
67. Sanyal, A.; Xavier, P.; André, M.; Marie, L.; Laustriat, M.D.; Fine, F. Rapeseed: Some Examples of Current French Research, Erucic Acid Rapeseed: Prospects of improvements. EDP Sciences, France. 2015. Available online: www.ocl-journal.org (accessed on 8 September 2020).
68. Choudhary, B.R.; Joshi, P.; Ramarao, S. Interspecific hybridization between *Brassica carinata* and *Brassica rapa*. *Plant Breed.* **2000**, *119*, 306–319. [CrossRef]
69. Cheung, K.W.; Razeq, F.M.; Sauder, C.A.; James, T.; Martin, S.L. Bidirectional but asymmetrical sexual hybridization between *Brassica carinata* and *Sinapis arvensis* (Brassicaceae). *J. Plant Res.* **2015**, *128*, 469–480. [CrossRef]
70. Kumar, S.; Seepaul, R.; Mulvaney, M.J.; Colvin, B.; George, S.; Marois, J.J.; Bennett, R.; Leond, R.; Wright, D.L.; Small, I.M. *Brassica carinata* genotypes demonstrate potential as a winter biofuel crop in South East United States. *Ind. Crops Prod.* **2020**, *150*, 112353. [CrossRef]
71. Jiang, Y.; Tian, E.; Li, R.; Chen, L.; Meng, J. Genetic diversity of *Brassica carinata* with emphasis on the interspecific crossability with *Brassica rapa*. *Plant Breed.* **2007**, *126*, 487–491. [CrossRef]
72. Xiao, Y.; Chen, L.; Zou, J.; Tian, E.; Xia, W.; Meng, J. Development of a population for substantial new type *Brassica napus* diversified at both A/C genomes. *Theor. Appl. Genet.* **2010**, *121*, 1141–1150. [CrossRef]
73. Thakur, A.K.; Singh, K.H.; Sharma, D.; Parmar, N.; Nanjundan, J. Breeding and genomics interventions in Ethiopian mustard (*Brassica carinata* A. Braun) improvement—A mini review. *S. Afr. J. Bot.* **2019**, *125*, 457–465. [CrossRef]
74. Zhang, W.; Hu, D.; Raman, R.; Guo, S.; Wei, Z.; Shen, X.; Meng, J.; Raman, H.; Zou, J. Investigation of the Genetic Diversity and Quantitative Trait Loci Accounting for Important Agronomic and Seed Quality Traits in Brassica carinata. *Front. Plant Sci.* **2017**, *8*, 1–17. [CrossRef] [PubMed]
75. Khedikar, Y.; Clarke, W.E.; Chen, L.; Higgins, E.E.; Kagale, S.; Koh, C.S.; Bennett, R.; Parkin, I.A.P. Narrow genetic base shapes population structure and linkage disequilibrium in an industrial oilseed crop, *Brassica carinata* A. Braun. *Sci. Rep.* **2020**, *10*, 12629. [CrossRef]
76. Wang, X.; Wang, H.; Wang, J.; Sun, R.; Wu, J.; Liu, S.; Bai, Y.; Mun, J.H.; Ian Bancroft, I.; Cheng, F.; et al. The genome of the mesopolyploid crop species *Brassica rapa*. *Nat. Genet.* **2011**, *43*, 1035–1040. [CrossRef] [PubMed]
77. Liu, S.; Liu, Y.; Yang, X.; Tong, C.; Edwards, D.; Parkin, I.A.; Zhao, M.; Ma, J.; Yu, J.; Huang, S.; et al. The *Brassica oleracea* genome reveals the asymmetrical evolution of polyploid genomes. *Nat. Commun.* **2014**, *5*, 1–11. [CrossRef] [PubMed]
78. Chalhoub, B.; Denoeud, F.; Liu, S.; Parkin, I.A.P.; Tang, H.; Wang, X.; Chiquet, J.; Belcram, H.; Tong, C.; Samans, B.; et al. Early allopolyploid evolution in the post-Neolithic *Brassica napus* oilseed genome. *Plant. Genet.* **2014**, *345*, 950–953. [CrossRef]
79. Zhang, X.; Yue, Z.; Mei, S.; Qiu, Y.; Yang, X.; Chen, X.; Cheng, F.; Wu, Z.Y.; Sun, Y.Y.; Jing, Y.; et al. A *de novo* Genome of a Chinese Radish Cultivar. *Hortic. Plant. J.* **2015**, *1*, 155–164.
80. Yang, J.; Liu, D.; Wang, X.; Ji, C.; Cheng, F.; Liu, B.; Hu, Z.; Chen, S.; Pental, D.; Ju, Y.; et al. The genome sequence of allopolyploid *Brassica juncea* and analysis of differential homoeolog gene expression influencing selection. *Nat. Genet.* **2016**, *48*, 1225–1232. [CrossRef]

© 2020 by the authors. Licensee MDPI, Basel, Switzerland. This article is an open access article distributed under the terms and conditions of the Creative Commons Attribution (CC BY) license (http://creativecommons.org/licenses/by/4.0/).

sustainability

MDPI

Article

Measurement of the Spatial Complexity and Its Influencing Factors of Agricultural Green Development in China

Hongpeng Guo, Shuang Xu and Chulin Pan *

Department of Agricultural and Forestry Economic Management, College of Biological and Agricultural Engineering, Jilin University, 5988 Renmin Street, Changchun 130022, China; ghp@jlu.edu.cn (H.G.); xushuang_jlu@163.com (S.X.)

* Correspondence: pancl@jlu.edu.cn

Received: 17 October 2020; Accepted: 5 November 2020; Published: 7 November 2020

Abstract: The article focuses on the spatial complexity of agricultural green development (AGD) in different regions. The article first constructs an evaluation index system for the level of AGD from four dimensions: Social development, economic benefits, resource input, and ecological environment. Then, the article uses an improved entropy weight method to evaluate the level of AGD with panel data of 31 provinces in China from 2007 to 2018. Finally, on the basis of Moran Index and the Spatial Durbin Model, the article analyzes the spatial heterogeneity of the factors that affect the green development of agriculture in China. The results show that: (1) From 2007 to 2018, the overall level of AGD shows a fluctuating upward trend in China, and there are differences among provinces. The level of AGD in the three major regions presents the characteristics of Eastern > Central > Western; (2) China's provincial AGD level has an obvious positive autocorrelation in spatial distribution, showing significant spatial agglomeration characteristics in space; (3) the four factors of urbanization level, agricultural mechanization level, scientific and technological R&D investment, and arable area, have different effects on the level of AGD in three major regions. This study provides a reference for understanding the status of China's agricultural green development level and policy recommendations on how to improve the level of agricultural green development. The results imply that some effective policy measures, such as prompting the integrated development of the three major industries and optimizing the industrial structure, should be taken to coordinate "green" with "development" from national and regional perspectives.

Keywords: agricultural green development; entropy weight method; spatial heterogeneity; spatial spillover effect; China

1. Introduction

Agriculture plays a central role in sustainable development, which determines the relationship between the global economy, society, and the natural world [1]. Since China implemented the reform and opening policy in 1978, China's agriculture has developed rapidly, producing considerable economic benefits and meeting the growing food demand of people [2,3]. China has done remarkable work in feeding 22% of the world's population with only 9% of the world's arable land, making an important contribution to international food security [4]. At the same time, China has also paid a huge environmental price. Environmental problems such as water eutrophication, soil acidification, air pollution, the reduction of biodiversity, etc., have become increasingly prominent, which greatly hinders the green and sustainable development of agriculture [5–7]. Therefore, China needs to coordinate the relationship between agricultural development and environmental protection urgently and promote the green transformation of agriculture.

In recent years, China has realized the importance of agricultural green development. Different ways have been adopted in different provinces in the process of agricultural green development, which have achieved varying degrees of effectiveness. However, how to realize the sustainable utilization of resources and sustainable development of agriculture while protecting the environment is still an urgent problem to be solved.

Agriculture Green Development is the future development direction of agriculture for China and the world [8–10]. In order to achieve green development in agriculture, it is necessary to make a quantitative evaluation of the current agricultural development level. In previous studies, in order to evaluate the sustainable agricultural development, many scholars have constructed different evaluation index systems for sustainable agricultural development around three dimensions, including society, economy, and ecology, evaluating the sustainable development of agriculture from the perspectives of farms, regions, and countries [11–16]. In recent years, under the background of green transformation of agriculture, scholars have studied from different perspectives on the construction of evaluation index system and evaluation methods of agricultural green development. For instance, Wei et al. selected 14 indicators from four dimensions of resource conservation, ecological conservation, environmental friendliness, and quality efficiency to construct China's agricultural green development index, then evaluating the level of agricultural green development in China [17]. Gong et al. selected 10 indicators from three dimensions of low-carbon economy, economic growth, and safe supply to construct a development index to evaluate the level of agricultural green development [18]. More scholars have used principal component analysis [19], entropy weight method [20–22], or analytic hierarchy process [23] to evaluate the level of agricultural green development in different provinces of China.

The overall objective of Agriculture Green Development is to coordinate "green" with "development" to realize the transformation of current agriculture with high resource consumption and high environmental costs into a green agriculture and countryside with high productivity, high resource use efficiency, and low environmental impact [10]. However, the resource and environment cost of agricultural production in China is too high. Irrational fertilization leads to serious acidification of croplands, which poses a threat to agricultural production and hinders China' s green development of agriculture seriously [5,24,25]. In areas with scarce water resources, the quality and safety of agricultural products cannot be guaranteed, which will also hinder the green development of agriculture [26]. Tu et al. pointed out that the green development of agriculture is not only reflected in the low agricultural non-point source pollution, but also in the high efficiency of agricultural production [27]. Inconsistent agricultural development conditions, different ecological environments, and different resource endowments in various regions have an impact on agricultural production and the level of agricultural development [28,29]. In addition, changes in institutions and policies, the level of urbanization, and the level of science and technology also have an impact on agricultural production efficiency [30]. Among many influencing factors, technological progress has been regarded as the main driving factor of agricultural production [31]. Meanwhile, the green development of agriculture puts forward higher requirements for technological progress, which requires the integration of multiple disciplines and technological innovation to realize the green development of agriculture [32,33].

Through reviewing literatures, since the concept of agricultural green development is relatively new, most studies mainly focus on two aspects: The first is the construction and level measurement of the evaluation index system of agricultural green development, and the second is the driving or restrictive factors that affect agricultural green development. The research on the spatial characteristics of agricultural green development and its influencing factors is still blank. Therefore, this article first evaluates the status of agricultural green development level in China through improved entropy weight method to measure the level of agricultural green development. Then, in order to study the spatial correlation and heterogeneity of agricultural green development in China, the article conducted an empirical analysis on the influencing factors of agricultural green development with panel data from 31 provinces in China. This study provides data support and empirical basis for the proposal and formulation of agricultural green development policy in different regions of China.

2. Materials and Methods

2.1. Materials

2.1.1. Construction of Evaluation Index System for Agricultural Green Development Level

According to the definition of the level of agricultural green development and the research of other scholars, this article selects 16 indicators from four perspectives, including social development, economic benefits, resource input, and ecological environment to construct an evaluation index system for the level of agricultural green development in China, as shown in Table 1. All data are derived from the *China Statistical Yearbook*, *China Rural Statistical Yearbook*, *China Environment Statistical Yearbook*, and *Provincial Statistical Yearbook* from 2008 to 2019.

Table 1. Evaluation index system of agricultural green development level.

Primary Indicators	Secondary Indicators	Units	Description	Source	Indicator Attribute
Social Development	Agricultural financial investment rate	%	Agricultural financial investment/total government expenditure	[21,34]	+
	Environmental protection investment rate	%	Environmental protection financial investment/total government expenditure	[35]	+
	Agricultural mechanization level	W/ha	Total power of agricultural machinery/sown area of crops	[22,34,36]	+
Economic Benefits	Irrigation level of farmland	%	Effective irrigation area/arable land area	[20,21,34,37]	+
	Land Multiple Cropping Index	%	Sown area of crops/arable land area	[21,22,38]	+
	Land yield rate	yuan/ha	Total agricultural output value/sown area of crops	[21,34,39]	+
	Rural Engel coefficient	%	Food consumption expenditure of rural residents/total consumption expenditure	[34,36]	−
	Per capita disposable income of rural residents	10^4 yuan/person	Per capita disposable income of rural residents	[20,34–36,38]	+
Resource Input	Pesticide application intensity	t/kha	Total pesticide input/total sown area	[20–22,34–36,38,39]	−
	Fertilizer application intensity	t/ha	Total fertilizer input/total sown area	[20–22,35,36,38,39]	−
	Agricultural plastic film application intensity	t/kha	Total agricultural plastic film input/total sown area	[20–22,34–36,38,39]	−
	The level of agricultural labor input	persons/ha	Total agricultural labor/arable land area	[34]	+
Ecological Environment	Harmless treatment rate of domestic garbage	%	The amount of domestic garbage treated in a harmless manner/the amount of domestic garbage generation	[34,37,38]	+
	Centralized treatment rate of sewage	%	The amount of centralized sewage treatment/total sewage generation	[34,37,38]	+
	Forest cover rate	%	Forest area/land area	[20,21,34–38,40,41]	+
	Disaster area rate	%	Disaster area/affected area	[20,21,39]	−

2.1.2. Influencing Factors of Agricultural Green Development

This paper used the level of agricultural green development in each province as the explained variable, and *urb*, *mech*, *rnd*, and *land* as the explanatory variables to analyze the impact of each variable on the green development of agriculture. The specific meaning is as follows:

urb: A variable that represents the level of urbanization, which is reflected by the proportion of urban population in the total population. On the one hand, urbanization can gather social resources in a certain space and accelerate the transformation and upgrading of industrial structure. On the other hand, the improvement of urbanization level can increase the emission of environmental pollution and inhibit the green development of agriculture [42,43].

mech: A variable that represents the level of mechanization, which is reflected by the total power of agricultural machinery. The improvement of the level of mechanization is conducive to improving the efficiency of agricultural production and has a positive impact on the green development of agriculture [27].

rnd: A variable that represents the investment in scientific and technological research and development, which is reflected by the scientific and technological expenditure. The investment in scientific research and development can promote technological progress, improve agricultural production efficiency, and promote the green development of agriculture [30,44].

land: A variable that represents the area of arable land, which is reflected by the arable land area. Arable land is an indispensable element in agricultural production. For one thing, arable land area represents the resource endowment of a certain area, which can increase crop yields and promote the green development of agriculture. For another, the area of arable land decreases year by year while the utilization and output rate of arable land are increased, which has a positive impact on the green development of agriculture.

2.2. Entropy Weight Method

In the application of comprehensive evaluation methods, there are multiple evaluation methods, which can be divided into subjective weighting method and objective weighting method when determining the weight of each index. Subjective weighting methods tend to rely on subjective judgments and lacks certain objectivity, which makes the evaluation result subjectively random. The objective method uses the value of indicators to calculate the weight without considering human factors. Entropy weight method determines each index's weight on the basis of amount of information. Among the objective methods, entropy weight method has been widely used in many fields, which will make the evaluation more accurate when dealing with large amounts of data.

Therefore, we selected the entropy weight method of objective weighting methods to determine the weight of various indicators through information entropy, and then made a comprehensive evaluation of the level of agricultural green development in China. Data selected were panel data containing both time series and cross-section. In order to guarantee the comparability among different years, time variable was introduced to improve the entropy weight method. The formula of improved entropy weight method is expressed as Equations (1)–(8) [45].

Step 1 Construct the original index matrix data: assume there are h years, n provinces, and m indicators, then the original index matrix is $X = \{x_{\lambda ij}\}_{h*n*m}(1 \leq \lambda \leq h, 1 \leq i \leq n, 1 \leq j \leq m)$, where $x_{\lambda ij}$ represents the index j of province n in year h. In this article, h, n and m are 12, 31, and 16.

Step 2 Range standardize and dimensionless processing of each indicator in the index system, as is shown in Equations (1) and (2).

Standardization of positive indicators:

$$Z_{\lambda ij} = \frac{x_{\lambda ij} - x_{\min}}{x_{\max} - x_{\min}} \tag{1}$$

Standardization of negative indicators:

$$Z_{\lambda ij} = \frac{x_{\max} - x_{\lambda ij}}{x_{\max} - x_{\min}} \tag{2}$$

Step 3 Normalization of indicators, as is shown in Equation (3):

$$P_{\lambda ij} = \frac{Z_{\lambda ij}}{\sum\limits_{\lambda=1}^{h}\sum\limits_{i=1}^{n} Z_{\lambda ij}} \tag{3}$$

Step 4 Calculate the entropy of each indicator, as is shown in Equations (4) and (5):

$$E_j = -k\sum_{\lambda=1}^{h}\sum_{i=1}^{n} P_{\lambda ij}\ln P_{\lambda ij} \tag{4}$$

$$k = \frac{1}{\ln(h \times n)} \tag{5}$$

Step 5 Calculate the redundancy of the entropy values of each indicator, as is shown in Equation (6):

$$D_j = 1 - E_j \tag{6}$$

Step 6 Calculate the weight of each indicator, as is shown in Equation (7):

$$W_j = \frac{D_j}{\sum\limits_{j=1}^{m} D_j} \tag{7}$$

Step 7 Calculate the level of agricultural green development in each province in each year, as is shown in Equation (8).

$$C_{\lambda i} = Z_{\lambda ij} \times W_j \tag{8}$$

2.3. The Spatial Autocorrelation Analysis Method

2.3.1. Global Spatial Autocorrelation

Before studying the spillover effects of agricultural green development level in China, it is necessary to test whether there exits statistical spatial correlation among the provincial agricultural green development level. It is common in the literature to measure the global spatial autocorrelation with Moran's I index. The formula of Moran's I is calculated in Equation (9) [46].

$$\text{Moran's } I = \frac{n}{\sum\limits_{i=1}^{n} (x_i - \overline{x})^2} \frac{\sum\limits_{i=1}^{n} \sum\limits_{j=1}^{n} w_{ij}(x_i - \overline{x})(x_j - \overline{x})}{\sum\limits_{i=1}^{n} \sum\limits_{j=1}^{n} w_{ij}} \tag{9}$$

where x_i denotes the observed value of the agricultural green development level in province i; x_j denotes the observed value of the agricultural green development level in province j; $\overline{x}$ denotes the mean of observed values of the agricultural green development level; n denotes the total provincial number; w_{ij} denotes the spatial weight matrix. We used 0-1 adjacent space weight matrix as the spatial weight matrix in this article. The definition is shown in Equation (10).

$$w_{ij} = \begin{cases} 1, \text{ province } i \text{ adjacent to province } j \\ 0, \text{ otherwise} \end{cases}, \; i = 1, 2, \cdots, n; j = 1, 2, \cdots, n \tag{10}$$

Moran's I index ranges from −1 to 1. The Moran's I index shows a positive spatial autocorrelation if it is between 0 and 1, which represents that the "high value" is adjacent to the "high value", and the "low value" is adjacent to the "low value". The Moran's I index shows a negative spatial autocorrelation if it is between −1 and 0, which represents that the "high value" is adjacent to the "low value". There is no spatial autocorrelation if the Moran's I index is 0 [47].

2.3.2. Local Spatial Autocorrelation

The meaning of the local Moran index is similar to that of the global Moran index, which is expressed as Equation (11). If Moran's $I_i > 0$, it means that the "high(low) value" in region i is surrounded by the "high(low) value". Otherwise, if Moran's $I_i < 0$, it means that the "high(low) value" in region i is surrounded by the "low(high) value" [47].

$$\text{Moran's } I_i = \frac{n^2}{\sum\limits_{i=1}^{n}(x_i-\overline{x})^2}\frac{(x_i-\overline{x})\sum\limits_{i=1}^{n}\sum\limits_{j=1}^{n}w_{ij}(x_j-\overline{x})}{\sum\limits_{i=1}^{n}\sum\limits_{j=1}^{n}w_{ij}} \tag{11}$$

2.4. Spatial Panel Data Model

Spatial econometric models mainly include Spatial Durbin Model (SDM), Spatial Autoregressive Model (SAR), and Spatial Error Model (SEM). The SDM model is more general, which is a general model including SAR model and SEM model. The model is expressed as Equation (12) [44].

$$Y = \rho WY + X\beta + \theta WX + \varepsilon \tag{12}$$

where ρ represents the spatial autocorrelation coefficient, W represents the spatial weight matrix, X represents the explanatory variables, WX represents the spatial lag of explanatory variables, β represents the regression coefficient of the explanatory variables, and ε represents the random error term.

When $\theta = 0$, the SDM model will be simplified to the SAR model. The model is expressed as Equation (13).

$$Y = \rho WY + X\beta + \varepsilon \tag{13}$$

When $\theta + \rho\beta = 0$, the SDM model will be simplified to the SEM model. The model is expressed as Equations (14) and (15).

$$Y = X\beta + u \tag{14}$$

$$u = \lambda Wu + \varepsilon \tag{15}$$

3. Empirical Results and Discussion

3.1. Calculation of Agricultural Green Development Level

According to the basic principle of entropy weight method, the article calculated the index weight of each index step by step, and then calculated the level of agricultural green development of 31 provinces (cities and municipalities) from 2007 to 2018 through weighted calculation. The results are shown in Table 2.

It can be seen from Table 2 that from 2007 to 2018, the overall level of agriculture green development in China shows a slow upward trend. The average national level of agricultural green development fluctuated from 0.429 to 0.468, with an average annual growth rate of 0.76%. Provincial level of the agricultural green development shows an upward trend, but there exist differences among provinces. Among the top 10 provinces in terms of average level are 8 provinces in the eastern region, while among the bottom 10 provinces in terms of average level are 7 provinces in the western region. It can be seen that the provinces with a higher level of agricultural green development in China are concentrated in the eastern area, and the provinces with a lower level of development are concentrated in the western region. The three major regions have significant regional differences in the level of agricultural green development, showing the spatial characteristics of Eastern > Central > Western on the whole. In addition, the range of provincial agricultural green development level decreased from 0.341 to 0.274, indicating that the gap of agricultural green development level among provinces in China is decreasing.

Table 2. The level of agricultural green development (AGD) evaluation results in China during 2007-2018.

Province	Year												Level	Ranking
	2007	2008	2009	2010	2011	2012	2013	2014	2015	2016	2017	2018		
Zhejiang	0.630	0.635	0.606	0.625	0.615	0.626	0.649	0.622	0.610	0.631	0.602	0.589	0.620	1
Fujian	0.564	0.599	0.581	0.564	0.594	0.607	0.637	0.622	0.616	0.648	0.594	0.596	0.602	2
Beijing	0.611	0.600	0.629	0.599	0.636	0.617	0.609	0.584	0.529	0.567	0.585	0.581	0.595	3
Guangdong	0.524	0.551	0.578	0.602	0.604	0.609	0.623	0.580	0.554	0.594	0.563	0.612	0.583	4
Shanghai	0.533	0.561	0.590	0.593	0.546	0.589	0.607	0.596	0.558	0.560	0.605	0.555	0.574	5
Jiangsu	0.508	0.498	0.522	0.501	0.528	0.543	0.569	0.555	0.545	0.570	0.521	0.543	0.534	7
Hebei	0.463	0.460	0.466	0.475	0.480	0.499	0.526	0.513	0.515	0.512	0.515	0.532	0.496	9
Shandong	0.493	0.496	0.481	0.477	0.504	0.517	0.526	0.492	0.481	0.478	0.492	0.486	0.494	10
Hainan	0.391	0.444	0.431	0.436	0.457	0.473	0.468	0.455	0.445	0.512	0.498	0.518	0.461	14
Tianjin	0.452	0.419	0.381	0.424	0.454	0.448	0.499	0.431	0.480	0.422	0.490	0.416	0.443	17
Liaoning	0.396	0.391	0.362	0.346	0.380	0.378	0.395	0.387	0.404	0.413	0.412	0.412	0.390	25
Eastern	**0.506**	**0.514**	**0.511**	**0.513**	**0.527**	**0.537**	**0.555**	**0.531**	**0.521**	**0.537**	**0.534**	**0.531**	**0.526**	
Hunan	0.510	0.518	0.540	0.544	0.545	0.574	0.597	0.596	0.587	0.611	0.581	0.585	0.566	6
Jiangxi	0.512	0.540	0.526	0.512	0.498	0.519	0.492	0.510	0.514	0.547	0.537	0.536	0.520	8
Henan	0.487	0.470	0.478	0.464	0.472	0.496	0.495	0.469	0.481	0.484	0.492	0.500	0.482	11
Hubei	0.408	0.418	0.444	0.465	0.465	0.472	0.510	0.489	0.475	0.480	0.485	0.510	0.468	12
Anhui	0.444	0.436	0.447	0.422	0.457	0.448	0.505	0.474	0.458	0.487	0.475	0.493	0.462	13
Heilongjiang	0.360	0.373	0.369	0.361	0.370	0.425	0.400	0.412	0.419	0.406	0.455	0.438	0.399	21
Jilin	0.336	0.392	0.362	0.350	0.403	0.420	0.421	0.406	0.392	0.379	0.351	0.383	0.383	28
Shanxi	0.334	0.360	0.351	0.342	0.337	0.375	0.413	0.391	0.373	0.369	0.357	0.398	0.367	29
Central	**0.424**	**0.438**	**0.440**	**0.432**	**0.443**	**0.466**	**0.479**	**0.468**	**0.462**	**0.470**	**0.467**	**0.480**	**0.456**	
Guangxi	0.409	0.426	0.443	0.445	0.445	0.456	0.498	0.476	0.478	0.485	0.481	0.473	0.460	15
Sichuan	0.463	0.452	0.452	0.432	0.434	0.469	0.493	0.467	0.460	0.446	0.451	0.461	0.457	16
Chongqing	0.427	0.420	0.428	0.451	0.439	0.453	0.472	0.432	0.431	0.438	0.443	0.442	0.440	18
Shaanxi	0.390	0.421	0.412	0.415	0.445	0.445	0.441	0.427	0.423	0.419	0.433	0.428	0.425	19
Yunnan	0.386	0.394	0.410	0.415	0.401	0.434	0.427	0.418	0.406	0.443	0.414	0.411	0.413	20
Guizhou	0.370	0.352	0.376	0.392	0.346	0.413	0.394	0.417	0.421	0.430	0.433	0.444	0.399	22
Qinghai	0.396	0.397	0.404	0.389	0.420	0.396	0.443	0.412	0.391	0.387	0.368	0.359	0.397	23
Xinjiang	0.397	0.387	0.356	0.396	0.378	0.412	0.409	0.409	0.392	0.399	0.380	0.400	0.393	24
Ningxia	0.350	0.395	0.355	0.405	0.391	0.394	0.403	0.397	0.363	0.359	0.394	0.427	0.386	26
Inner Mongolia	0.372	0.421	0.360	0.360	0.367	0.374	0.390	0.390	0.388	0.380	0.392	0.412	0.384	27
Tibet	0.392	0.355	0.345	0.348	0.384	0.371	0.352	0.337	0.381	0.330	0.333	0.368	0.358	30
Gansu	0.289	0.296	0.309	0.314	0.290	0.310	0.305	0.296	0.295	0.277	0.331	0.338	0.304	31
Western	**0.387**	**0.393**	**0.387**	**0.397**	**0.395**	**0.411**	**0.419**	**0.407**	**0.402**	**0.399**	**0.404**	**0.414**	**0.401**	
Nation	**0.429**	**0.440**	**0.436**	**0.434**	**0.446**	**0.463**	**0.474**	**0.460**	**0.456**	**0.461**	**0.461**	**0.468**	**0.452**	

3.2. *Spatial Correlation Analysis*

3.2.1. Global Autocorrelation Test

The results of the Moran's *I* index are shown in Table 3. It shows that the Moran's *I* index of provincial agricultural green development level in China is positive at the significance of 5% from 2007 to 2018. It indicates that provincial agricultural green development has an obvious positive autocorrelation in the spatial distribution. The level of provincial agricultural green development presents a spatial agglomeration effect. A positive spatial correlation means that the characteristics of adjacent provinces are similar, that is, provinces with higher level tend to be close to those with higher level, or provinces with lower level tend to be relatively close to those with lower level. Therefore, there is a spatial correlation between the provincial level of agricultural green development in China on the whole, which means that the level of agricultural green development shows significant spatial agglomeration characteristics in space.

Table 3. Global Moran's *I* of AGD level during 2007–2018.

Year	I	E (I)	Sd (I)	Z	*p*-Value
2007	0.501	−0.033	0.107	4.422	0.000
2008	0.482	−0.033	0.107	4.223	0.000
2009	0.505	−0.033	0.108	4.529	0.000
2010	0.536	−0.033	0.107	4.686	0.000
2011	0.504	−0.033	0.107	4.479	0.000
2012	0.531	−0.033	0.107	4.689	0.000
2013	0.553	−0.033	0.107	4.760	0.000
2014	0.544	−0.033	0.107	4.823	0.000
2015	0.610	−0.033	0.107	5.407	0.000
2016	0.642	−0.033	0.107	5.735	0.000
2017	0.636	−0.033	0.108	5.456	0.000
2018	0.612	−0.033	0.108	5.371	0.000

3.2.2. Local Correlation Test

In order to further test the spatial relationship of agricultural green development in China and investigate the distribution characteristics of agricultural green development in local areas, the article draws the local Moran index scatter plot of some years (2007, 2011, 2015, 2018) based on the 0–1 adjacent spatial weight matrix, as is shown in Figure 1a–d. The presentation of Moran scatter plot is to divide the space into four quadrant units, and different quadrant units represent the relationship between the local spatial unit and its adjacent spatial unit. Specifically, the first quadrant is a high-high (H-H) agglomeration area, which means that the region with a high level is surrounded by regions with high level. The second quadrant is a low-high (L-H) agglomeration area, which means that the region with a low level is surrounded by regions with high level. The third quadrant is a low-low (L-L) agglomeration area, which means that the regions with a low level is surrounded by regions with low level. The fourth quadrant is a high-low (H-L) agglomeration area, which means that the region with a high level is surrounded by regions with low level.

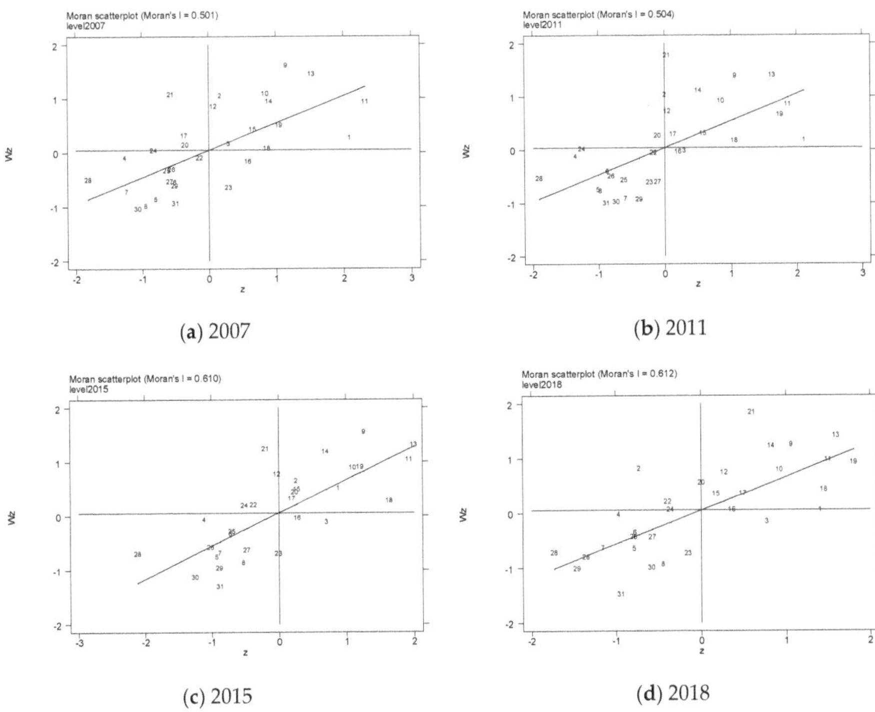

Figure 1. Local Moran index scatter plot of AGD level.

In order to understand the agglomeration or dispersion of provinces more intuitively, the data of each quadrant were summarized and presented in the form shown in Table 4. From the perspective of time dynamics, the level of green agricultural development in various provinces has changed, but in the selected four years, most of the provinces are located in the H-H and L-L agglomeration areas. From a long-term perspective, the provinces located in each quadrant have relatively little change. Provinces located in the H-H agglomeration area for a long time are mostly provinces in the eastern coastal area, while the provinces located in the L-L agglomeration area are mostly provinces in the western region. It indicates that agricultural green development in China exists as a local spatial correlation, and the level of agricultural green development in different provinces exists spatially heterogeneously.

Table 4. Local Moran index province distribution of AGD level.

	2007	2011	2015	2018
H-H agglomeration (the first quadrant)	Beijing, Tianjin, Hebei, Shanghai, Jiangsu, Zhejiang, Anhui, Fujian, Jiangxi, Shandong, Hunan, Guangdong	Beijing, Shanghai, Jiangsu, Zhejiang, Anhui, Fujian, Jiangxi, Shandong, Hubei, Hunan, Guangdong, Hainan	Beijing, Tianjin, Shanghai, Jiangsu, Zhejiang, Fujian, Jiangxi, Shandong, Hubei, Hunan, Guangdong, Guangxi	Beijing, Shanghai, Jiangsu, Zhejiang, Anhui, Fujian, Jiangxi, Shandong, Henan, Hubei, Hunan, Guangdong, Guangxi, Hainan
L-H agglomeration (the second quadrant)	Hubei, Guangxi, Hainan, Guizhou	Tianjin, Guangxi	Anhui, Hainan, Chongqing, Guizhou	Tianjin, Chongqing, Guizhou

Table 4. *Cont.*

	2007	2011	2015	2018
L-L agglomeration the third quadrant)	Shanxi, Inner Mongolia, Liaoning, Jilin, Heilongjiang, Chongqing, Yunnan, Tibet, Shaanxi, Gansu, Qinghai, Ningxia, Xinjiang	Shanxi, Inner Mongolia, Liaoning, Jilin, Heilongjiang, Chongqing, Sichuan, Guizhou, Yunnan, Tibet, Shaanxi, Gansu, Qinghai, Ningxia, Xinjiang	Shanxi, Inner Mongolia, Liaoning, Jilin, Heilongjiang, Sichuan, Yunnan, Tibet, Shaanxi, Gansu, Qinghai, Ningxia, Xinjiang	Shanxi, Inner Mongolia, Liaoning, Jilin, Heilongjiang, Sichuan, Yunnan, Tibet, Shaanxi, Gansu, Qinghai, Ningxia, Xinjiang
H-L agglomeration (the fourth quadrant)	Henan, Sichuan	Hebei, Henan	Hebei, Henan	Hebei

3.3. *Spatial Econometric Analysis*

3.3.1. Model Test and Selection

Before the spatial econometric regression, it is necessary to analyze the panel data with the ordinary least squares (OLS) model. First, we used F test to compare the fixed effect model with the mixed effect model without considering spatial correlation, and the result showed that the fixed effect model was better than the mixed effect model. Then, we used the Breusch and Pagan (B&P) test to compare the random effect model and the mixed effect model, and the result showed that the random effect model was better than the mixed effect model. Finally, we used Hausman test to compare the random effect model and the fixed effect model, and the result showed that the fixed effect model was better than the random effect model. The test results are shown in Table 5. Through the above three tests, we found that the fixed effect model is more suitable. Therefore, the fixed effect model was used for further spatial econometric model selection to analyze the influencing factors of agricultural green development level in China.

Table 5. Panel data related effect diagnosis results.

Test	F	Chi2	*p*-Value
F test	56.10		0.0000
B&P test		1023.21	0.0000
Hausman test		20.14	0.0012

Considering the spatial correlation, we used LM test and Robust LM test to compare the Spatial Autoregressive Model (SAR) with the Spatial Error Model (SEM), as is shown in Table 6. It can be seen from the LM results that the LM statistics of both the SAR model and the SEM model are at the significance level of 5%, rejecting the null hypothesis that there is no spatial lag term and spatial error term. The result indicates that the two models can be used to analyze the influencing factors of agricultural green development. From the perspective of robust LM test results, the SAR model failed to pass the significance test, indicating that the SEM model is better than the SAR model.

Table 6. LM test results.

Test	Statistics	Df	*p*-Value
Spatial error:			
Lagrange multiplier	229.836	1	0.000
Robust Lagrange multiplier	213.301	1	0.000
Spatial lag:			
Lagrange multiplier	17.272	1	0.000
Robust Lagrange multiplier	0.736	1	0.391

In order to make the estimation results more accurate, based on the fixed effect model estimation results of Spatial Durbin Model (SDM), we used Wald test and likelihood ratio (LR) test to determine whether the SDM model with fixed effect needs to be simplified to SAR model or SEM model. The test results are shown in Table 7, where the null hypothesis of test 1 is $\theta = 0$, and the null hypothesis of test 2 is $\theta = -\rho\beta$. From the test results, we can see that the statistics of the two tests are both at the significance of 1% and reject the null hypothesis. The results mean that there may be bias when only using the SAR model or the SEM model to analyze the spatial spillover effect of influencing factors of agricultural green development, and the SDM model is the suitable spatial econometric model.

Table 7. Model comparison test results.

	Chi2 (4)	Prob > Chi2
Test 1	21.85	0.0002
Test 2	18.40	0.0010

In this article, we listed the regression results of the fixed effect model under the SDM model, the SAR model, the SEM model, and the OLS model for comparative analysis, and the results are shown in Table 8. According to the results in Table 8, compared with the SAR model, the SEM model, and the OLS model, the SDM model considering spatial lag effect terms and spatial error effect terms has the best fitting effect, indicating that spatial geographical factors cannot be ignored when analyzing the influencing factors of agricultural green development level. After comparing the significance of each variable, the significance of spatial effect, and the Log-likelihood, it is found that the spatial effect of the three spatial econometric models is at the significance of 1%. However, the independent variable coefficient of the SDM model is more significant, and the Log-likelihood value is the largest, which further confirms that the SDM model is the optimal estimation model.

In order to clarify the type of the selected model, we made a further test on the fixed effect model under the Spatial Durbin model, exploring the choice of spatial fixed effect model, time fixed effect model, or spatial and time double fixed effect model, as is shown in Table 9. According to the results in Table 9, although the fitting effect under the time fixed effect is lower than that under the spatial fixed effect, the coefficient of each variable of the estimated result under the time fixed effect is significant. Therefore, this article selected the time fixed effect model under the Spatial Durbin Model for analysis.

Table 8. Model comparison estimation results.

	SDM	SAR	SEM	OLS
lnurb	0.2325 *** (0.0719)	0.0407 (0.0535)	0.1093 * (0.0602)	0.0388 (0.0574)
lnmech	0.0901 *** (0.0196)	0.0803 *** (0.0157)	0.0892 *** (0.0171)	0.0853 *** (0.0168)
lnrnd	0.1925 * (0.0111)	0.0181 ** (0.0087)	0.0162 * (0.0093)	0.0263 *** (0.0090)
lnland	−0.2373 *** (0.0747)	−0.1407 *** (0.0513)	-0.1649 *** (0.0589)	−0.1499 *** (0.0550)
W×*lnurb*	−0.4328 *** (0.1002)			
W×*lnmech*	−0.0333 (0.0324)			
W×*lnrnd*	0.0366 ** (0.0144)			
W×*lnland*	0.2180 ** (0.1039)			
ρ	0.2574 *** (0.0644)	0.2252 *** (0.0633)		
λ			0.2706 *** (0.0676)	
$sigma^2$	0.0018 *** (0.0001)	0.0020 *** (0.0001)	0.0019 *** (0.0001)	
R^2	0.2920	0.2539	0.2540	0.2573
Log-likelihood	639.8792	629.3011	630.7409	

Note: Standard errors in parentheses * $p < 0.10$, ** $p < 0.05$, *** $p < 0.01$.

Table 9. Estimation of fixed effect types results.

	Spatial Fixed Effect	Time Fixed Effect	Spatial and Time Fixed Effect
lnurb	0.2325 *** (0.0719)	0.2051 *** (0.0361)	0.2597 *** (0.0701)
lnmech	0.0901 *** (0.0196)	0.1682 *** (0.0145)	0.0870 *** (0.0188)
lnrnd	0.1925 * (0.0111)	0.0700 *** (0.0081)	0.0086 (0.0111)
lnland	−0.2373 *** (0.0747)	−0.1652 *** (0.0146)	−0.2338 *** (0.0718)
W×*lnurb*	−0.4328 *** (0.1002)	−0.2738 *** (0.0572)	−0.5617 *** (0.1156)
W×*lnmech*	−0.0333 (0.0324)	−0.2203 *** (0.0286)	−0.0159 (0.0369)
W×*lnrnd*	0.0366 ** (0.0144)	0.0793 *** (0.0147)	−0.0034 (0.0179)
W×*lnland*	0.2180 ** (0.1039)	0.1218 *** (0.0270)	0.1269 (0.1119)
ρ	0.2574 *** (0.0644)	0.2795 *** (0.0645)	0.0755 (0.0727)
$sigma^2$	0.0018 *** (0.0001)	0.0076 *** (0.0006)	0.0017 *** (0.0001)
R^2	0.2920	0.2292	0.0199
Log-likelihood	639.8792	377.5491	659.0086

Note: Standard errors in parentheses * $p < 0.10$, ** $p < 0.05$, *** $p < 0.01$.

3.3.2. Results and Discussion

Due to the vast territory of China, there are obvious differences in agricultural resource endowments and supporting policies in different provinces. Therefore, there are obvious regional differences in the factors that affecting the level of agricultural green development. In order to analyze the spatial characteristics and spatial spillover effect of the factors affecting the level of agricultural green development in different regions, China can be divided into three areas based on the geographical and administrative location, i.e., the eastern, central, and western areas, as is shown in Table 10 [48]. In this study, 31 provinces (municipalities, autonomous regions) were considered exclusive of Hong Kong, Macao, and Taiwan for the data absence.

Table 10. The regional classification in China.

Regions	Provinces (Municipalities, Autonomous Regions)
Eastern area	Beijing, Tianjin, Hebei, Liaoning, Shanghai, Jiangsu, Zhejiang, Fujian, Shandong, Guangdong, Hainan
Central area	Shanxi, Jilin, Heilongjiang, Anhui, Jiangxi, Henan, Hubei, Hunan
Western area	Mongolia, Guangxi, Chongqing, Sichuan, Guizhou, Yunnan, Tibet, Shaanxi, Gansu, Qinghai, Ningxia, Xinjiang

The article constructs the Spatial Durbin Model with time fixed effect on the influencing factors of agricultural green development in the nation, the eastern area, the central area, and the western area for analysis, and the estimated results are shown in Table 11.

Table 11. Estimation of the Spatial Durbin Model (SDM) model of influencing factors of AGD level.

	Nation	Eastern	Central	Western
lnurb	0.2051 ***	−0.4919 ***	−0.3015 ***	0.1775 ***
	(0.0361)	(0.1299)	(0.1044)	(0.0374)
lnmech	0.1682 ***	0.0800 **	0.0315	0.1821 ***
	(0.0145)	(0.0359)	(0.0266)	(0.0319)
lnrnd	0.0700 ***	0.1225 ***	0.0249	0.1447 ***
	(0.0081)	(0.0157)	(0.0164)	(0.0152)
lnland	−0.1652 ***	−0.1719 ***	−0.2492 ***	−0.2242 ***
	(0.0146)	(0.0447)	(0.0282)	(0.0247)
W×*lnurb*	−0.2738 ***	−0.0964	−0.2603	−0.3382 ***
	(0.0572)	(0.2098)	(0.1980)	(0.0592)
W×*lnmech*	−0.2203 ***	−0.1627 **	−0.2768 ***	−0.0309
	(0.0286)	(0.0658)	(0.0522)	(0.0756)
W×*lnrnd*	0.0793 ***	0.0315	−0.0459 **	0.0439
	(0.0147)	(0.0244)	(0.0216)	(0.0438)
W×*lnland*	0.1218 ***	0.1223	−0.0931 **	0.0316
	(0.0270)	(0.0991)	(0.0375)	(0.0570)
ρ	0.2795 ***	0.1203	−0.1041	−0.0661
	(0.0645)	(0.0968)	(0.1007)	(0.1322)
sigma^2	0.0076 ***	0.0062 ***	0.0021 ***	0.0033 ***
	(0.0006)	(0.0008)	(0.0003)	(0.0004)
R^2	0.2292	0.1181	0.3336	0.1681
Log-likelihood	377.5491	148.1407	160.2003	207.1059

Note: Standard errors in parentheses ** $p < 0.05$, *** $p < 0.01$.

From the national perspective, the four influencing factors of urbanization level, agricultural mechanization level, scientific and technological R&D investment, and arable land area have significant impacts on the level of agricultural green development in China at the significance level of 1%. Among the four factors, urbanization level, agricultural mechanization level, and scientific and technological R&D investment have a positive impact on the level of agricultural green development, while arable land area has a negative impact on the level of agricultural green development. Meanwhile, the factors in the adjacent regions have a significant impact on the provincial agricultural green development level. Among the four factors, the urbanization level and agricultural mechanization level in the adjacent provinces have a significant negative impact on the provincial agricultural green development level, while the scientific and technological R&D investment and arable land area in the adjacent provinces have a significant positive impact on the provincial agricultural green development level.

From the regional level, the impact of various factors on the level of agricultural green development in different regions is different, and the impact direction is also different. Therefore, we adopted the horizontal comparison method to analyze each variable in the eastern, central, and western areas to illustrate the spatial heterogeneity.

Considering the impact of the variable "*lnurb*" on the level of agricultural green development, the urbanization level has a significant impact on the green development of agriculture in the eastern, central, and western areas at a significance level of 1%, but there are differences in the impact direction. The coefficient of "*lnurb*" is significantly negative in the eastern and central areas and is significantly positive in the western area. The eastern and western areas have superior geographical location and are economically developed. The improvement of urbanization level releases the surplus labor force in rural areas to engage in the secondary and tertiary industries, which increases the emission of environmental pollution, thus restraining the green development of agriculture. The urbanization level in the western area is in the initial stage, and the improvement of urbanization level gathers some social resources, which is conducive to accelerating the promotion of the green development of agriculture.

Considering the impact of the variable "*lnmech*" on the level of agricultural green development, the impact of agricultural mechanization level on the green development of agriculture in the eastern and western areas is positive at the significance level of 5% and 1%, respectively. In the process of agricultural production, agricultural machinery can improve agricultural production efficiency, improve land productivity, save labor, and promote the green development of agriculture.

Regarding the impact of the variable "*lnrnd*" on the level of agricultural green development, the impact of scientific and technological R&D investment on the green development of agriculture in the eastern and western areas is positive at the significance level of 1%. Investment in scientific and technological research and development can effectively promote technological progress, and technological progress provides guarantee for the entire process of agricultural production, as well as improving the agricultural production efficiency in the meantime, which can promote the green development of agriculture.

With regards to the impact of the variable "*lnland*" on the level of agricultural green development, the impact of arable land area on the green development of agriculture in the eastern, central, and western areas is negative at the significance level of 1%. In recent years, with the acceleration of industrialization and urbanization, the arable land area has a trend of continuous decrease, but the increase of the utilization rate and output rate of arable land has played a significant positive role in the green development of agriculture.

Considering the impact of the variable "$W \times$*lnurb*" on the level of agricultural green development, the urbanization level in adjacent provinces is significantly negative in the western area at a significance level of 1%, indicating that the regional spillover effect of urbanization on agricultural green development is negative. The improvement of urbanization level in adjacent provinces has a restraining effect on the local green development of agriculture for the western area. The possible reason is that the development of urbanization in the western area can attract agricultural surplus labor in adjacent provinces, which has led to a decline in the level of agricultural development in the local area.

Concerning the impact of the variable "$W \times$*lnmech*" on the level of agricultural green development, the agricultural mechanization level in adjacent provinces is significantly negative in the eastern and central areas at a significance level of 5% and 1%, respectively, indicating that the regional spillover effect of agricultural mechanization level on agricultural green development is negative. The improvement of agricultural mechanization level in adjacent provinces has a restraining effect on the local green development of agriculture for the eastern and central area.

Regarding the impact of the variable "*W*×*lnrnd*" on the level of agricultural green development, the scientific and technological R&D investment in adjacent provinces is significantly negative in the central area at a significance level of 5%, indicating that the regional spillover effect of scientific and technological R&D investment on agricultural green development is negative. The increase of scientific and technological R&D investment in adjacent provinces has a restraining effect on the local green development of agriculture for the central area.

Finally, with regards to the impact of the variable "*W*×*lnland*" on the level of agricultural green development, the arable land area in adjacent provinces is significantly negative in the central area at a significance level of 5%, indicating that the regional spillover effect of arable land area on agricultural green development is negative. The increase of arable land area in adjacent provinces has a restraining effect on the local green development of agriculture for the central area.

4. Conclusions and Policy Implications

4.1. Conclusions

Firstly, we evaluated the level of agricultural green development in China from 2007 to 2018, and the results showed that the level of agricultural green development in China presented a fluctuating upward trend, and there are differences among provinces. On the whole, the level of agricultural green development in the three major regions presented the characteristics of Eastern > Central > Western.

Secondly, we explored the spatial relationship of the level of agricultural green development among provinces in China. The results showed that China's provincial agricultural green development (AGD) level has an obvious positive autocorrelation in spatial distribution, showing the significant characteristics of spatial agglomeration in space

Finally, the Spatial Durbin Model was used to analyze the influencing factors of agricultural green development in China. The results showed that, from the national perspective, the urbanization level, agricultural mechanization level, scientific and technological R&D investment, and the arable land area had significant impact on the level of agricultural green development in the local and adjacent areas. From the regional perspective, in the eastern area, the four factors had a significant impact on the local green development agriculture, and the agricultural mechanization level in adjacent areas had a negative significant impact on the local green development of agriculture. In the central area, the urbanization level and arable land area had a significant impact on the local green development of agriculture, and the mechanization level, scientific and technological R&D investment, and arable land area in adjacent areas had a negative significant impact on the local green development of agriculture. In the western area, the four factors had a significant impact on the local green development of agriculture, and the urbanization level in adjacent areas had a significant negative impact on the local green development of agriculture.

4.2. Policy Implications

On the basis of the above results, several relevant policy implications can be derived, as follows:

First, agricultural green development is a systematic and complex issue. Therefore, China should introduce an authoritative and scientific agricultural green development level evaluation system and strengthen the monitoring of various data such as agricultural resources and the environment. In this way, China can make dynamic evaluation of agricultural green development in various provinces, which is helpful for the timely issue policies to promote the green transformation of agriculture.

Second, China should accelerate the integration of urban and rural areas, promote the integrated development of the three major industries, and optimize the structure; in addition, paying more attention to the preference for agriculture, so as to realize green efficiency in agricultural production and stable growth of farmers' income.

Third, China should increase the scientific and technological R&D investment in the whole process of agricultural production, including agricultural machinery, agricultural intelligent equipment, and water-saving and energy-saving equipment, etc.

Fourth, the stability of the amount of arable land resources is the most basic element to ensure national food security. It is necessary to establish a strict protection system to protect the red line of 1.8 billion mu of arable land. While ensuring the quantity of arable land resources, the quality of soil should also be improved to ensure the stability and quality of arable land.

5. Limitations and Future Work

There are some limitations in this article. First, this study focuses on the estimation of the agricultural green development and its influencing factors in China based on the static Spatial Durbin Model, but the long-term dynamic association may be different or more complex. Therefore, the dynamic Spatial Durbin Model with time series is also worth discussing. Second, the spatial heterogeneity of China's agricultural green development is only divided into three regions of the eastern, the central, and the west, therefore there may be a spatial spillover effect bias problem in this article. If the geographical regions of China are divided into more regions, the influence of various factors on different regions may be changed.

In the future research, additional aspects should be considered. First, we will incorporate the time series factor into the model and discuss the time and spatial heterogeneity of agricultural green development in China with the dynamic panel data under the consideration of time and space. Second, we will subdivide the geographical regions of China from three major regions into six geographical regions to make the research conclusions more accurate.

Author Contributions: Conceptualization, H.G. and C.P.; methodology, H.G. and S.X.; software, H.G. and S.X.; validation, H.G. and S.X.; formal analysis, H.G.; resources, H.G.; data curation, C.P.; writing—original draft preparation, H.G.; writing—review and editing, S.X. and C.P.; visualization, H.G.; supervision, C.P.; project administration, H.G.; funding acquisition, H.G. All authors have read and agreed to the published version of the manuscript.

Funding: This research was funded by Social Science Fund Project of Jilin Province, China, grant No.2017B032; MOE (Ministry of Education in China) Project of Humanities and Social Sciences, grant/award No.18YJC630128; Social Science Fund Project of Jilin Province, China, grant No.2018BS33; and Social Science Fund Project of "the 13th Five-Year" of Education Department of Jilin Province, China, grant No.JJKH20190736SK.

Conflicts of Interest: The authors declare no conflict of interest.

References

1. Kanter, D.R.; Musumba, M.; Wood, S.L.R.; Palm, C.; Antle, J.; Balvanera, P.; Dale, V.H.; Havlik, P.; Kline, K.L.; Scholes, R.J.; et al. Evaluating agricultural trade-offs in the age of sustainable development. *Agric. Syst.* **2018**, *163*, 73–88. [CrossRef]
2. Ma, L.; Zhang, W.F.; Ma, W.Q.; Velthof, G.L.; Oenema, O.; Zhang, F.S. An Analysis of Developments and Challenges in Nutrient Management in China. *J. Environ. Qual.* **2013**, *42*, 951–961. [CrossRef] [PubMed]
3. Yang, H.S. Resource management, soil fertility and sustainable crop production: Experiences of China. *Agric. Ecosyst. Environ.* **2006**, *116*, 27–33. [CrossRef]
4. Liu, X.B.; Zhang, X.Y.; Herbert, S.J. Feeding China's growing needs for grain. *Nature* **2010**, *465*, 420. [CrossRef] [PubMed]
5. Guo, J.H.; Liu, X.J.; Zhang, Y.; Shen, J.L.; Han, W.X.; Zhang, W.F.; Christie, P.; Goulding, K.W.T.; Vitousek, P.M.; Zhang, F.S. Significant Acidification in Major Chinese Croplands. *Science* **2010**, *327*, 1008–1010. [CrossRef] [PubMed]
6. Liu, X.J.; Zhang, Y.; Han, W.X.; Tang, A.H.; Shen, J.L.; Cui, Z.L.; Vitousek, P.; Erisman, J.W.; Goulding, K.; Christie, P.; et al. Enhanced nitrogen deposition over China. *Nature* **2013**, *494*, 459–462. [CrossRef] [PubMed]

7. Stoll-Kleemann, S.; Schmidt, U.J. Reducing meat consumption in developed and transition countries to counter climate change and biodiversity loss: A review of influence factors. *Reg. Environ. Chang.* **2017**, *17*, 1261–1277. [CrossRef]
8. Firbank, L.G. Towards the sustainable intensification of agriculture-a systems approach to policy formulation. *Front. Agric. Sci. Eng.* **2020**, *7*, 81–89. [CrossRef]
9. Schmidt-Traub, G. The urgency of Agriculture Green Development. *Front. Agric. Sci. Eng.* **2020**, *7*, 108–109. [CrossRef]
10. Shen, J.B.; Zhu, Q.C.; Jiao, X.Q.; Ying, H.; Wang, H.L.; Wen, X.; Xu, W.; Li, T.Y.; Cong, W.F.; Liu, X.J.; et al. Agriculture Green Development: A model for China and the world. *Front. Agric. Sci Eng.* **2020**, *7*, 5–13. [CrossRef]
11. Nambiar, K.K.M.; Gupta, A.P.; Fu, Q.L.; Li, S. Biophysical, chemical and socio-economic indicators for assessing agricultural sustainability in the Chinese coastal zone. *Agric. Ecosyst. Environ.* **2001**, *87*, 209–214. [CrossRef]
12. Quintero-Angel, M.; Gonzalez-Acevedo, A. Tendencies and challenges for the assessment of agricultural sustainability. *Agric. Ecosyst. Environ.* **2018**, *254*, 273–281. [CrossRef]
13. Rigby, D.; Caceres, D. Organic farming and the sustainability of agricultural systems. *Agric. Syst.* **2001**, *68*, 21–40. [CrossRef]
14. Taylor, D.; Mohamed, Z.; Shamsudin, M.N.; Mohayidin, M.G.; Chiew, E. Creating a farmer sustainability index: A Malaysian case study. *Am. J. Altern. Agric.* **1993**, *8*, 175–184. [CrossRef]
15. Van Calker, K.J.; Berentsen, P.B.M.; Romero, C.; Giesen, G.W.J.; Huirne, R.B.M. Development and application of a multi-attribute sustainability function for Dutch dairy farming systems. *Ecol. Econ.* **2006**, *57*, 640–658. [CrossRef]
16. Zhen, L.; Routray, J.K. Operational indicators for measuring agricultural sustainability in developing countries. *Environ. Manag.* **2003**, *32*, 34–46. [CrossRef]
17. Wei, Q.; Zhang, B.; Jin, S. A Study on Construction and Regional Comparison of Agricultural Green Development Index in China. *Issues Agric. Econ.* **2018**, *11*, 11–20.
18. Gong, Q.; Li, X. Construction and Measurement of Agricultural Green Development Index: 2005–2018. *Reform* **2020**, *1*, 133–145.
19. Gong, X.; Luo, R. The Evaluation of Agricultural Green Development Ability in Western Regions from the Perspective of Precise Poverty Alleviation. *Ecol. Econ.* **2018**, *34*, 128–132.
20. Jia, Y.; Zhao, B.; He, Z.; Zhang, Z.; Zhang, R. Research on evaluation and direction of the green development of agriculture in Henan province. *J. Henan Agric. Univ.* **2019**, *53*, 823–830.
21. Jin, S. Evaluation of provincial agricultural green development level and regional differences in China. *Seeker* **2019**, 89–95. [CrossRef]
22. Zhao, H.; Yu, F. Evaluation of Agricultural Green Development Level in Main Grain Producing Areas based on Entropy Method. *Reform* **2019**, *11*, 136–146.
23. Zhou, L. Research on the Green Development of Tibetan Agriculture under the Background of Rural Revitalization. *Northwestern J. Ethnol.* **2019**, 116–127. [CrossRef]
24. Carson, H.L. Silent Spring. *Coll. Univ.* **1963**, *38*, 294–296.
25. Davies, W.J.; Shen, J.B. Reducing the environmental footprint of food and farming with Agriculture Green Development. *Front. Agric. Sci. Eng.* **2020**, *7*, 1–4. [CrossRef]
26. Godfray, H.C.J.; Beddington, J.R.; Crute, I.R.; Haddad, L.; Lawrence, D.; Muir, J.F.; Pretty, J.; Robinson, S.; Thomas, S.M.; Toulmin, C. Food Security: The Challenge of Feeding 9 Billion People. *Science* **2010**, *327*, 812–818. [CrossRef]
27. Tu, Z.; Gan, T. Research on the Regional Difference and Dynamic Mechanism of Agricultural Green Development in China. *Wuhan Univ. J. (Philos. Soc. Sci.)* **2019**, *72*, 165–178.
28. Koohafkan, P.; Altieri, M.A.; Gimenez, E.H. Green Agriculture: Foundations for biodiverse, resilient and productive agricultural systems. *Int. J. Agric. Sustain.* **2012**, *10*, 61–75. [CrossRef]
29. Leimona, B.; Amaruzaman, S.; Arifin, B.; Yasmin, F.; Hasan, F.; Agusta, H.; Sprang, P.; Jaffee, S.; Frias, J. Indonesia's 'Green Agriculture' Strategies and Policies: Closing the Gap Between Aspirations and Application. In *Bogor, Indonesia: World Agroforestry Centre and World Bank*; World Agroforestry Centre: Nairobi, Kenya, 2015.
30. Yu, J.L.; Wu, J. The Sustainability of Agricultural Development in China: The Agriculture-Environment Nexus. *Sustain-Basel* **2018**, *10*, 1776. [CrossRef]

31. Jin, S.Q.; Ma, H.Y.; Huang, J.K.; Hu, R.F.; Rozelle, S. Productivity, efficiency and technical change: Measuring the performance of China's transforming agriculture. *J. Prod. Anal.* **2010**, *33*, 191–207. [CrossRef]
32. Yu, X.R. Promoting Agriculture Green Development to realize the great rejuvenation of the Chinese nation. *Front. Agric. Sci. Eng.* **2020**, *7*, 112–113.
33. Zhang, F.S.; Dong, W.C.; Li, J.Q. Creating a new university-education system to promote Agriculture Green Development. *Front. Agric. Sci. Eng.* **2020**, *7*, 114–116. [CrossRef]
34. Zhang, J.J.; Cui, S.L.; Ma, L.; Meng, F.L.; Song, C.Y.; Li, Y.M.; Ma, W.Q. Construction of a green development index system for agriculture in China and examples. *Chin. J. Eco-Agric.* **2020**, *28*, 1113–1126.
35. Jiang, Y. *Level of Development of Hunan Agricultural Green Measure and Development Path Selection Research*; Hunan Agricultural University: Changsha, China, 2018.
36. Ren, H.J.; Liu, Y.F.; Ding, S.X. Research on the Measurement of Agricultural Green Development Level in Qinghai Province. *IOP Conf. Ser. Earth Environ.* **2019**, 267. [CrossRef]
37. Cheng, Y. *Study on Evaluation of Agricultural Green Development in Shanxi Province*; Shanxi University of Finance and Economics: Taiyuan, China, 2019.
38. Yang, X. *Study on the Measurement and Promotion Path of Agricultural Green Development Level in China*; Hebei University of Economics and Business: Shijiazhuang, China, 2019.
39. Guo, M. *Reasearch of Agricultural Green Development and A Preliminary study on the Evaluation System*; Beijing Forestry University: Beijing, China, 2011.
40. Jing, P.; Zhao, W.; Zheng, Q.; Zhang, D. Study On Dynamic Prediction and Path of Green Development Level of Agriculture in Anhui Province. *Chin. J. Agric. Resour. Regional. Plan.* **2018**, *39*, 51–56.
41. Zhang, N.; Zhang, L.; Zhao, H.; Han, Y.; Duan, Y. Construction and Application of Evaluation Index System for Agricultural Green Development. *Ecol. Econ.* **2018**, *34*, 21–24+46.
42. Lin, X.; Wang, D.; Ren, W.; Liu, Y. Research on the mechanism of urbanization to economic increase in China. *Geogr. Res.* **2013**, *32*, 691–700.
43. Wang, B.; Tang, W.; Wu, Y.; Zhang, N. Does Urbanization Increase China's Green Development Efficiency? *Econ. Rev.* **2014**, *4*, 38–49.
44. Xue, L.; Shen, Y.; Xu, C. A Research on Spillover Effects of Agricultural Agglomeration on Agricultural Green Development Efficiency. *Econ. Surv.* **2020**, *37*, 45–53.
45. Yang, L.; Sun, Z. The Development of Western New-type Urbanization Level Evaluation Based on Entropy Method. *Econ. Probl.* **2015**, 115–119. [CrossRef]
46. Moran, P.A.P. The Interpretation of Statistical Maps. *J. R. Stat. Soc. Ser. B Stat. Methodol.* **1948**, *37*, 243–251. [CrossRef]
47. Anselin, L. Local Indicators of Spatial Association-Lisa. *Geogr. Anal.* **1995**, *27*, 93–115. [CrossRef]
48. Xu, S.; Li, Y.; Tao, Y.; Wang, Y.; Li, Y. Regional Differences in the Spatial Characteristics and Dynamic Convergence of Environmental Efficiency in China. *Sustain.-Basel* **2020**, *12*, 7423. [CrossRef]

Publisher's Note: MDPI stays neutral with regard to jurisdictional claims in published maps and institutional affiliations.

© 2020 by the authors. Licensee MDPI, Basel, Switzerland. This article is an open access article distributed under the terms and conditions of the Creative Commons Attribution (CC BY) license (http://creativecommons.org/licenses/by/4.0/).

sustainability

MDPI

Article

Impacts of Climate and Phenology on the Yields of Early Mature Rice in China

Yahui Guo [1], Wenxiang Wu [2,3,*], Yumei Liu [4], Zhaofei Wu [1], Xiaojun Geng [1], Yaru Zhang [1], Christopher Robin Bryant [5,6] and Yongshuo Fu [1,7,*]

1 College of Water Sciences, Beijing Normal University, Beijing 100875, China; guoyh@mail.bnu.edu.cn (Y.G.); 202031470020@mail.bnu.edu.cn (Z.W.); xgeng@mail.bnu.edu.cn (X.G.); 201921470030@mail.bnu.edu.cn (Y.Z.)
2 Key Laboratory of Land Surface Pattern and Simulation, Institute of Geographic Sciences and Natural Resources Research, Chinese Academy of Sciences, Beijing 100101, China
3 CAS Center for Excellence in Tibetan Plateau Earth Sciences, Chinese Academy of Sciences (CAS), Beijing 100101, China
4 College of Agricultural and Biological Engineering, Heze University, Heze 274015, China; liuyumei02@caas.cn
5 School of Environmental Design and Rural Development, University of Guelph, Guelph, ON N1G2W5, Canada; chris.r.bryant@umontreal.ca
6 Département de Géographie, Université de Montréal, Montréal, QC H2V2B8, Canada
7 Plants and Ecosystems Research Group (PLECO), University of Antwerp, B-2106 Antwerp, Belgium
* Correspondence: wuwx@igsnrr.ac.cn (W.W.); yfu@bnu.edu.cn (Y.F.)

Received: 16 October 2020; Accepted: 1 December 2020; Published: 4 December 2020

Abstract: Phenological variables are closely correlated with rice (*Oryza sativa* L.) yields as they play important roles in influencing and controlling the carbon allocations between plant organs, but their impacts on rice yields and their relative importance compared with climatic variables are not yet well investigated. In this study, the impacts and the relative importance of climatic and phenological variables on the yields of early mature rice were assessed using the trial data from 75 agricultural stations across China, spanning from 1981–2010. We found that both daily maximum (Tmax) and daily minimum (Tmin) temperatures during the growing season (from transplanting to maturity) increased significantly, while sunshine duration (SD) and precipitation (Prep) did not change significantly. The average transplanting date was advanced by 3.18 days/decade, and the heading (maturity) dates were delayed by 2.47 (4.55) days/decade, with yields significantly increased by 9.65 g/m^3 per decade across all sites. Partial correlation coefficients between most phenological variables and rice yields were negative, whereas most of the climatic variables were positively correlated with rice yields. The average of partial correlation coefficients between transplanting, heading, and maturity dates and rice yields were −0.10, −0.15, and −0.01, respectively, and the average of coefficients between Tmax, Tmin, SD, and Prep and rice yields were 0.08, 0.02, 0.12, and −0.05, respectively. Interestingly enough, phenological variables were the dominating influencing factors on rice yields at 63% of the sites, suggesting that the relative importance of phenology to rice yields may be even higher than that of climate. The climatic variables were closely correlated with rice yields as they are fundamental growth materials for crops, and phenological variables strongly influenced the growth and development of rice. Our results highlight that phenology should be precisely evaluated in crop models to improve the accuracy of simulating their response to climate change. Furthermore, due to limited understanding of phenological processes, manipulative experiments are urgently needed to comprehensively improve our understanding of rice phenology and rice yield response to ongoing climate change.

Keywords: rice yields; climate change; phenology; relative contribution; partial correlation

1. Introduction

The average global temperature has increased by roughly 1 °C since 1950, and such warming and associated climatic changes have been demonstrated to influence agricultural growth, limiting potential production and threatening global food availability and stability [1–4]. The Intergovernmental Panel on Climate Change (IPCC) has released a special report on the impacts of global warming of 1.5 °C above preindustrial levels between 2030 and 2050 [5–9]. Many studies have shown that a warming climate could reduce agricultural yields by accelerating development and shortening the growth duration of crops [10–13]. Meanwhile, phenological shifts are also important ecological indicators that influence crop growth and determine agricultural yields [14–16]. Phenological variables are very important indicators for rice yields as they occupy fundamental roles in dominating and controlling the carbon allocation between plant organs [17–20]. However, the impacts of phenological variables on crop yields and their relative importance compared with climatic variables are seldomly evaluated or investigated. Therefore, it is essential to quantitatively study the response of crop phenology, yield, and agricultural productivity to climate change and to quantify the relative importance of phenology and climate on crop yields. Meanwhile, it is of great importance to improve our understanding of how to acknowledge crop development, optimize field management, and take adaptive measures to mitigate the negative effects under an ongoing changing climate.

The impacts of climate change on agricultural production may be more apparent in developing countries because of their greater reliance on agriculture and the lack of advanced management practices [21–23]. Guo et al. simulated the yields of rice from 2016 to 2099 in Zhejiang province, China, and the results implied that the rice yields would be inevitably reduced by the warming climate [24]. Xu et al. modeled the spatial and temporal change of rice yields in the Sichuan Basin, China, and the results indicated that adaptive measures should be adopted in the near future to countermeasure the negative impacts of the warming climate [25]. Peng et al. investigated the trends of arable land, consumer demand, and climate change in China during the past decades and found that new technology should be adopted to improve rice production to meet future food demand in the 2030s [26]. Liu et al. found that the change of phenology of spring and winter wheat in China was closely influenced by climate change [27]. As the largest developing economy, China feeds 22% of the world's population with only 9% of the world's arable land [28]. A warming climate has been detected here and is expected to continue for decades; thus, investigating and understanding the mechanisms of climate change and phenology on agricultural production in China should be a high priority [29–33].

Studies based on large-scale analyses have tried to investigate the mechanisms of climate change on crop growth and yields using process-based simulation models. However, these models are commonly limited by the uncertainties in predictions of the overall conditions of crop growth, and they cannot incorporate all circumstances, such as pest control and extreme weather [34–41]. In addition, crop models can simulate the impacts of climate change only with defined cultivar characteristics, management practices, soil properties, and initial conditions. Moreover, the crop models can hardly simulate all circumstances under the changing climate, such as changes in sowing dates, crop varieties, and management practices. Crop trial data usually contain crop cultivars, phenology dates, yields, and management practices, as well as weather information [42]. Over the past few decades, statistical approaches, integrated with the long time series of high-quality crop trial data, have been applied to understand the impacts of climate change variability on crop growth and yields [43,44]. The statistical approaches have been quite suitable for investigating how both climate change and phenology have affected crop growth and yield by removing the impact of other influencing variables [45]. Thus, the high-quality data, combined with the statistical approaches, have provided a chance to explore the inner mechanism of impacts of climate and phenology on crops. Tao et al. assessed the impacts of climate change on rice using partial correlation analysis and found that growth duration was prolonged by changing rice cultivars and the rice yields were significantly influenced by climate change [46]. Wang et al. calculated growing degree day (GDD) data and evaluated the impacts of climate change on rice; then, the linear regression method was conducted to explore the constraints factors to rice yields [47]. Pirjo et al. explored the coincidence of changes in yields of rice and variations of climate in Europe using statistical methods, including linear regression analysis and random coefficient regression [48]. Although previous studies have explored the impacts of climate change on rice growth and rice yield using statistical approaches with long-time trial data, their research only focused on the impacts of climate on agricultural production. The influences from phenology were ignored, and, therefore, the results can hardly reveal the real mechanisms of climate change or phenology impacts on agricultural production. In agriculture, phenology represents an important indicator that controls crop growth, which will further impact agricultural production [49–51]. However, almost no studies have calculated the relative importance of phenological variables on rice yields rather than that of climatic variables. Therefore, some concerns need to be explored, such as the impacts of phenology on rice yields, and the relative importance of phenological variables (transplanting, heading, and maturity dates) to rice yields compared with climatic variables (temperature, precipitation, sunshine durations). Since the phenology impacts on agricultural production are unknown and the relative importance between phenology and climate is unknown, it is of vital importance to improve the understanding of the potential correlation relationships between phenology and agricultural production.

In this study, long-term trial data, including climate, phenology, and yields of early mature rice spanning the years from 1981 to 2010 from 75 agricultural meteorological stations across China, were adopted for data analysis. The stations are evenly distributed across two major rice cultivation regions in southern China. Climatic data from weather stations included the temperature, sunshine duration, and precipitation. Phenology data contained the transplanting, heading, and maturity dates. The yield of early mature rice was observed and recorded in the format of kg/ha by professional technical workers. We aim here to (1) explore spatial and temporal changes and trends in phenology, climate, and yield of early mature rice over the past three decades; (2) quantify the relative importance of phenological and climatic variables to yields of early mature rice using partial correlation analysis.

2. Materials and Methods

2.1. Rice Trial Data and Climate

The study area included two major early-mature rice production regions in China, of which one is in the middle and lower reaches of the Yangtze River (YZ) and the other is in the Pearl River Delta (PRD). The YZ region contains Anhui, Jiangxi, Zhejiang, Hubei, and Hunan provinces (49 sites), and the PRD region contains the Guangdong and Guangxi provinces (26 sites; Figure 1). These agro-meteorological experimental stations are maintained by professional agricultural technicians following standardized observation criteria and strictly prescribed procedures under the control of the Chinese Meteorological Administration (CMA) [52].

The long trial data of early mature rice contains the information of rice cultivars, transplanting dates, heading dates, maturity dates, yields, and management practices at each site. Therefore, these data provided us with a unique chance to match weather variables with farm-specific planting and harvesting dates and rice growth phases for every year. At the same time, the data were all very valuable for investigating the impacting mechanisms from climate change and phenology on rice yields. The average transplanting and maturity dates of the sites were, respectively, 114 and 175 (day of year, DOY) for early mature rice. The rice was well irrigated and fertilized during the growth periods, with strict pest control. The detailed phenological records of transplanting, heading, and maturity dates in DOY format at 75 sites from 1981 to 2010 were selected as the phenological events to be investigated for determining the relationships between phenology and the yields of early mature rice. The climate data containing daily maximum (Tmax) and daily minimum (Tmin) temperatures, sunshine duration (SD), and precipitation (Prep) of each site during the growing seasons from 1981 to 2010 were obtained from the China Meteorological Data Sharing Service System (http://data.cma.cn/). The climatic data were applied for investigating the impacts of climate on rice yields and its relative importance compared with phenology.

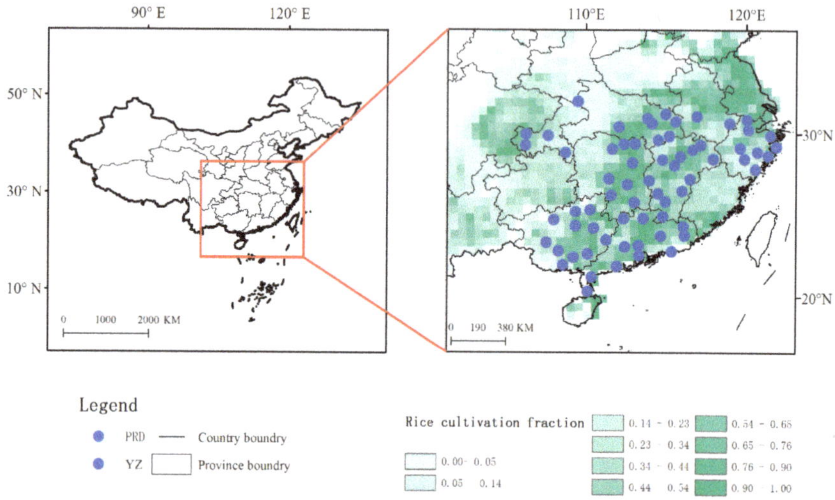

Figure 1. Spatial distributions of agro-meteorological experimental stations across China. Note: The blue dots refer to agro-meteorological experimental stations; the thick lines represent the boundary of the nation and the thin lines represent the boundaries of the provinces; the shaded areas indicate the proportion of rice cultivation.

2.2. Data Analysis Method

The agro-meteorological experimental stations with more than five years of records of crop growth and weather parameters were selected for further data analysis. The rice growth period (RGP) was defined as the days between transplanting dates and maturity dates of early mature rice; the average values of the climatic variables (Tmax, Tmin, SD, and Prep) during RGPs of all sites were calculated. Then, the linear regression method (LRM) was applied to investigate the change of climatic variables during RGPs from 1981 to 2010 [53–55]. Similarly, to identify the temporal changes in phenology and yields of early mature rice during the past decades, the average dates of the phenological variables (transplanting dates, heading dates, and maturity dates) and yields of all sites were calculated as well. Meanwhile, LRM was also applied to show the temporal changes.

To investigate the correlations between phenological variables and climatic variables during RGPs and annual yields across all sites, the yields and phenological variables were linearly detrended to counter the effects that were due to the differences in rice cultivars and management practices. The climatic variables of Tmax and Tmin were accumulated during RGPs at each site over the study period, and the climatic variables of SD and Prep were averaged. Partial correlation analysis was applied to investigate the relationship between the detrended yields and the three detrended phenological variables (transplanting, heading, and maturity dates) and the four detrended climatic variables (Tmax, Tmin, SD, and Prep) during the period 1981–2010, respectively. The absolute values of coefficients of the partial correlation analysis were calculated, and the coefficients (phenological or climatic variable) with the highest absolute value were confirmed as the main influencing factor for each site. For each site, there was only one dominating influencing factor that was most correlated with rice yields, and the spatial distribution of the main influencing factor for each site was obtained. Then, the relative importance of phenological and climatic variables was calculated using the percentage of the number of total dominating phenological variables (climatic) of all sites.

3. Results

3.1. Change and Trend of Climatic Variables during Rice Growth from 1981–2010

The average values of the climatic variables (Tmax, Tmin, SD, and Prep) at all sites during RGPs from 1981–2010 are shown in Figure 2. The ratios of the equation using LRM were 0.052 and 0.040 for Tmax ($p < 0.01$) and Tmin ($p < 0.01$), which meant that the average daily temperatures increased significantly during RGPs over the study period, and the increasing trend of average daily Tmax was a slightly higher than Tmin. The increasing temperature may indicate that a warming climate is occurring in most of the sites in China, which may have influenced the yields of early mature rice. In contrast, decreasing trends in SD and Prep were detected using LRM, but the *p*-values were not significant.

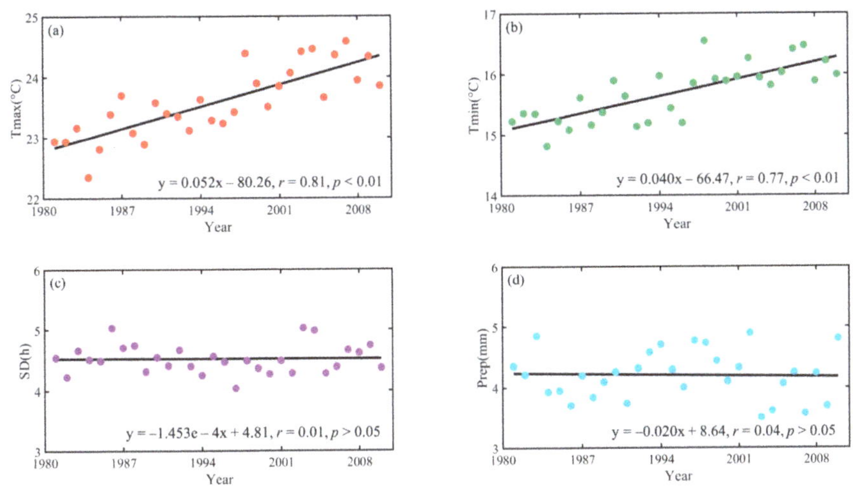

Figure 2. Trends of daily maximum temperature (Tmax), daily minimum temperature (Tmin), sunshine duration (SD), and precipitation (Prep) across all sites during the study period 1981–2010. Note: (**a–d**) represents Tmax, Tmin, SD, and Prep, respectively.

3.2. Averages and Trends of Phenological Variables of Rice

The average dates of the phenological variables (transplanting, heading, and maturity dates) and yields of early mature rice at each site during the study period were obtained (Figure 3). The average transplanting date was 114.94, 110.19, and 123.63 day of year (DOY) for all sites, YZ, and PRD, respectively. Similarly, the average heading date was 175.38, 172.94, and 181.26 DOY for all sites, YZ, and PRD, respectively, with the average maturity date being 197.89, 193.74, and 205.20 DOY, respectively. It can be concluded that the phenological variables of all sites in PRD were commonly later than those in YZ. The average value of yields was 910.59, 806.78, and 580.61 g/m^3 for all sites, YZ, and PRD, respectively. The average yield at sites in YZ was generally higher than those in PRD.

Using LRM, the trends of phenological variables and yields of all sites during the study period were calculated (Figure 4). The average trends of transplanting dates for all sites, YZ, and PRD were −3.18, −3.11, and −3.15 days/decade, respectively. The average trend of heading dates for all sites, YZ, and PRD were 0.58, 0.76, and 0.78 days/decade, respectively. The average trend of maturity dates for all sites, YZ, and PRD, were 2.47, 5.39, and −0.63 days/decade, respectively. The average trends of yields for all sites, YZ, and PRD were 4.55, 3.90, and 9.65 g/m^3 per/decade, respectively. The spatial results in Figure 4 indicate that transplanting dates at most sites had advanced, while all heading dates and maturity dates were delayed over the study period.

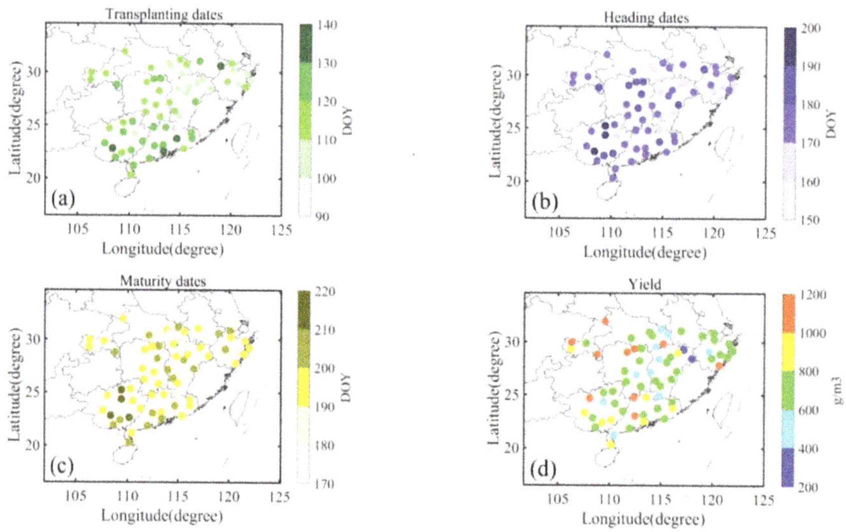

Figure 3. Average transplanting dates, heading dates, maturity dates, and yields across all sites during 1981–2010. Note: (**a**) represents the transplanting dates, (**b**) represents the heading dates, (**c**) represents the maturity dates, and (**d**) represents the rice yield.

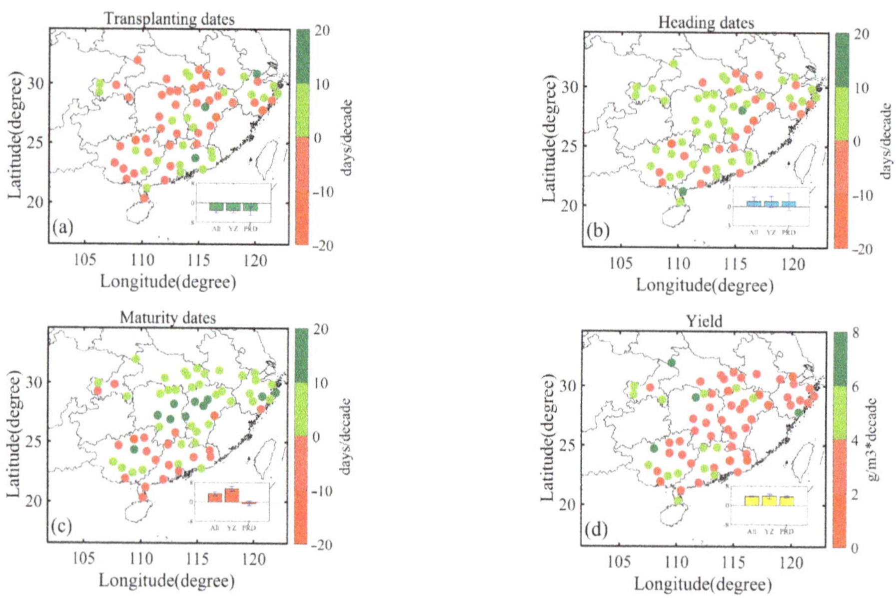

Figure 4. Spatial trends of transplanting dates, heading dates, maturity dates, and yields across sites during the period 1981–2010. Note: (**a**) represents the transplanting dates, (**b**) represents the heading dates, (**c**) represents the maturity dates, and (**d**) represents the rice yield respectively.

3.3. Partial Correlation Analysis

Since the impacts of three phenological variables (transplanting, heading, and maturity dates) and four climatic variables (Tmax, Tmin, SD, and Prep) were relatively large, these seven variables were selected to confirm the main influencing factors using the method introduced in the Methods section. The detailed results are shown in Figure 5, in which the dominating phenological (climatic) variables are shown in green (red) to show the different main influencing factors, respectively. Since there had been misalignments of the phenological variables, climatic variables, and yields, only 44 sites were chosen for the final results in this analysis. The phenological variables painted with green are the dominating influencing factors compared with climatic variables painted with red, which correlate with the yields of early mature rice. Thus, the phenological variables (transplanting, heading, and maturity dates) played important roles in influencing rice yields at most sites in China during the study period. When undertaking a deeper investigation, sites in PRD were found to be most impacted by phenological variables, while sites in YZ were impacted by both phenological and climatic variables.

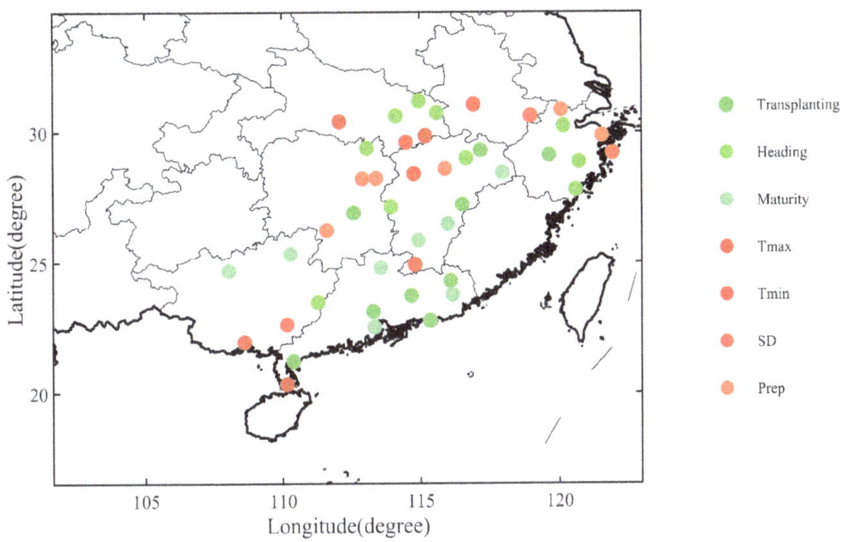

Figure 5. Spatial distribution of partial correlation coefficients between phenological and climatic variables and rice yields.

The average partial correlation coefficients between phenological variables, climatic variables, and yields were calculated (Figure 6). For all sites, the calculated coefficients of transplanting, heading, and maturity dates were −0.10, −0.15, and −0.01, respectively. The transplanting date had advanced at most sites, and the negative relationship may indicate that the transplanting date would have increased rice production. Meanwhile, the delayed phenological variables of heading and maturity dates and their negative relationships imply that they would have reduced rice production. In contrast, the average coefficients of Tmax, Tmin, SD, and Prep were 0.08, 0.02, 0.12, and −0.05, respectively, which implied that most climatic variables positively influenced rice yields and increased rice production. The average value of the coefficients was different for the two regions, and the phenological variables were all negative for YZ; the heading date was the only factor that was negatively correlated with yields for PRD. Thus, the influencing mechanisms of phenological variables on rice yield appear to be different for YZ and PRD. The influencing mechanisms of climatic variables on rice yield were also different, and the impacts of Tmax, SD, and Prep were all positive to rice yields, while Tmin was in a negative relationship with rice yields. Tmax was in a positive relationship with rice yields, perhaps because Tmax had not reached the highest temperature threshold for rice growth and development. SD and Prep promoted rice production, while Tmin may have reduced the yield of early mature rice by putting stress on sites in YZ. Tmin and SD were both in positive relationships with rice yields in PRD, indicating they had increased rice yields, while the negative Tmax may indicate the yield here was affected by stress from high temperatures and lack of water. In total, the phenological indicators played important roles, which were even greater than those of climatic variables, in affecting rice yields.

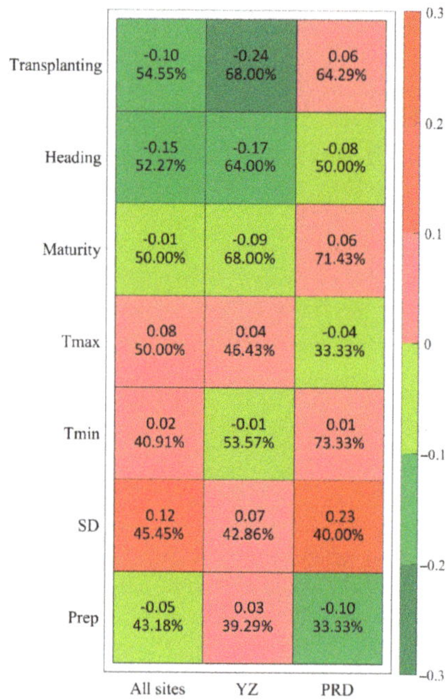

Figure 6. Average partial correlation coefficients between phenological and climatic variables and rice yields across all sites, Yangtze River (YZ), and Pearl River Delta (PRD), respectively.

To better investigate and confirm the dominating influencing factor on rice yields at each site, the absolute values of the coefficients of all variables were calculated and their relative importance was obtained. In order to better present relative importance, the results were classified into two categories: phenology and climate (Figure 7). The green color represents the phenological variables, while the red color represents the climatic variables. It can be clearly acknowledged that the sum of phenological variables was almost equal to the sum of climatic variables for all sites. However, for the two regions, the situations were significantly related to different influencing mechanisms. Phenological variables were the dominating influencing factors that actually played more important roles than climatic variables in YZ, while climatic variables were the dominating influencing factors that were more important than phenological variables in PRD.

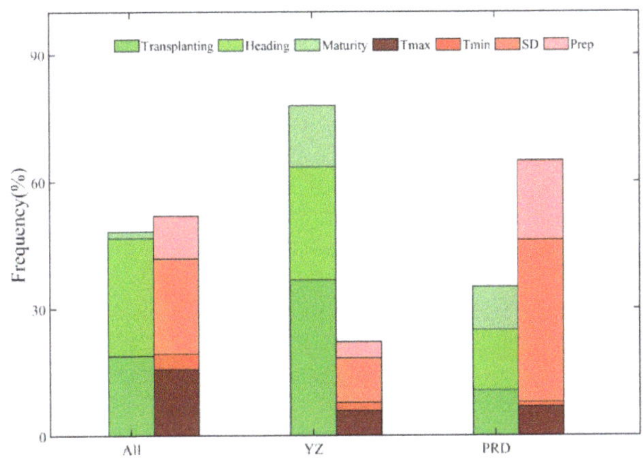

Figure 7. Relative importance of phenological variables and climatic variables on the yields of early mature rice.

4. Discussion

4.1. Spatial Changes of Phenological Variables and Impacts on Early Mature Rice during 1981–2010

Detailed trial data of early mature rice from the past three decades have been analyzed to investigate the impacts of multiple phenological variables on yields during these past decades under the changing climate. In this study, the rice at all sites were all transplanted after seeding at most sites; thus, the phenological variables have played important roles in affecting rice growth and rice yields. The change of phenological variables at 75 sites were used for the analysis, and it was found that the average transplanting date had advanced by 3.18 days/decade for early mature rice. This finding was consistent with previous studies that had also detected the advance of transplanting dates [49]. However, delayed dates in heading and maturity for early mature rice were also detected in this research; these dates were not consistent with previous studies. The difference may be related to the inconsistent total number of sites and the inconsistent length of study periods. The advancement of transplanting dates may primarily reflect manual intervention, and the delayed heading and maturity dates may be due to the adoption of high-yielding rice varieties of rice cultivars, which rely more on the cumulative effect of growing degree days (GDDs) compared to the previous commonly adopted rice cultivars.

The partial correlation coefficients between phenological variables and rice yields showed that phenological variables had relatively important impacts on the yields of early mature rice. Phenological variables were the dominating influencing factors for the yields of early mature rice in nearly 63% of the sites. The results of this study are in line with previous studies, where the phenological variables are important indicators that control the carbon allocation between plant organs, and the phenological events are different forms that represent vegetative growth and reproductive growth periods [56]. Phenological variables divided the whole growth period of rice into several different stages, and the carbon allocation of total organic matter to organs varied at different phenological durations. The negative impacts of earlier phenological variables on canopy radiation capture, biomass production, and grain yield for early mature rice were found, which were consistent with previous studies [57,58]. We further investigated the impacts of three important phenological events on rice yields and found that the phenological variables were negatively correlated with yields at most sites. The transplanting dates had advanced; in other words, the transplanting dates were earlier than in normal practice. Thus, it was likely that the earlier transplanting dates increased the potential risk of cold damage during rice reproduction stages, and the rice may have suffered from cooler temperatures during the growing periods of early mature rice. This result is consistent with a previous study on northern Japan, where cold stress on rice was found and negative impacts were also detected [59]. The delayed heading and maturity dates meant that these two phenological dates were mostly later than in normal practice. On the one hand, both Tmax and Tmin had increased significantly ($p < 0.05$), whereas the heading and maturity dates were prolonged rather than advanced. This may be due to an improvement of management practice and a change to new and advanced rice cultivars that have longer development periods [60]. The delayed heading and maturity dates may indicate that the rice may suffer from high-temperature stress during the important growth stages of early mature rice.

Even though partial correlation analysis was applied to exclude the effects of climatic variables when analyzing phenological impacts on rice yields, there were still uncertainties regarding management practice, weather variability, insects, and disease that could modify the trend and correlation analyses. It was quite difficult to isolate the influence of a single factor from the phenological variables that have influenced rice yields. For example, crop yields may have increased because of the increased usage of modern cultivars and technology during the study period. If so, the estimated decreases (increases) in crop productivity may be larger (smaller) than our estimates. Although there were limitations, we found that the phenological variables were all negatively correlated with the rice yields of early mature rice and the phenological variables had significantly affected the growth and yields of early mature rice. Since uncertainties still remain as to the method of statistics, we suggest that control experiments should be conducted to better investigate the causal mechanisms between climate, phenology, and rice yields in future related studies.

4.2. Spatial Changes of Climatic Variables and Impacts on Early Mature Rice during 1981–2010

Tmax and Tmin were significantly increased during the rice growth periods at all sites, and the detected ratio of Tmax was about 1.3 times of the corresponding Tmin during 1981–2010. Our finding was consistent with previous studies, where the increased rate of minimum temperature was found to be less than twice the increased rate of the corresponding maximum temperature during 1981–2000 and general warming trends were found during all growth periods across most sites [61]. Partial correlation analysis using data from all sites showed that both Tmax and Tmin were positively correlated with rice yields for all sites. Figure 1 shows the detailed locations of YZ and PRD, where YZ is in the subtropical region, while PRD is in the tropical region. The meteorological information shown in Figures A1 and A2 implies that there were different basic climatic variables and different trends in those two regions (YZ and PRD) during the study period. The results displayed in Figure 6 showed the Tmax was positively correlated with rice yields in YZ, while it was negatively correlated in PRD. This can be explained by the fact that the most suitable temperature interval for rice growth is from 27 to 32 °C, whereas the average Tmax in YZ was 22.07 °C; thus, Tmax was far from the threshold of maximum temperature for rice growth. Since rice is a thermophilic crop, a suitable temperature in daylight will benefit rice growth and accumulate dry matter [62]. For Tmin, there is evidence that when Tmin of pollen mother cells in the meiosis stage (microspore stage and meiosis fine line stage) is lower than 15–17 °C, it causes the degeneration of spikelets, inducing an increase in false grains and resulting in the delay of heading. The average Tmin of YZ (14.17 °C) was below the minimum threshold of temperature that can maintain the growth of early mature rice. The mechanism of temperature influence on rice yields was totally different in PRD because the basic temperature of PRD was about 5 °C higher than that in YZ; Tmax here was 26.15 °C and probably higher than the threshold of maximum temperature that rice can stand during the whole growth period of one single year. The rice in YZ was mostly influenced by the cold stress of Tmin, while the rice in PRD was mostly influenced by the heat stress of Tmax. This is consistent with a previous study, where the negative impacts on rice yields from cold stress were detected despite the consideration of CO_2 fertilization effects [63]. The mechanisms of Tmax and Tmin in influencing rice growth and rice yields are totally different. Further investigations of the effects of temperature (Tmax and Tmin) on the physiological processes governing rice growth and the impact on rice yields should be adopted; further analysis of adaptive measures should also be equally developed.

Similar to temperature, the average and trends of SD and precipitation for YZ and PRD were also different. Sun-loving rice is a C3 plant, and it does not have a critical light length for panicles, which requires higher light conditions than other crops, but its photosynthesis with an increase in illumination is not as obvious as that of C4 corn. The results of this study show that the average SD was 4.52 and 4.48 (hours/day) for YZ and PRD, respectively, and the SDs were in positive relationships with rice yields in both YZ and PRD. Even though the trend of SD of YZ in Figure A1 shows that the length of SD was decreasing, there was no doubt that SD was promoting rice growth and yields of early mature rice. As to the increasing trend of SD in PRD, shown in Figure A2, this means that SD will constantly be conducive to rice development. There is no doubt that the SD was promoting rice growth and rice yields in both YZ and PRD. Paddy rice wis the dominant agricultural crop in most sites, and the average precipitation was 3.93 and 4.62 (mm/day) in YZ and PRD, respectively. Precipitation was positively correlated with rice yields at sites in YZ because the rice here had suffered from water stress for a short period. However, PRD is full of water, and it was not likely that the rice had a lack of water there. More precipitation means less sunshine length at sites in this region. Thus, the impacts of precipitation on rice yields were also different in YZ and PRD. In this study, the climatic variables were each analyzed using partial correlation with yields; the climatic variables were highly intercorrelated with each other. In other words, Tmax and Tmin were correlated, and SD and precipitation were also correlated. We recommend the performance of integrated research instead of the impacts of single factors on rice yields when analyzing agricultural applications.

5. Conclusions

In this study, a significantly increasing trend of temperature was detected at most sites, whereas the not-significant decreasing trends of sunshine and precipitation were detected. Phenological variables, including transplanting, heading, and maturity dates, had significantly influenced the growth and yields of early mature rice in China. The phenological variables were actually controlling the carbon allocation between plant organs, which further influenced rice growth. Since the phenological variables of agriculture are controlled by both the changing climate and the improvement of management practices, the adaptation of better transplanting dates, which are mainly controlled by farmers, should be further explored. Climatic variables are also important as they are fundamental materials for maintaining the growth of rice. Heat stress was detected in PRD and cold stress was detected in YZ. SD increased the yields at most sites, and precipitation was positively correlated with rice yields in YZ but negatively correlated in PRD. The mechanisms were different for YZ and PRD because their basic climatic variables and trends were different. Although there were limitations and uncertainties arising from management practices, rice cultivars, and pests, this study provides empirical evidence that phenological variables (transplanting, heading, and maturity dates) are relatively important compared with climatic variables as they are responsible for carbon allocation. We also propose that more advanced statistical techniques, such as machine learning and deep learning, integrated with mechanism approaches such as crop models and combined with additional techniques such as remote sensing, should be used for the deeper investigation of impacts from both phenological and climatic variables as we are supposed to raise agricultural production by adjusting management practices and rice cultivars under the ongoing warming climate changes in temperature and precipitation.

Author Contributions: Conceptualization, Y.G. and W.W.; methodology, Y.G. and Y.F.; software, Y.G.; validation, Y.G., Y.L., Z.W. and X.G.; formal analysis, Y.G., X.G., Y.Z. and C.R.B.; investigation, Y.G., Z.W., X.G. and Y.Z.; resources, Y.G., W.W. and Y.F.; data curation, Y.G., W.W. and Y.L.; writing—original draft preparation, Y.G.; writing—review and editing, Y.G., W.W. and Y.F.; visualization, Y.G., Z.W., X.G. and Y.Z.; supervision, W.W. and Y.F.; project administration, W.W. and Y.F.; funding acquisition, W.W. and Y.F. All authors have read and agreed to the published version of the manuscript.

Funding: This study was supported by the General Program of the National Nature Science Foundation of China (Grant No. 31770516), the National Key Research and Development Program of China (2017YFA06036001), and the 111 Project (B18006) and Fundamental Research Funds for the Central Universities (2018EYT05).

Acknowledgments: We would like to express our deepest gratitude to Hongyong Sun, who has provided us with valuable assistance in the writing of the original manuscript.

Conflicts of Interest: The authors declare no conflict of interest.

Appendix A

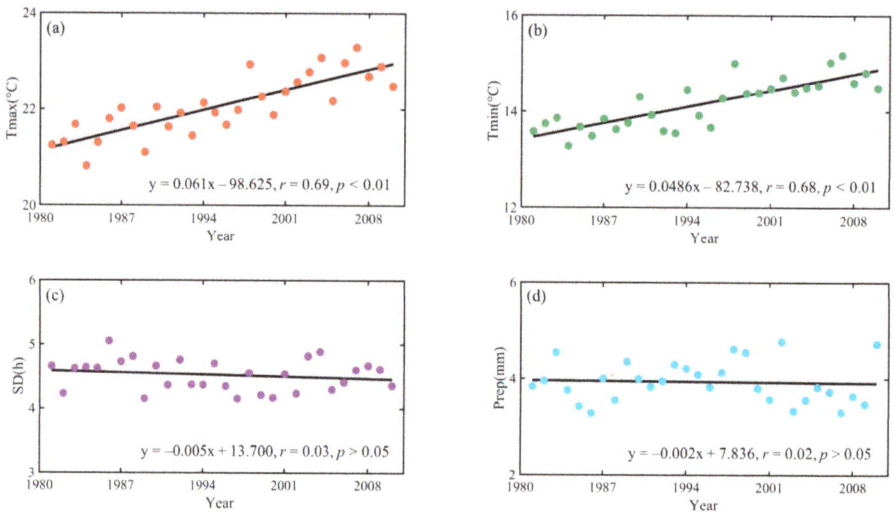

Figure A1. Trends of Tmax, Tmin, SD, and Prep across sites in YZ during the period 1981–2010. Note: (**a–d**) represents Tmax, Tmin, SD, and Prep, respectively.

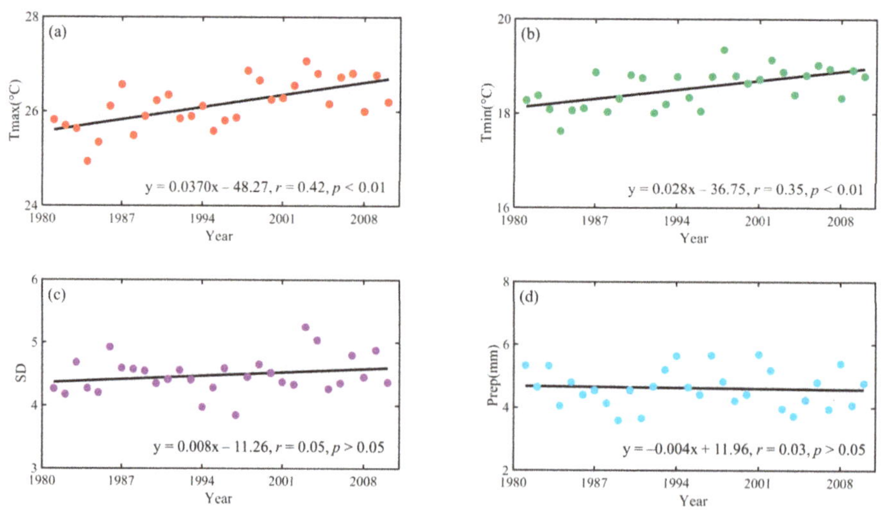

Figure A2. Trends of Tmax, Tmin, SD, and Prep across sites in PRD during the period 1981–2010. Note: (**a–d**) represents Tmax, Tmin, SD, and Prep, respectively.

References

1. Piao, S.; Ciais, P.; Huang, Y.; Shen, Z.; Peng, S.; Li, J.; Zhou, L.; Liu, H.; Ma, Y.; Ding, Y. The impacts of climate change on water resources and agriculture in China. *Nature* **2010**, *467*, 43–51. [CrossRef] [PubMed]
2. Lobell, D.B.; Schlenker, W.; Costa-Roberts, J. Climate trends and global crop production since 1980. *Science* **2011**, *333*, 616–620. [CrossRef] [PubMed]
3. Peng, S.; Huang, J.; Sheehy, J.E.; Laza, R.C.; Visperas, R.M.; Zhong, X.; Centeno, G.S.; Khush, G.S.; Cassman, K.G. Rice yields decline with higher night temperature from global warming. *Proc. Natl. Acad. Sci. USA* **2004**, *101*, 9971–9975. [CrossRef] [PubMed]
4. Fu, Y.H.; Zhao, H.; Piao, S.; Peaucelle, M.; Peng, S.; Zhou, G.; Ciais, P.; Huang, M.; Menzel, A.; Peñuelas, J. Declining global warming effects on the phenology of spring leaf unfolding. *Nature* **2015**, *526*, 104–107. [CrossRef]
5. Hoegh-Guldberg, O.; Jacob, D.; Taylor, M.; Bindi, M.; Brown, S.; Camilloni, I.; Diedhiou, A.; Djalante, R.; Ebi, K.L.; Engelbrecht, F. Impacts of 1.5 °C global warming on natural and human systems. In *Global Warming of 1.5 °C: An IPCC Special Report on the Impacts of Global Warming of 1.5 °C Above Pre-Industrial Levels and Related Global Greenhouse Gas Emission Pathways, in the Context of Strengthening the Global Response to the Threat of Climate Change, Sustainable Development, and Efforts to Eradicate Poverty*; Intergovernmental Panel on Climate Change: Geneva, Switzerland, 2018.
6. Masson-Delmotte, V. *Global Warming of 1.5 °C: An IPCC Special Report on the Impacts of Global Warming of 1.5 C above Pre-Industrial Levels and Related Global Greenhouse Gas Emission Pathways, in the Context of Strengthening the Global Response to the Threat of Climate Change, Sustainable Development, and Efforts to Eradicate Poverty*; World Meteorological Organization: Geneva, Switzerland, 2018.
7. You, Q.; Wu, F.; Shen, L.; Pepin, N.; Jiang, Z.; Kang, S. Tibetan Plateau amplification of climate extremes under global warming of 1.5 °C, 2 °C and 3 °C. *Glob. Planet. Chang.* **2020**, *192*, 103261. [CrossRef]
8. Shkolnik, I.; Pavlova, T.; Efimov, S.; Zhuravlev, S. Future changes in peak river flows across northern Eurasia as inferred from an ensemble of regional climate projections under the IPCC RCP8. 5 scenario. *Clim. Dyn.* **2018**, *50*, 215–230. [CrossRef]
9. Xu, Y.; Ramanathan, V.; Victor, D.G. *Global Warming Will Happen Faster Than We Think*; Nature Publishing Group: Berlin, Germany, 2018.
10. Zhang, T.; Huang, Y.; Yang, X. Climate warming over the past three decades has shortened rice growth duration in China and cultivar shifts have further accelerated the process for late rice. *Glob. Chang. Biol.* **2013**, *19*, 563–570. [CrossRef]
11. Hu, X.; Huang, Y.; Sun, W.; Yu, L. Shifts in cultivar and planting date have regulated rice growth duration under climate warming in China since the early 1980s. *Agric. For. Meteorol.* **2017**, *247*, 34–41. [CrossRef]
12. He, L.; Asseng, S.; Zhao, G.; Wu, D.; Yang, X.; Zhuang, W.; Jin, N.; Yu, Q. Impacts of recent climate warming, cultivar changes, and crop management on winter wheat phenology across the Loess Plateau of China. *Agric. For. Meteorol.* **2015**, *200*, 135–143. [CrossRef]
13. Ye, T.; Zong, S.; Kleidon, A.; Yuan, W.; Wang, Y.; Shi, P. Impacts of climate warming, cultivar shifts, and phenological dates on rice growth period length in China after correction for seasonal shift effects. *Clim. Chang.* **2019**, *155*, 127–143. [CrossRef]
14. ur Rehman, H.; Nawaz, Q.; Basra, S.M.A.; Afzal, I.; Yasmeen, A. Seed priming influence on early crop growth, phenological development and yield performance of linola (Linum usitatissimum L.). *J. Integr. Agric.* **2014**, *13*, 990–996. [CrossRef]
15. Xiao, D.; Moiwo, J.P.; Tao, F.; Yang, Y.; Shen, Y.; Xu, Q.; Liu, J.; Zhang, H.; Liu, F. Spatiotemporal variability of winter wheat phenology in response to weather and climate variability in China. *Mitig. Adapt. Strateg. Glob. Chang.* **2015**, *20*, 1191–1202. [CrossRef]
16. Sanchez, B.; Rasmussen, A.; Porter, J.R. Temperatures and the growth and development of maize and rice: A review. *Glob. Chang. Biol.* **2014**, *20*, 408–417. [CrossRef] [PubMed]
17. Sacks, W.J.; Kucharik, C.J. Crop management and phenology trends in the US Corn Belt: Impacts on yields, evapotranspiration and energy balance. *Agric. For. Meteorol.* **2011**, *151*, 882–894. [CrossRef]
18. Brüggemann, N.; Gessler, A.; Kayler, Z.E.; Keel, S.; Badeck, F.W.; Barthel, M.; Boeckx, P.; Buchmann, N.; Brugnoli, E.; Esperschütz, J. Carbon allocation and carbon isotope fluxes in the plant-soil-atmosphere continuum: A review. *Biogeosci. Discuss.* **2011**, *8*, 3619–3695. [CrossRef]

19. Litton, C.M.; Raich, J.W.; Ryan, M.G. Carbon allocation in forest ecosystems. *Glob. Chang. Biol.* **2007**, *13*, 2089–2109. [CrossRef]
20. Guo, Y.; Fu, Y.; Hao, F.; Zhang, X.; Wu, W.; Jin, X.; Robin Bryant, C.; Senthilnath, J. Integrated phenology and climate in rice yields prediction using machine learning methods. *Ecol. Indic.* **2021**, *120*, 106935. [CrossRef]
21. Mirza, M.M.Q. Climate change and extreme weather events: Can developing countries adapt? *Clim. Policy* **2003**, *3*, 233–248. [CrossRef]
22. Mertz, O.; Halsnæs, K.; Olesen, J.E.; Rasmussen, K. Adaptation to climate change in developing countries. *Environ. Manag.* **2009**, *43*, 743–752. [CrossRef]
23. Müller, C.; Cramer, W.; Hare, W.L.; Lotze-Campen, H. Climate change risks for African agriculture. *Proc. Natl. Acad. Sci. USA* **2011**, *108*, 4313–4315. [CrossRef]
24. Guo, Y.; Wu, W.; Du, M.; Bryant, C.R.; Li, Y.; Wang, Y.; Huang, H. Assessing potential climate change impacts and adaptive measures on rice yields: The case of zhejiang province in China. *Sustainability* **2019**, *11*, 2372. [CrossRef]
25. Xu, C.-C.; Wu, W.-X.; Ge, Q.-S.; Zhou, Y.; Lin, Y.-M.; Li, Y.-M. Simulating climate change impacts and potential adaptations on rice yields in the Sichuan Basin, China. *Mitig. Adapt. Strateg. Glob. Chang.* **2017**, *22*, 565–594. [CrossRef]
26. Peng, S.; Tang, Q.; Zou, Y. Current status and challenges of rice production in China. *Plant Prod. Sci.* **2009**, *12*, 3–8. [CrossRef]
27. Liu, Y.; Chen, Q.; Ge, Q.; Dai, J.; Qin, Y.; Dai, L.; Zou, X.; Chen, J. Modelling the impacts of climate change and crop management on phenological trends of spring and winter wheat in China. *Agric. For. Meteorol.* **2018**, *248*, 518–526. [CrossRef]
28. Li, G.; Zhao, Y.; Cui, S. Effects of urbanization on arable land requirements in China, based on food consumption patterns. *Food Secur.* **2013**, *5*, 439–449. [CrossRef]
29. Li, Q.; Dong, W.; Li, W.; Gao, X.; Jones, P.; Kennedy, J.; Parker, D. Assessment of the uncertainties in temperature change in China during the last century. *Chin. Sci. Bull.* **2010**, *55*, 1974–1982. [CrossRef]
30. Peaucelle, M.; Janssens, I.A.; Stocker, B.D.; Ferrando, A.D.; Fu, Y.H.; Molowny-Horas, R.; Ciais, P.; Peñuelas, J. Spatial variance of spring phenology in temperate deciduous forests is constrained by background climatic conditions. *Nat. Commun.* **2019**, *10*, 1–10. [CrossRef]
31. Liu, Q.; Fu, Y.H.; Zhu, Z.; Liu, Y.; Liu, Z.; Huang, M.; Janssens, I.A.; Piao, S. Delayed autumn phenology in the Northern Hemisphere is related to change in both climate and spring phenology. *Glob. Chang. Biol.* **2016**, *22*, 3702–3711. [CrossRef]
32. Bi, L.; Zhang, B.; Liu, G.; Li, Z.; Liu, Y.; Ye, C.; Yu, X.; Lai, T.; Zhang, J.; Yin, J.; et al. Long-term effects of organic amendments on the rice yields for double rice cropping systems in subtropical China. *Agric. Ecosyst. Environ.* **2009**, *129*, 534–541. [CrossRef]
33. Liu, L.; Wang, E.; Zhu, Y.; Tang, L.; Cao, W. Effects of warming and autonomous breeding on the phenological development and grain yield of double-rice systems in China. *Agric. Ecosyst. Environ.* **2013**, *165*, 28–38. [CrossRef]
34. Zhang, S.; Tao, F. Modeling the response of rice phenology to climate change and variability in different climatic zones: Comparisons of five models. *Eur. J. Agron.* **2013**, *45*, 165–176. [CrossRef]
35. Xu, C.; Wu, W.; Ge, Q. Impact assessment of climate change on rice yields using the ORYZA model in the Sichuan Basin, China. *Int. J. Climatol.* **2018**, *38*, 2922–2939. [CrossRef]
36. Zhao, J.; Yang, X. Distribution of high-yield and high-yield-stability zones for maize yield potential in the main growing regions in China. *Agric. For. Meteorol.* **2018**, *248*, 511–517. [CrossRef]
37. Li, T.; Hasegawa, T.; Yin, X.; Zhu, Y.; Boote, K.; Adam, M.; Bregaglio, S.; Buis, S.; Confalonieri, R.; Fumoto, T.; et al. Uncertainties in predicting rice yield by current crop models under a wide range of climatic conditions. *Glob. Chang. Biol.* **2015**, *21*, 1328–1341. [CrossRef] [PubMed]
38. Yuan, S.; Peng, S.; Li, T. Evaluation and application of the ORYZA rice model under different crop managements with high-yielding rice cultivars in central China. *Field Crop. Res.* **2017**, *212*, 115–125. [CrossRef]
39. Zhang, T.; Yang, X.; Wang, H.; Li, Y.; Ye, Q. Climatic and technological ceilings for Chinese rice stagnation based on yield gaps and yield trend pattern analysis. *Glob. Chang. Biol.* **2014**, *20*, 1289–1298. [CrossRef] [PubMed]

40. Chun, J.A.; Li, S.; Wang, Q.; Lee, W.-S.; Lee, E.-J.; Horstmann, N.; Park, H.; Veasna, T.; Vanndy, L.; Pros, K.; et al. Assessing rice productivity and adaptation strategies for Southeast Asia under climate change through multi-scale crop modeling. *Agric. Syst.* **2016**, *143*, 14–21. [CrossRef]
41. Gaydon, D.S.; Probert, M.E.; Buresh, R.J.; Meinke, H.; Suriadi, A.; Dobermann, A.; Bouman, B.; Timsina, J. Rice in cropping systems—Modelling transitions between flooded and non-flooded soil environments. *Eur. J. Agron.* **2012**, *39*, 9–24. [CrossRef]
42. Zhang, S.; Tao, F.; Zhang, Z. Changes in extreme temperatures and their impacts on rice yields in southern China from 1981 to 2009. *Field Crop. Res.* **2016**, *189*, 43–50. [CrossRef]
43. Lobell, D.B.; Burke, M.B. On the use of statistical models to predict crop yield responses to climate change. *Agric. For. Meteorol.* **2010**, *150*, 1443–1452. [CrossRef]
44. Zhang, T.; Zhu, J.; Yang, X.; Zhang, X. Correlation changes between rice yields in North and Northwest China and ENSO from 1960 to 2004. *Agric. For. Meteorol.* **2008**, *148*, 1021–1033. [CrossRef]
45. Schlenker, W.; Roberts, M.J. Nonlinear temperature effects indicate severe damages to US crop yields under climate change. *Proc. Natl. Acad. Sci. USA* **2009**, *106*, 15594–15598. [CrossRef] [PubMed]
46. Tao, F.; Zhang, Z.; Shi, W.; Liu, Y.; Xiao, D.; Zhang, S.; Zhu, Z.; Wang, M.; Liu, F. Single rice growth period was prolonged by cultivars shifts, but yield was damaged by climate change during 1981–2009 in China, and late rice was just opposite. *Glob. Chang. Biol.* **2013**, *19*, 3200–3209. [CrossRef] [PubMed]
47. Wang, X.; Li, T.; Yang, X.; Zhang, T.; Liu, Z.; Guo, E.; Liu, Z.; Qu, H.; Chen, X.; Wang, L.; et al. Rice yield potential, gaps and constraints during the past three decades in a climate-changing Northeast China. *Agric. For. Meteorol.* **2018**, *259*, 173–183. [CrossRef]
48. Peltonen-Sainio, P.; Jauhiainen, L.; Trnka, M.; Olesen, J.E.; Calanca, P.; Eckersten, H.; Eitzinger, J.; Gobin, A.; Kersebaum, K.C.; Kozyra, J. Coincidence of variation in yield and climate in Europe. *Agric. Ecosyst. Environ.* **2010**, *139*, 483–489. [CrossRef]
49. Zhao, H.; Fu, Y.H.; Wang, X.; Zhao, C.; Zeng, Z.; Piao, S. Timing of rice maturity in China is affected more by transplanting date than by climate change. *Agric. For. Meteorol.* **2016**, *216*, 215–220. [CrossRef]
50. Chmielewski, F.-M. Phenology in agriculture and horticulture. In *Phenology: An Integrative Environmental Science*; Springer: Berlin, Germany, 2013; pp. 539–561.
51. Sadras, V.O.; Monzon, J.P. Modelled wheat phenology captures rising temperature trends: Shortened time to flowering and maturity in Australia and Argentina. *Field Crop. Res.* **2006**, *99*, 136–146. [CrossRef]
52. Ying, M.; Zhang, W.; Yu, H.; Lu, X.; Feng, J.; Fan, Y.; Zhu, Y.; Chen, D. An Overview of the China Meteorological Administration Tropical Cyclone Database. *J. Atmos. Ocean. Technol.* **2014**, *31*, 287–301. [CrossRef]
53. Kumaresan, M.; Riyazuddin, P. Factor analysis and linear regression model (LRM) of metal speciation and physico-chemical characters of groundwater samples. *Environ. Monit. Assess.* **2008**, *138*, 65–79. [CrossRef]
54. Kumar, P.S.; Delson, P.D.; Vernon, J.; James, E. A linear regression model (LRM) for groundwater chemistry in and around the Vaniyambadi industrial area, Tamil Nadu, India. *Chin. J. Geochem.* **2013**, *32*, 19–26. [CrossRef]
55. Mukherjee, A.; Ramachandran, P. Prediction of GWL with the help of GRACE TWS for unevenly spaced time series data in India: Analysis of comparative performances of SVR, ANN and LRM. *J. Hydrol.* **2018**, *558*, 647–658. [CrossRef]
56. Huang, G.; Li, C.H.; Li, Y. Phenological responses to nitrogen and water addition are linked to plant growth patterns in a desert herbaceous community. *Ecol. Evol.* **2018**, *8*, 5139–5152. [CrossRef] [PubMed]
57. Farrell, T.; Fukai, S.; Williams, R. Minimising cold damage during reproductive development among temperate rice genotypes. I. Avoiding low temperature with the use of appropriate sowing time and photoperiod-sensitive varieties. *Aust. J. Agric. Res.* **2006**, *57*, 75–88. [CrossRef]
58. Tao, F.; Yokozawa, M.; Xu, Y.; Hayashi, Y.; Zhang, Z. Climate changes and trends in phenology and yields of field crops in China, 1981–2000. *Agric. For. Meteorol.* **2006**, *138*, 82–92. [CrossRef]
59. Shimono, H. Earlier rice phenology as a result of climate change can increase the risk of cold damage during reproductive growth in northern Japan. *Agric. Ecosyst. Environ.* **2011**, *144*, 201–207. [CrossRef]
60. Jiang, Y.; van Groenigen, K.J.; Huang, S.; Hungate, B.A.; van Kessel, C.; Hu, S.; Zhang, J.; Wu, L.; Yan, X.; Wang, L. Higher yields and lower methane emissions with new rice cultivars. *Glob. Chang. Biol.* **2017**, *23*, 4728–4738. [CrossRef] [PubMed]
61. Li, X.; Jiang, L.; Meng, F.; Wang, S.; Niu, H.; Iler, A.M.; Duan, J.; Zhang, Z.; Luo, C.; Cui, S. Responses of sequential and hierarchical phenological events to warming and cooling in alpine meadows. *Nat. Commun.* **2016**, *7*, 1–8. [CrossRef] [PubMed]

62. Krishnan, P.; Swain, D.K.; Chandra Bhaskar, B.; Nayak, S.K.; Dash, R.N. Impact of elevated CO_2 and temperature on rice yield and methods of adaptation as evaluated by crop simulation studies. *Agric. Ecosyst. Environ.* **2007**, *122*, 233–242. [CrossRef]
63. Shimono, H.; Okada, M.; Yamakawa, Y.; Nakamura, H.; Kobayashi, K.; Hasegawa, T. Rice yield enhancement by elevated CO_2 is reduced in cool weather. *Glob. Chang. Biol.* **2007**, *14*, 276–284. [CrossRef]

Publisher's Note: MDPI stays neutral with regard to jurisdictional claims in published maps and institutional affiliations.

© 2020 by the authors. Licensee MDPI, Basel, Switzerland. This article is an open access article distributed under the terms and conditions of the Creative Commons Attribution (CC BY) license (http://creativecommons.org/licenses/by/4.0/).

sustainability
MDPI

Article

Seed Security Factors Driving Farmer Decisions on Uptake of Tissue Culture Banana Seed in Central Uganda

Lucy Mulugo [1,*], Florence Birungi Kyazze [1], Paul Kibwika [1], Bonaventure Aman Omondi [2] and Enoch Mutebi Kikulwe [3]

1 Department of Extension and Innovation Studies, College of Agricultural and Environmental Sciences, Makerere University, P.O. Box 7062, Kampala, Uganda; fbirungikyazze@gmail.com (F.B.K.); pkibwika@caes.mak.ac.ug (P.K.)
2 Bioversity International C/O IITA/Benin Research Station, 08 Boite Postale, Cotonou 0932, Benin; b.a.omondi@cgiar.org
3 Bioversity International, Katalima Road, Naguru, P.O. Box 24384 Plot 106, Kampala, Uganda; e.kikulwe@cgiar.org
* Correspondence: mulugo@caes.mak.ac.ug; Tel.: +25-(670)-280-1136

Received: 16 October 2020; Accepted: 19 November 2020; Published: 7 December 2020

Abstract: Despite the promotion of tissue culture (TC) banana to curb the spread of diseases, farmer use of such quality planting material remains low. This study utilizes the Double-Hurdle model on cross-sectional data of 174 banana farmers in Central Uganda to analyze the drivers for uptake of TC banana plant materials. Results show acceptability ($\beta = 0.74$; $p < 0.01$), adaptability ($\beta = 0.69$; $p < 0.01$) and availability for farmer use ($\beta = 1.04$; $p < 0.01$) along with social influence, farmer competences and socioeconomic factors positively influence farmer uptake of the TC banana plantlets. For uptake intensity, the main drivers include acceptability ($\beta = 0.39$; $p < 0.05$), accessibility ($\beta = 0.39$; $p < 0.01$) and farmer competences. This study demonstrates that seed security factors with farmer competencies, social influence and socioeconomic factors influence farmer decisions on uptake of TC technology for banana production. Findings emphasize the need for more involvement of extension services and research institutions in the education and promotion of TC plants in farming communities. We recommend that banana TC developers and promoters focus attention on banana varieties that are acceptable and adaptable to farmer environmental conditions.

Keywords: seed security; banana tissue culture planting material; uptake; banana farmers; central Uganda

1. Introduction

The per capita food output in Sub-Saharan Africa (SSA) has considerably declined, thus, contributing to increased food and income insecurity [1,2]. At the center of the debate on food and income, insecurity is the inability of the smallholder farmer to use quality plant material void of pests and diseases [3]. In the SSA region, the propagation system is characterized by both formal and informal plant material supplies. For banana (*Musa* spp.), the majority of smallholder farmers depend on the informal supply (including home-saved material from previous season harvests) [4]. For example, over 90% of banana farmers in East and Central Africa rely on suckers sourced from friends, neighbors, relatives and/or their own fields to either expand or establish new banana plantations [5,6]. The high prevalence of pests and disease in the home-saved plant material, however, has necessitated research and development of expert practitioners to increase the use of high-quality, formal supplies of plant material [7]. Such an approach, embedded in the farmers' social, cultural environment, guarantees quality banana propagation material [8]. One such effort is the development of tissue culture (TC) planting materials (TC banana planting materials, TC banana plantlets and TC seed are used interchangeably throughout the manuscript), which are always free of pests and diseases.

The development of banana tissue culture is a response strategy for addressing the dual challenges of (1) supplying disease-free plant material and (2) enhancing farm-level yields in banana production. First, banana is a major staple for more than half of Uganda's population, and it provides a wide range of products (animal feeds, charcoal briskets, crafts, construction materials, etc.) which significantly contribute to food and income security of the populace and consequently to national development [9]. Despite the value and benefits derived from bananas, diseases such as banana *Xanthomonas* wilt (BXW) threaten its survival in the country. Between 2002 and 2005, BXW caused losses equivalent to 61.1 million US dollars to the country, mainly associated with the East African highland banana (EAHB) "Matooke" (AAA-EAHB genome) and the "Kayinja" beer banana (ABB genome) [10]. As such, farmer use of quality banana material (e.g., tissue culture plantlets) is considered a vital component for increasing banana survival and boosting agricultural productivity in the country.

Tissue culture (TC) generated banana plants are presumed to be free of BXW and recommended for the establishment of clean banana plantations [7]. However, the use of TC banana plantlets remains low [11], at less than 7% of the total banana production in the country [7]. For instance, a study by Akankwasa et al. [12] reveals that only two hundred and fifty mother gardens had been established in Uganda, and 40,000 tissue-cultured plantlets were distributed to banana farmers. Further, results indicate that merely 6% of banana farmers are willing to use TC banana planting material. Existing studies [13,14] have mainly focused on extrinsic (mainly socioeconomic) factors to explain the uptake of TC banana planting materials. However, studies that examine the role of security factors (availability, accessibility, acceptability and adaptability) in light of the farmer environment (competence, social influence and farmer's socioeconomic factors still contribute to the lack of TC plant use and necessitate further investigation.

Research is well established in root and tuber crops [15,16]. Studies on the farmer environment to explain farmer behavior in the uptake of agricultural technologies [17–19] have been conducted, but most of the existing research is qualitative. Research is needed to understand quantitative factors influencing uptake of agricultural technologies such as banana TC plantlets. Particularly, quantitative data are needed on how TC plant availability, accessibility, acceptability and adaptability combine with the farmer environment to influence the uptake of TC banana technology. The objective of this study is to explore the role of seed security factors and the farmer environment in the uptake of TC banana planting materials. Results will be important in designing interventions that ensure the sustainability of the banana crop in central Uganda.

2. Conceptual Framework, Study Area and Methods

2.1. Conceptual Framework

The uptake of agricultural technologies like TC banana planting materials is a complex, nonlinear process influenced by multiple factors. As such, the use of a single theory to analyze farmer decision-making in using tissue culture banana planting materials does not provide a holistic picture of the uptake process. Thus, we developed a conceptual framework (Figure 1) that encompasses seed security factors and the farmer environment to analyze the uptake of TC banana plantlets.

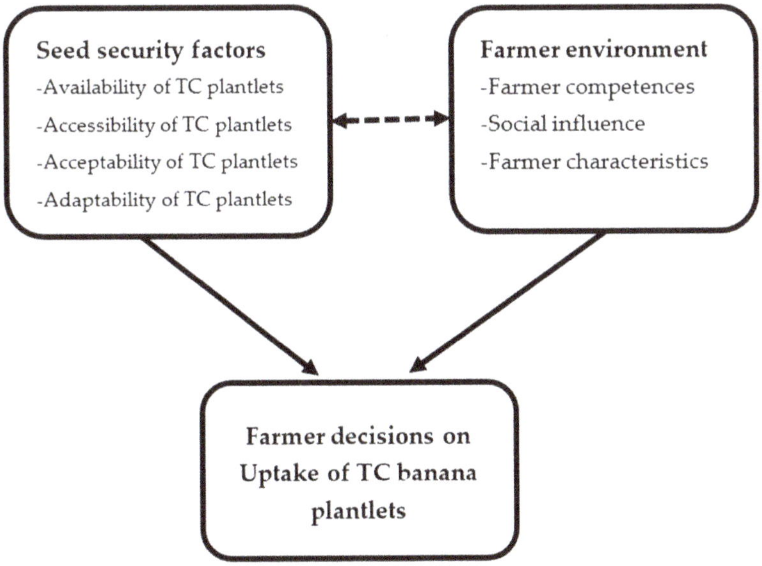

Figure 1. Seed security factors and farmer environment influencing uptake of TC banana plantlets (source: developed by authors).

The seed security framework with a focus on root tuber and banana (RTB) crops [20–22] explains that for sustained uptake of certified planting materials in a seed system (a seed system is the network of stakeholders involved in producing and planting seed (including vegetative planting material) of a particular crop in a certain area [15,20]), the factors:—availability, accessibility and quality are of relevance. It has been argued that the focus of seed security factors stimulates farmer confidence in the uptake of technologies and is of utmost relevance in identifying why uptake efforts fail (or succeed), thus aiding in the more effective design of future efforts [21]. Studies on seed security factors exist for potato [23,24], yam [25], cassava [26,27] banana and plantain [28,29] and sweet potato [30,31]. These studies reiterate the relevance of seed security factors in guiding programs that encourage the uptake of newly introduced technologies.

Availability indicates a sufficient and timely supply of propagation material from existing and functional sources to farmers [20]. Accessibility pertains to farmers' ease of acquiring TC banana planting material. This refers to whether farmers have the financial capital to purchase the plantlets and the feasibility of transporting the banana planting materials from the TC sources to destined localities [20]. Provision of information pertaining to seed technologies has also been found to be key in aiding farmers to access seed [15,23,24]. Acceptability relates to the provision of preferred and desirable seed varieties that are acceptable to meet farmers' tastes and preferences [22,32]. In accordance, Mulugo et al. [33] attest to banana farmers in central Uganda having desirable banana varieties preferred based on taste, aroma, color and texture. Similarly, Akankwasa et al. [12] found that banana taste, flavor, texture and color were key in determining consumers' likelihood of purchasing hybrid banana varieties in the four regions of Uganda.

Adaptability refers to the ability of TC banana plantlets to perform well in newly introduced environments and farming conditions. Kilwinger et al. [5] cite prolonged drought effects on banana productivity in central Uganda; Nyombi [34] mentions low soil fertility in the same region. However, previous studies tend to exclude TC established plantations, whose plantlets are characteristically fragile and sensitive to harsh environmental conditions [35]. Sinja et al. [36] confirm that farmers will only take up a technology that is best adapted to their environment. Thus, it is imperative that

TC planting materials adapt to the prevailing soil and environmental conditions in the study region. For this study, adaptability measures drought tolerance capabilities of TC established.

Theories on competence development posit knowledge, attitude and skill to be important factors that determine individual capabilities [37,38]. Competences consist of integrated pieces of knowledge, skills and attitudes that can be used to perform a task successfully, such as the uptake of TC plantlets [39]. Scholars [39,40] have shown skills to be interwoven with knowledge and viewed conjointly as doing or acting in practice. Meijer et al. [19] refer to knowledge as factual information and understanding of how a new agricultural technology works and what benefits the farmers can derive from it. Ugochukwu and Phillips [41], Kuehne et al. [42] and Meijer et al. [19] indicate that farmers' knowledge about the existence of a new technology extends to how to apply it and what the outcomes are in terms of products, yield, potential benefits, risks and costs. In this regard, knowledge is operationalized in this study to include elements of skill and technical knowledge that a farmer needs to grow TC plant materials, as well as the application of local technical knowledge to control BXW. In accordance with the tripartite theory, attitudes have three components- affective, behavioral and evaluative [43,44], which are key for the uptake of agricultural technologies. The affective domain refers to the emotional response of liking/disliking an object [45]. In this context, the TC plantlets. The behavioral component is a verbal or overt tendency by an individual [46] consisting of actions or observable responses that are the result of the object. The evaluative component, on the other hand, constitutes an individual's opinion of either belief or disbelief about a technology. For this study, attitude captures the affective and evaluative domains that include farmers' passion for growing TC plants and their perception of how the use of TC plants may reflect their personality to other people in their community.

Social influence refers to the degree to which a farmer perceives that relevant people believe that he or she should use agricultural technology [17]. Studies [47,48] attest to social influence triggering individuals' behavioral intentions to use new technologies. Thus, for this study, social influence captures the elements of farmer persuasion by members of farmer groups, faith-based leaders and community elders/leaders to grow bananas using TC planting material.

Previous research [12,49,50] shows the importance of farmer socioeconomic factors or characteristics (sex, age, education, farming experience, farm size, membership to farmer groups, and accessibility to agricultural extension services) on farmer uptake of agricultural technologies. Thus, these factors are incorporated into this study to evaluate whether they are critical in the uptake of TC banana plants. It is assumed that these factors in the presence of seed security factors could differently be influencing the uptake of TC banana seed.

2.2. Study Area

The study was conducted in Luweero and Mukono districts (Figure 2) in Central Uganda, where TC banana planting materials have been promoted for more than a decade. The two districts experience a high prevalence of BXW [51] despite numerous interventions to curb the disease. In each district, villages that hosted the community-based TC nurseries—Gonve and Nambi villages were selected on the assumption that proximity to TC nurseries enhances farmer physical access to TC banana planting materials. Farmers in the two villages were linked to TC laboratories (from whence they obtained TC plantlets) through farmer-managed community-based TC nurseries.

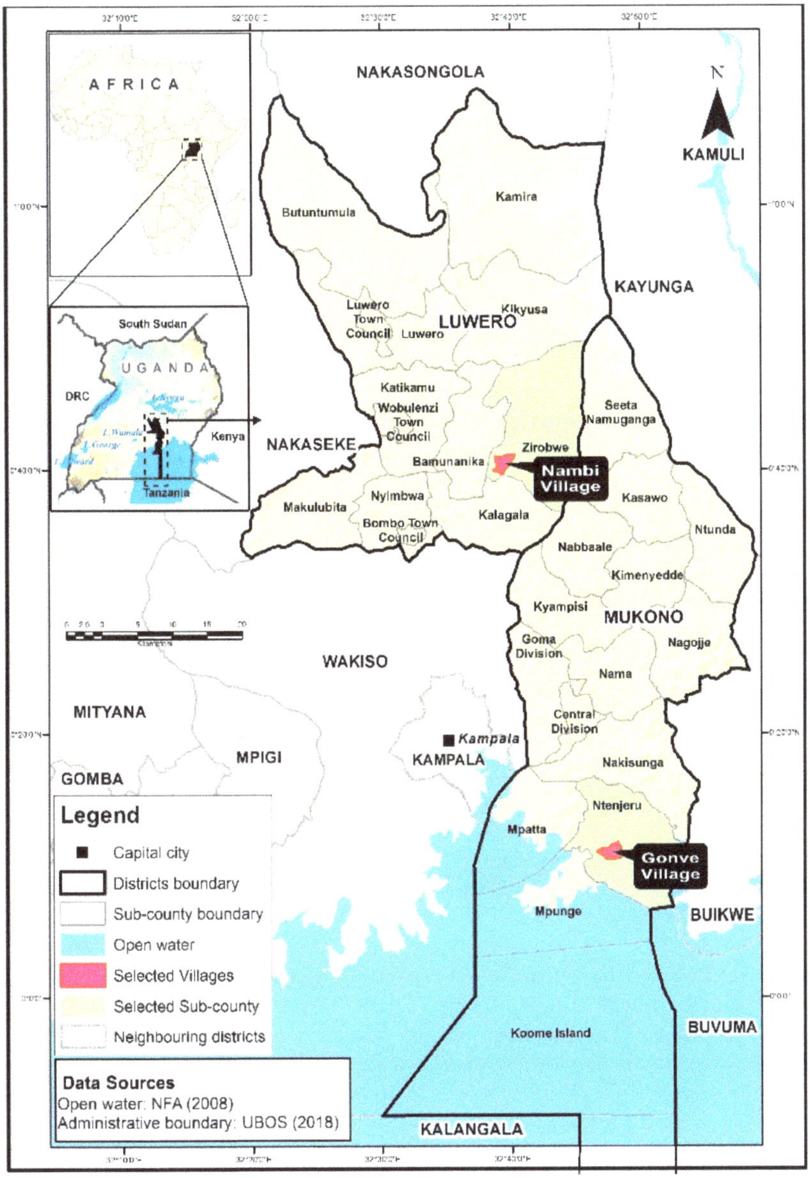

Figure 2. Map showing the study areas.

2.3. Study Design and Sample Selection

This study employed a quantitative research design. Data were obtained through a cross-sectional survey conducted in April and May 2018. The unit of analysis was the individual banana farmer. A questionnaire was administered to every respondent to generate data on variables of interest.

With the help of local council village chairpersons, names of all banana farmers from the selected villages that had been trained by The International Institute for Tropical Agriculture (IITA) between 2008 and 2011 (comprising adopters and potential adopters of the TC banana technology) were compiled to generate the sampling frame with a total of 340 banana farmers (Table 1). Unlike Mulugo et al. [17], who sampled users and non-users of the TC banana plantlets, here we selected adopters and potential adopters of the TC planting materials. The farmers had been trained on how to grow and manage TC banana plantlets, business and marketing and the establishment of farmer cooperatives linked to the TC community nurseries. From a predetermined sample size of 174 banana farmers, estimated using the method suggested by Krejcie and Morgan [52], respondents were proportionately drawn from each of the study villages, as shown in Table 1 below.

Table 1. Number of farmers selected for the study.

Location	Total Number of Farmers Listed	Number of Farmers Selected
Village: Nambi Luwero district	120	68
Village: Gonve Mukono district	220	106
Total	340	174

2.4. Data Collection

Primary data were collected from the 174 randomly selected banana farmers using a semi-structured questionnaire. The questionnaire was pretested for reliability and suitability and then modified for clarity and sequencing of questions based on the pretest experiences and results. Data were collected through face-to-face interviews with the selected banana farmers. Data were collected on seed security factors, farmer competencies and farmer characteristics related to uptake of the TC banana planting materials.

Measures

Availability was measured based on statements (e.g., "TC is available in sufficient quantities"). Accessibility was measured based on 3 items (e.g., "The TC nursery operator provides information on how to plant TC plantlet"), acceptability was measured based on 3 items (e.g., "TC banana plantlets is of desirable banana varieties"), and adaptability was measured based on 2 items (e.g., "TC banana plantlets thrive in all soil types"). To measure farmer competencies, 8 statements were used. Knowledge was measured with 5 items (e.g., "I have sufficient technical knowledge to grow TC banana plants"), and attitude was measured with 3 items (e.g., "I have enough passion for growing TC banana plants"). The social influence variable had 3 items. A sample item reads as follows: "If I am informed about TC by a community leader, then I can use it as banana seed". Each item was measured on a five-point rating scale from 1 = least and 5 = highest.

In addition, the questionnaire included demographic characteristics of the farmers: These included age in years, the highest level of formal education in years, the gender of the farmer, farm size in acres, experience in banana farming in years, access to agricultural extension services, membership to farmer groups, receipt of agricultural information, land apportioned for crop production and banana cultivation.

2.5. Analytical Framework

Preceding the regression analysis, principal component analysis (PCA) was carried out for data reduction and extraction of variables. Based on the criterion of eigenvalues being greater than 1 [53] in the PCA, underlying dimensions among the farmers' perceptions of the TC plantlets' attributes were generated.

This study breaks the banana farmer decision-making into two stages. In the first stage, the banana farmer's decision is either uptake of the banana TC technology or not. In the second stage, the farmer's decision is on how much of the TC plant material to use (uptake intensity), which in this study is represented by the size of land allocated to the banana TC plantlets. Data were analyzed using Cragg's double-hurdle model [54] consisting of a two-stage regression in which the first stage is a probit model to analyze factors influencing a binary decision on uptake of technology and the second stage is a truncated model which analyzes factors that affect uptake intensity. Farmer uptake of banana TC plantlets was analyzed in the following equation:

$$Y = \beta_0 + \beta_i[Seed\_Sec]_i + \beta_j[Farm\_Envt]_j + \varepsilon \quad (1)$$

where Y represents farmer's decision-making for the uptake of banana TC plantlets uptake = 1; no uptake = 0) and in the second stage, Y equals the uptake intensity. $[Seed\_Sec]$ is a vector of plant security factors, which includes perceived acceptability, accessibility, adaptability and availability of the plantlets. $[Farm\_Envt]$ comprises three components: (1) social influence, (2) farmer competencies, and (3) farmer socioeconomic factors or characteristics. β_0 is the constant while β_i represent the various coefficients of the security factors ranging 1–4. β_j are the coefficients of the farmer environment factors ranging from 1 to 9 and ε is the error term. The a priori hypothesized signs of the coefficients are shown in Table 2.

Table 2. Apriori signs of explanatory variables used in the study.

Variable	Apriori Sign	Reference
Availability of TC banana plantlets	(+)	[15,16]
Accessibility of TC banana plantlets	(+)	[15,16]
Acceptability of TC banana plantlets	(+)	[22,32]
Adaptability of TC banana plantlets	(+)	[22,32]
Farmer Knowledge on TC plantlets	(+)	[41,42]
Social Influence	(+)	[17,47,48]
Farmer Attitude towards TC plantlets	(+)	[19,43]
Gender	(+)	[12,50]
Age	(+/−)	[12]
Education	(+)	[12,49]
Farm size	(+)	[12,50]
Land allocated to other crops	(-)	[12]
Group membership	(+)	[50]
Extension services	(+)	[12,50]

3. Results and Discussion

Table 3 presents a summary of key descriptive statistics of study respondents. Results show an equal number of male and female farmers were interviewed. About 33% of respondents have adopted the TC banana planting materials. The average age of banana farmers in the study communities was 43 years, with an average of eight (8) years of formal education. The average farm size for all farmers was 4.3 acres, with an average of about 2.6 acres allocated to crops and about 1.1 acres to banana production.

Table 3. Description and summary statistics of respondent characteristics.

Variable	Mean (SD)	Min	Max	Percentage (%)
Sex (1 = male, 0 = female)	0.53 (0.50)	0	1	-
Age of farmer (years)	42.83 (13.63)	18	93	-
Formal education (years)	7.74 (3.45)	1	16	-
Banana farming experience (Years)	18.10 (13.10)	0	70	-
Farm size (acres)	4.27 (5.75)	0.50	69	-
Land allocated to crops (acres)	2.60 (2.41)	0.42	22.75	-
Land under banana cultivation (acres)	1.05 (1.08)	0.13	10	-
Availability of banana TC plantlets	3.07 (1.166)	1	5	-
Accessibility of banana TC plantlets	2.94 (1.027)	1	5	-
Acceptability of TC banana plantlets	3.60 (0.946)	1	5	-
Adaptability of TC banana plantlets	2.74 (0.971)	1	5	-
Farmer knowledge about TC plantlets	2.45 (1.365)	1	5	-
Farmer attitude towards TC plantlets	3.13 (1.211)	1	5	-
Social influence	3.83 (0.845)	1	5	-
Farmer attitude towards TC plantlets	3.13 (1.211)	1	5	-
Banana TC users (Yes = 1)	-	-	-	33.3
Access to extension services (Yes = 1)	-	-	-	38.5
Receipt of agricultural information (Yes = 1)	-	-	-	15.5
Membership to groups (Yes = 1)	-	-	-	39.7

Standard deviation (SD) is in parenthesis.

Results of the Kaiser–Meyer–Olkin (KMO) measure (0.815) and Bartlett's test of sphericity ($\chi = 3804.143$; $p < 0.001$) (Table 4) indicate sampling adequacy and suitability of the data for factor analysis [55]. Principal component results show that seven (7) extracted factors explained 76.2% of the total variance in the principal components. Specifically, as presented in Table 4, the variance extracted ranged from 32.4% (for farmer knowledge) to 4.4% (TC plant adaptability). In addition, the factor loadings for the extracted variables ranged from 0.508 to 0.925, and thus, convergent validity was confirmed [56]. Lastly, the Cronbach's alpha values (Table 4) range from 0.600 to 0.992, signifying adequacy of internal consistency, and thus, confirmation of measurement validity [57] in this study.

Table 4. Loadings of perception and attitude factors for uptake of TC banana planting materials (n = 174).

Item Description	Cronbach's Alpha	Factor Loadings *						
		Factor 1 Knowledge	Factor 2 Availability	Factor 3 Social Influence	Factor 4 Attitude	Factor 5 Accessibility	Factor 6 Acceptability	Factor 7 Adaptability
Apply organic fertilizer to TC plantlets	0.992	0.925						
Apply organic pesticide to TC plantlets		0.921						
Use LTK in BXW management with TC plantlets		0.883						
Sufficiency in technical knowledge to grow TC plantlets		0.538						
Skilled to grow TC plantlets		0.601						
Existence of TC nurseries/sources	0.915		0.883					
Availability of functional TC nurseries/sources			0.881					
Availability of TC plantlets in sufficient quantities			0.790					
Availability of TC plantlets in time			0.710					
If informed about TC plantlets by a community leader, then I can use it as planting material	0.939			0.912				
If informed about TC plantlets by a faith-based leader, then I can use it				0.906				
If informed about TC plantlets by a member of a farmer group, then I can use it as planting material				0.844				
According to people that are important to me, I should use TC plantlets	0.834				0.618			
I have enough passion for planting TC plantlets					0.658			

Table 4. *Cont.*

Item Description	Cronbach's Alpha	Factor Loadings *						
		Factor 1 Knowledge	Factor 2 Availability	Factor 3 Social Influence	Factor 4 Attitude	Factor 5 Accessibility	Factor 6 Acceptability	Factor 7 Adaptability
Use of TC plantlets reflects my personality to other farmers					0.751			
Access to information on how to grow TC plantlets	0.839					0.851		
Access to information on how to manage TC plantlets						0.864		
Affordability of TC plantlets						0.666		
Farmer desirability of banana varieties supplied as TC plantlets	0.651						0.786	
Appropriateness of size of TC plantlets							0.664	
Acceptability of taste of food from TC plantlets							0.681	
Drought tolerance capability of TC plantlets	0.600							0.725
Capability of TC established plantations to last long								0.744
Ability of TC plantlets to thrive in all soil types								0.508
Eigenvalues		8.098	3.080	2.448	1.660	1.483	1.886	1.091
% of variance explained		32.393	12.321	9.793	6.639	5.934	4.743	4.363

Kaiser–Meyer–Olkin measure of sampling adequacy = 0.815; approx. chi-squared = 3804.143. Bartlett's sphericity test: df = 300; $p < 0.001$.

3.1. Factors Associated with the Uptake Decision for Farmer Use of TC Banana Plantlets

Results of the first stage analysis (probit regression) show that farmer perceived acceptability ($\beta = 0.74$; $p < 0.01$) has a positive and significant influence on farmer decisions for the uptake of the banana TC planting material (Table 5). Similarly, perceived adaptability ($\beta = 0.69$; $p < 0.01$) and perceived availability ($\beta = 1.04$; $p < 0.01$) had a positive and significant influence on uptake. These findings are in agreement with other studies regarding the influence of security factors on the uptake of improved plant materials [15,20,21]. In terms of marginal effects, TC plant acceptability (0.062) implies that, on average, a 1% increment in farmer acceptance of banana TC plants increases the probability of farmer uptake by 6.2%. This is likely because farmers tend to prefer introduced varieties that are comparable to their local varieties with regard to desirable attributes. Similar results are also reported by Mulugo et al. [33] and Akankwasa et al. [12] for the case of Uganda.

Table 5. Factors that influence farmers' decisions regarding adoption of tissue culture banana planting materials: results of the probit model.

Variable	Coefficient	Robust Std. Error	$p > z$	Average Marginal Effects (dy/dx)
Seed security factors				
Perceived acceptability of TC plantlets	0.737	0.276	0.008 **	0.062
Perceived accessibility of TC plantlets	0.302	0.229	0.187	0.025
Perceived adaptability of TC plantlets	0.688	0.263	0.009 **	0.058
Perceived availability of TC plantlets	1.037	0.340	0.002 **	0.087
Farmer competences				
Knowledge about TC plantlets	2.155	0.434	<0.00 ***	0.180
Attitude to TC plantlets	1.000	0.337	0.003 **	0.084
Farmer socioeconomic factors				
Social influence	0.810	0.372	0.030 *	0.068
Sex of farmer	−1.377	0.562	0.014 *	−0.115
Age of farmer	−0.088	0.297	0.767	−0.007
Education of farmer	0.059	0.062	0.347	0.005
Group membership	−0.496	0.551	0.369	−0.042
Access to extension services	−0.344	0.525	0.513	−0.029
Land allocated to crops	0.341	0.154	0.027 *	0.029
Experience in banana farming	−0.013	0.025	0.624	−0.001
Constant	−1.016	1.158	0.381	
Number of observations	174			
Log-likelihood ratio	−25.98			
Wald χ^2	169.56 ***			
Mean VIF	1.8			

* $p < 0.05$, ** $p < 0.01$, *** $p < 0.001$; dependent variable: uptake measured as a binary.

For farmer perceived adaptability of TC plantlets (marginal effect = 0.058), an additional increase in adaptability is associated with a probability of a 5.8% increase in uptake. Notably, the importance of TC plant adaptability to drought and poor soil fertility is emphasized [5,34]. Lastly, the average marginal effect of TC plant availability (0.087) implies that an additional TC nursery is associated with a likelihood of an 8.7% increase in farmer uptake of TC plantlets. Related studies like Okechukwu and Kumar [27] on availing disease-resistant varieties in Africa and Kromann et al. [23] on the provision of quality seed in Ecuador have reported sufficient and timely availability of quality plant material to increase farmer uptake of plant technologies.

Farmer knowledge about TC plant cultivation ($\beta = 2.16$; $p < 0.001$) has a positive and significant effect on uptake. Similarly, farmer attitude ($\beta = 1.10$; $p < 0.01$) posits a positive and significant effect on uptake. These results concur with the theory that farmer competencies predict uptake [19,41,42].

In terms of marginal effects, the result on farmer knowledge about TC plants (0.180) implies that, on average, an increment in farmer skill and technical knowledge on how to grow TC plantlets increases the likelihood of farmer uptake by 18%. This finding validates the study by Atieno and Schulte-Geldermann [24] on public–private partnerships in the multiplication of plant materials in Kenya. Specifically, knowledge sharing and training by extension agents increase farmer uptake of new technology.

Similarly, our findings substantiate previous research about farmer attitudes towards the uptake of introduced agricultural technologies [58,59]. Particularly, Mekoya et al. [58] found Ethiopian farmers' positive attitude towards multipurpose fodder trees (for their feed value and contribution to soil conservation) enhanced farmer uptake of the agroforestry technology.

Social influence ($\beta = 0.81$; $p < 0.05$) also had a positive and significant influence on farmer decisions for uptake of the TC banana plants [19,60]. The average marginal effect (0.068) infers that having an influential person inform farmers about the importance of banana TC plants increases the likelihood of farmer uptake by 6.8%, suggesting that locally institutionalized mechanisms need to be promoted. Similar results are also reported by Mulugo et al. [17] and Wauters et al. [61] for Uganda and Belgium, respectively. Specific for Uganda, using faith-based leaders, political and community leaders urge farmers to use TC banana plantlets is a significant predictor of farmer intentions to use TC planting materials.

Contrary to most studies [12,62,63], the model estimates show that the gender of the farmer ($\beta = -1.38$; $p < 0.05$) has a significant effect on the uptake decision, indicating that women were more likely to use TC banana planting materials compared to men. A possible explanation for women's interest in TC banana planting material could be attributed to the distinct role that women play in ensuring food security for their families. Food security in central Uganda is culturally viewed as a women's role, and women must ensure that their homesteads are food secure (Sanya et al. [64]). The study finding is in line with Sanya et al. [64], who found female farmers to be 12.4% more likely to adopt hybrid banana varieties compared to male farmers. Nevertheless, with recent developments by the government being skewed to banana-based processing enterprises, it is likely that men will increasingly grow bananas to boost their sources of income.

The amount of land allocated to crop production ($\beta = 0.34$; $p < 0.05$) has a positive influence on the likelihood of farmer uptake of the TC banana technology. Based on the average marginal effect (0.029), increasing the land size by one acre increases the likelihood that farmers would try the TC planting materials by 2.9%. A possible explanation for this finding could be that farmers with larger farms may be more willing to take risks and devote portions of their land to growing bananas using TC plantlets, compared with those with smaller land areas. This is in line with most adoption studies [50,65,66] that found that farmers with larger farm sizes have more land to allocate to new agricultural technologies.

3.2. Factors Associated with Intensity of Farmer Uptake of TC Banana Plantlets

Results of second stage analysis (truncated regression) show that farmer perceived acceptability ($\beta = 0.39$; $p < 0.05$) and accessibility ($\beta = 0.39$; $p < 0.01$) of TC plants posit a positive and significant influence on farmer decisions (Table 6). Essentially, the result specifies that if farmers perceive that the varieties being promoted through TC match their preferred food attributes, then they are more likely to expand their banana plantations using such a technology. Previous studies [12,33] attest to this finding.

Similarly, farmer accessibility (marginal effect = 0.058) indicates that farmers with fairly priced plant material coupled with information on how to grow it promotes plantation expansion by 5.8%. This finding is in tandem with studies [15,23,24] confirming accessibility in terms of affordability and awareness creation to be crucial for use and extent of uptake of introduced plant technologies. Specifically, access to quality potato planting material in Kenya increased farmer uptake of the planting materials by 30–40%, leading to increased yields (5184 t/year of potatoes) and profits ($777,600) [24].

Table 6. Factors that influence the intensity of uptake of the TC banana plantlets: results of the truncated regression model.

Variable	Coefficient	Robust Std. Error	$p > z$	Average Marginal Effects (dy/dx)
Seed security factors				
Perceived acceptability of TC plantlets	0.390	0.179	0.029 *	0.018
Perceived accessibility of TC plantlets	0.392	0.119	<0.001 **	0.058
Perceived adaptability of TC plantlets	0.019	0.084	0.822	0.003
Perceived availability of TC plantlets	0.049	0.086	0.566	0.007
Farmer competences				
Knowledge about TC plantlets	1.461	0.197	<0.00 ***	0.067
Attitude to TC plantlets	0.524	0.185	0.005 **	0.024
Farmer socioeconomic factors				
Social influence	−0.189	0.098	0.055 *	−0.028
Sex of farmer	−0.874	0.347	0.012 *	−0.040
Age of farmer	−0.001	0.058	0.927	−0.001
Education of farmer	0.010	0.020	0.623	0.002
Group membership	−0.119	0.175	0.497	−0.018
Access to extension services	0.166	0.440	0.706	0.008
Land allocated to crops	0.218	0.031	<0.00 ***	0.032
Farm size	0.119	0.058	0.041 *	0.006
Experience in banana farming	0.003	0.008	0.678	0.678
Constant	−0.814	0.365	0.026	
Number of observations	58			
Log-likelihood ratio	−34.39			
Wald chi-squared	187.44 ***			
Mean VIF	1.88			

* $p< 0.05$, ** $p < 0.01$, *** $p < 0.001$; dependent variable: proportion of land planted with banana TC seed.

Farmer knowledge about TC plant materials ($\beta = 1.46$; $p < 0.001$) and farmer attitude towards TC technology ($\beta = 0.52$; $p < 0.01$) both postulate a positive and significant influence on farmer decisions to allocate more of their land to TC banana production. The findings concur with previous research about farmer knowledge on introduced agricultural technologies to influence uptake and extent of use of improved crop varieties [19,24,41,42]. Similarly, farmer attitude result resonates with findings by [67] in Kenya, showing farmer attitudes to be key indicators in predicting uptake of aquaculture technologies among smallholder fish farmers.

Surprisingly, social influence ($\beta = -0.19$; $p < 0.05$) had a negative influence on farmer decisions to apportion more of their land to TC banana plantlets. This varied from the results of earlier research [47,68], which reported positive effects of social influence on farmer uptake of agricultural technologies. This probably is attributable to the cost implication of expanding banana cultivation using TC plant materials, inevitably requiring farmers to purchase plantlets, inputs and follow through with the recommended cultivation instructions that often are labor-intensive [14]. As such, it may be important that a related study on social influence is designed in a different context to confirm the actual impact of social influence on farmer intensity of use of agricultural technologies.

Although not a significant predictor of uptake, farm size ($\beta = 0.12$; $p < 0.05$) had a positive influence on uptake intensity of TC banana planting materials.

4. Conclusions

This study shows that farmers' decisions for the uptake of banana TC plants are positively and significantly influenced by seed security factors. From a practical perspective, the study contributes to results that show the importance of developers in the seed system in focusing on farmer desired crop attributes. Further, the study emphasizes the need for more involvement of extension services and research institutions in education about cultivation and promotion of TC planting materials in the banana farming communities. This involvement could incorporate the use of community role models since social influence plays a pivotal role in increasing uptake. We recommend seed security factors (acceptability, accessibility, adaptability and availability), social influence and farmer competence (knowledge, skill and attitude) as variables to be considered in programs aimed at increasing farmer uptake for seed system technologies.

Author Contributions: L.M., F.B.K., P.K., B.A.O. and E.M.K. conceived and designed the study. L.M., F.B.K. and P.K. developed the data collection tools. L.M., F.B.K. and P.K. participated in field research. L.M. and F.B.K. analyzed and interpreted the data, E.M.K., and L.M. acquired funding. All authors participated in writing, reviewing and editing the article. All authors have read and agreed to the published version of the manuscript.

Funding: This research was funded by the CGIAR Research Program (CRP) Roots Tubers and Bananas (RTB) through the "RTB Seed Systems Project" grant number 96202-500-A1050-B100461-C100078-UGAND and the Regional Universities Forum for Capacity Building in Africa (RUFORUM) with funding from the Carnegie Corporation of New York grant number RU/2016/Carnegie/DRG/009.

Acknowledgments: We are grateful to the banana farmers in the study communities who participated in this study. Our appreciation goes to Emmanuel Ngolobe, Eva Tereka, Ruth Nakintu, Ssebunya Mohammed Ali and Ssenyonjo Michael for their participation in data collection. Justine Kyobe, Sserunkuma Margaret, Nakafu Margaret Kyasa (Mukono district) and Kiggundu (Luweero district) are appreciated for their role in mobilizing the respondents. This research was undertaken as part of the CGIAR Research Program on Roots, Tubers and Bananas (RTB).

Conflicts of Interest: The authors declare no conflict of interest.

References

1. FAO; IFAD; UNICEF; WFP; WHO. *The State of Food Security and Nutrition in the World 2019*; Safeguarding against economic slowdowns and downturns; FAO: Rome, Italy, 2019.
2. Reynolds, T.W.; Waddington, S.R.; Anderson, C.L.; Chew, A.; True, Z.; Cullen, A. Environmental impacts and constraints associated with the production of major food crops in Sub-Saharan Africa and South Asia. *Food Secur.* **2015**, *7*, 795–822. [CrossRef]
3. Kusena, K.; Wynberg, R.; Mujaju, C. Do smallholder farmer-led seed systems have the capacity to supply good-quality, fungal-free sorghum seed? *Agric. Food Secur.* **2017**, *6*, 52. [CrossRef]
4. McGuire, S.; Sperling, L. Seed systems smallholder farmers use. *Food Secur.* **2016**, *8*, 179–195. [CrossRef]
5. Kilwinger, F.B.M.; Rietveld, A.M.; Groot, J.C.J.; Almekinders, C.J.M. Culturally embedded practices of managing banana diversity and planting material in central Uganda. *J. Crop Improv.* **2019**, *33*, 456–477. [CrossRef]
6. Lwandasa, H.; Kagezi, G.H.; Akol, A.M.; Mulumba, J.W.; Nankya, R.; Fadda, C.; Jarvis, D.J. Assessment of Farmers Knowledge and Preferences for Planting Materials to Fill-Gaps in Banana Plantations in Southwestern Uganda. *Uganda J. Agric. Sci.* **2014**, *15*, 165–178.
7. Dubois, T.; Dusabe, Y.; Lule, M.; Van Asten, P.; Coyne, D.; Hobayo, J.C.; Nkurunziza, S.; Ouma, E.; Kabunga, N.; Qaim, M.; et al. Tissue culture banana (Musa spp.) for smallholder farmers: Lessons learnt from East Africa. *Acta Hort.* **2013**, *986*, 51–59. [CrossRef]
8. Almekinders, C.J.; Walsh, S.; Jacobsen, K.S.; Andrade-Piedra, J.L.; McEwan, M.A.; de Haan, S.; Kumar, L.; Staver, C. Why interventions in the seed systems of roots, tubers and bananas crops do not reach their full potential. *Food Secur.* **2019**, *11*, 23–42. [CrossRef]
9. Kikulwe, E.; Okurut, S.; Ajambo, S.; Nowakunda, K.; Stoian, D.; Naziri, D. Postharvest losses and their determinants: A challenge to creating a sustainable cooking banana value chain in Uganda. *Sustainability* **2018**, *10*, 2381. [CrossRef]
10. Tushemereirwe, W.K.; Kubiriba, J.; Nankinga, C.; Masanza, M.; Muhangi, J.; Ssekiwoko, F.; Karamura, E. Management of banana xanthomonas wilt in Uganda. In Proceedings of the Workshop on Review of the Strategy for the Management of Banana Xanthomonas Wilt, Kigali, Rwanda, 23–27 July 2007; pp. 25–35.
11. Jogo, W.; Karamura, E.; Tinzaara, W.; Kubiriba, J.; Rietveld, A. Determinants of farm-level adoption of cultural practices for banana Xanthomonas wilt control in Uganda. *J. Agric. Sci.* **2013**, *5*, 70–81. [CrossRef]

12. Akankwasa, K.; Ortman, G.F.; Tushemereirwe, W.K.; Wale, E. Early-Stage Adoption of Improved Banana Matooke" Hybrids in Uganda: A Count Data Analysis Based on Farmers' Perceptions. *Int. J. Innov. Technol. Manag.* **2016**, *13*, 60–86. [CrossRef]
13. Murongo, M.; Wangai, K.J.; Mwine, T.J.; Ayuke, O.F. Farmer-based dynamics in tissue culture banana technology adoption: A socio-economic perspective among small holder farmers in Uganda. *Afr. J. Agric. Res.* **2018**, *13*, 2836–2854. [CrossRef]
14. Kabunga, N.S.; Dubois, T.; Qaim, M. Heterogenous information exposure and technology adoption: The case of tissue culture bananas in Kenya. *Agric. Econ.* **2012**, *43*, 473–485. [CrossRef]
15. Bentley, J.W.; Andrade-Piedra, J.; Demo, P.; Dzomeku, B.; Jacobsen, K.; Kikulwe, E.; Kromann, P.; Kumar, P.L.; McEwan, M.; Mudege, N.; et al. Understanding root, tuber, and banana seed systems and coordination breakdown: A multi-stakeholder framework. *J. Crop Improv.* **2018**, *32*, 599–621. [CrossRef]
16. Andrade-Piedra, J.L.; Bentley, J.W.; Almekinders, C.J.M.; Jacobsen, K.; Walsh, S.; Thiele, G. (Eds.) *Case Studies of Root, Tuber and Banana Seed Systems*; 244. CGIAR Research Program on Roots, Tubers and Bananas (RTB), RTB Working Paper No. 2016-3; RTB: Lima, Peru, 2016; ISSN 2309-6586.
17. Mulugo, L.; Kyazze, B.L.; Kibwika, P.; Kikulwe, E.; Omondi, A.B.; Ajambo, S. Unravelling technology-acceptance factors influencing farmer use of banana tissue culture planting materials in Central Uganda. *Afr. J. Sci. Technol. Innov. Dev.* **2019**, *12*, 453–465. [CrossRef]
18. Zhou, D.; Abdulla. The Acceptance of Solar Water Pump Technology among Rural Farmers of Northern Pakistan: A Structural Equation Model. *Cogent Food Agric.* **2017**, *3*, 1280882. [CrossRef]
19. Meijer, S.S.; Catacutan, D.; Ajayi, O.C.; Sileshi, G.W.; Nieuwenhuis, M. The role of knowledge, attitudes and perceptions in the uptake of agricultural and agroforestry innovations among smallholder farmers in sub-Saharan Africa. *Int. J. Agric. Sustain.* **2015**, *13*, 40–54. [CrossRef]
20. RTB (Research Program in Roots, Tubers and Bananas). *User's Guide to the Multi-Stakeholder Framework for Intervening in RTB Seed Systems*; RTB: Lima, Peru, 2016.
21. Sperling, L.; Ortiz, O.; Thiele, G. *RTB Seed Systems: Conceptual Frameworks for Guiding Practical Interventions*; Draft 1—RTB Working Paper 2013-1. CGIAR Research Program on Roots, Tubers and Bananas; RTB: Lima, Peru, 2013.
22. Sperling, L.; Cooper, H.D.; Remington, T. Moving towards more effective seed aid. *J. Dev. Stud.* **2008**, *44*, 586–612. [CrossRef]
23. Kromann, P.; Montesdeoca, F.; Andrade-Piedra, J. Integrating formal and informal potato seed systems in Ecuador. In *Case Studies of Roots, Tubers and Bananas Seed Systems*; RTB Working Paper No. 2016-3; RTB: Lima, Peru, 2016; Chapter 2.
24. Atieno, E.; Schulte-Geldermann, E. Public-private partnerships to multiply seed potato in Kenya. In *Case Studies of Roots, Tubers and Bananas Seed Systems*; RTB Working Paper No. 2016-3; RTB: Lima, Peru, 2016; Chapter 8.
25. Odu, B.O.; Coyne, D.; Kumar, P.L. Adapting a yam seed technique to meet farmers' criteria. In *Case Studies of Roots, Tubers and Bananas Seed Systems*; RTB Working Paper No. 2016-3; RTB: Lima, Peru, 2016; Chapter 4.
26. Ospina, B. Research reawakens in Nicaragua. In *Case Studies of Roots, Tubers and Bananas Seed Systems*; RTB Working Paper No. 2016-3; RTB: Lima, Peru, 2016; Chapter 9.
27. Okechukwu, R.; Kumar, P.L. Releasing disease-resistant varieties of cassava in Africa. In *Case Studies of Roots, Tubers and Bananas Seed Systems*; RTB Working Paper No. 2016-3; RTB: Lima, Peru, 2016; Chapter 11.
28. Kikulwe, E. Banana tissue culture: Community nurseries for African farmers. In *Case Studies of Roots, Tubers and Bananas Seed Systems*; RTB Working Paper No. 2016-3; RTB: Lima, Peru, 2016; Chapter 12.
29. Jacobsen, K.; Dzomeku, B. Bananas and plantains in Ghana. In *Case Studies of Roots, Tubers and Bananas Seed Systems*; RTB Working Paper No. 2016-3; RTB: Lima, Peru, 2016; Chapter 5.
30. Ogero, K.; McEwan, M.; Pamba, N. Clean vines for smallholder farmers in Tanzania. In *Case Studies of Roots, Tubers and Bananas Seed Systems*; RTB Working Paper No. 2016-3; RTB: Lima, Peru, 2016; Chapter 6.
31. Nshimiyimana, J.C.; Ndirigwe, J.; Sindi, K.; Uwase, V.; McEwan, M.; Low, J. Delivering clean sweetpotato vines in Rwanda. In *Case Studies of Roots, Tubers and Bananas Seed Systems*; RTB Working Paper No. 2016-3; RTB: Lima, Peru, 2016; Chapter 7.
32. Remington, T.; Maroko, J.; Walsh, S.; Omanga, P.; Charles, E. Getting off the seeds–and–tools treadmill with CRS seed vouchers and fairs. *Disasters* **2002**, *26*, 316–328. [CrossRef]

33. Mulugo, L.; Kibwika, P.; Kyazze, B.F.; Omondi, A.B.; Kikulwe, E. The Contestations of Diversity, Culture and Commercialization: Why Tissue Culture Technology will not solve the BXW disease in Central Uganda. *J. Agric. Hum. Values* **2020**, forthcoming.
34. Nyombi, K. Towards sustainable highland banana production in Uganda: Opportunities and challenges. *Afr. J. Food Agric. Nutr. Dev.* **2013**, *13*, 7544–7561.
35. Kavoo-Mwangi, A.M.; Kahangi, E.M.; Ateka, E.; Onguso, J.; Jefwa, J.M. Integration of commercial microbiological products into soil fertility practices as a potential option for acclimatization and growth of TC banana in Kenya. *Open J. Soil Sci.* **2014**, *4*, 259. [CrossRef]
36. Sinja, J.; Karugia, J.T.; Baltenweck, I.; Waithaka, M.M.; Miano, M.D.; Nyikal, R.A.; Romney, D. Farmer perception of technology and its impact on technology uptake: The case of fodder legume in Central Kenya highlands. In Proceedings of the Inaugural Symposium of African Association of Agricultural Economists, Nairobi, Kenya, 6–8 December 2004.
37. Eraut, M. *Developing Professional Knowledge and Competence*; Routledge Falmer: London, UK; New York, NY, USA, 1994.
38. Kaslow, N.J.; Rubin, N.J.; Bebeau, M.J.; Leigh, I.W.; Lichtenberg, J.W.; Nelson, P.D.; Portnoy, S.M.; Smith, I.L. Guiding principles and recommendations for the assessment of competence. *Prof. Psychol. Res. Pract.* **2007**, *38*, 441. [CrossRef]
39. Baartman, L.K.; De Bruijn, E. Integrating knowledge, skills and attitudes: Conceptualising learning processes towards vocational competence. *Educ. Res. Rev.* **2011**, *6*, 125–134. [CrossRef]
40. Morrison, G.R.; Ross, S.M.; Kemp, J.E. *Designing Effective Instruction*; John Wiley: New York, NY, USA, 2001.
41. Ugochukwu, A.I.; Phillips, P.W. Technology Adoption by Agricultural Producers: A Review of the Literature. In *From Agriscience to Agribusiness*; Springer: Berlin/Heidelberg, Germany, 2018; pp. 361–377.
42. Kuehne, G.; Llewellyn, R.; Pannell, D.J.; Wilkinson, R.; Dolling, P.; Ouzman, J.; Ewing, M. Predicting farmer uptake of new agricultural practices: A tool for research, extension and policy. *Agric. Syst.* **2017**, *156*, 115–125. [CrossRef]
43. Fabrigar, L.R.; MacDonald, T.K.; Wegener, D.T. The structure of attitudes. In *The Handbook of Attitudes*; Routledge: London, UK, 2005; p. 80.
44. Rosenberg, M.J.; Hovland, C. Cognitive, affective, and behavioral components of attitudes. In *Attitude Organization and Change*; Rosenberg, M., McGuire, C.H.W., Abelson, R., Brehm, J., Eds.; Yale University Press: New Haven, CT, USA, 1960.
45. Agarwal, J.; Malhotra, N.K. An integrated model of attitude and affect. *J. Bus. Res.* **2005**, *58*, 483–493. [CrossRef]
46. Wicker, A. Attitudes versus actions: The relationship of verbal and overt behavioural responses to attitude objects. *J. Soc. Issues* **1969**, *25*, 41–78. [CrossRef]
47. Adam, R.I.; Badstue, L.; Sindi, K. The dynamics of smallholder farmers' acquisition and distribution of sweetpotato vines in the Lake Victoria Zone Region, Tanzania. *Food Secur.* **2018**, *10*, 339–350. [CrossRef]
48. Coomes, O.T.; McGuire, S.J.; Garine, E.; Caillon, S.; McKey, D.; Demeulenaere, E.; Jarvis, D.; Aistara, G.; Barnaud, A.; Clouvel, P.; et al. Farmer seed networks make a limited contribution to agriculture? Four common misconceptions. *Food Policy* **2015**, *56*, 41–50. [CrossRef]
49. Gibreel, T.M. Crop commercialization and adoption of gum-arabic agroforestry and their effect on farming system in western Sudan. *Agrofor. Syst.* **2013**, *87*, 311–318. [CrossRef]
50. Barungi, M.; Edriss, A.; Mugisha, J.; Waithaka, M.; Tukahirwa, J. Factors influencing the adoption of soil erosion control technologies by farmers along the slopes of Mt. Elgon in eastern Uganda. *J. Sustain. Dev.* **2013**, *6*, 9. [CrossRef]
51. Ocimati, W.; Nakato, G.V.; Fiaboe, K.M.; Beed, F.; Blomme, G. Incomplete systemic movement of Xanthomonas campestris pv. musacearum and the occurrence of latent infections in xanthomonas wilt-infected banana mats. *Plant Pathol.* **2015**, *64*, 81–90. [CrossRef]
52. Krejcie, R.V.; Morgan, D.W. Determining sample size for research activities. *Educ. Psychol. Meas.* **1970**, *30*, 607–610. [CrossRef]
53. Kaiser, H.F. A note on Guttman's lower bound for the number of common factors. *Br. J. Stat. Psychol.* **1961**, *14*, 1–2. [CrossRef]
54. Cragg, J.G. Some statistical models for limited dependent variable with application to the demand for durable goods. *Econometrica* **1971**, *39*, 829–844. [CrossRef]

55. Leech, N.L.; Barrett, K.C.; Morgan, G.A. *Spss for Intermediate Statistics: Use and Interpretation*; Psychology Press: East Sussex, UK, 2005.
56. Hair, J.F.; Black, W.C.; Babin, B.J.; Anderson, R.E.; Tatham, R.L. *Multivariate Data Analysis*, 6th ed.; Prentice-Hall International: Upper saddle River, NJ, USA, 2006.
57. Taber, K.S. The use of Cronbach's alpha when developing and reporting research instruments in science education. *Res. Sci. Educ.* **2018**, *48*, 1273–1296. [CrossRef]
58. Mekoya, A.; Oosting, S.J.; Fernandez-Rivera, S.; Van der Zijpp, A.J. Farmers' perceptions about exotic multipurpose fodder trees and constraints to their adoption. *Agrofor. Syst.* **2008**, *73*, 141–153. [CrossRef]
59. Sileshi, G.W.; Kuntashula, E.; Matakala, P.; Nkunika, P.O. Farmers' perceptions of tree mortality, pests and pest management practices in agroforestry in Malawi, Mozambique and Zambia. *Agrofor. Syst.* **2008**, *72*, 87–101. [CrossRef]
60. Dessart, F.J.; Barreiro-Hurlé, J.; van Bavel, R. Behavioural factors affecting the adoption of sustainable farming practices: A policy-oriented review. *Eur. Rev. Agric. Econ.* **2019**, *46*, 417–471. [CrossRef]
61. Wauters, E.; Bielders, C.; Poesen, J.; Govers, G.; Mathijs, E. Adoption of soil conservation practices in Belgium: An examination of the theory of planned behavior in the agri-environmental domain. *Land Use Policy* **2010**, *27*, 86–94. [CrossRef]
62. Awotide, B.A.; Abdoulaye, T.; Alene, A.; Manyong, V.M. Assessing the extent and determinants of adoption of improved banana varieties in South-western Nigeria. *J. Dev. Agric. Econ.* **2014**, *6*, 376–385.
63. Matata, P.Z.; Ajay, O.C.; Oduol, P.A.; Agumya, A. Socio-economic factors influencing adoption of improved fallow practices among smallholder farmers in western Tanzania. *Afr. J. Agric. Res.* **2010**, *5*, 818–823.
64. Sanya, L.N.; Sseguya, H.; Kyazze, F.B.; Diiro, G.M.; Nakazi, F. The role of variety attributes in the uptake of new hybrid bananas among smallholder rural farmers in central Uganda. *Agric. Food Secur.* **2020**, *9*, 1–13. [CrossRef]
65. Lunduka, R.; Fisher, M.; Snapp, S. Could farmer interest in a diversity of seed attributes explain adoption plateaus for modern maize varieties in Malawi? *Food Policy* **2012**, *37*, 504–510. [CrossRef]
66. Wubeneh, N.G.; Sanders, J.H. Farm-level adoption of sorghum technologies in Tigray, Ethiopia. *Agric. Syst.* **2006**, *91*, 122–134. [CrossRef]
67. Obiero, K.O.; Waidbacher, H.; Nyawanda, B.O.; Munguti, J.M.; Manyala, J.O.; Kaunda-Arara, B. Predicting uptake of aquaculture technologies among smallholder fish farmers in Kenya. *Aquac. Int.* **2019**, *27*, 1689–1707. [CrossRef]
68. Ramirez, A. The influence of social networks on agricultural technology adoption. *Procedia-Social Behav. Sci.* **2013**, *79*, 101–116. [CrossRef]

Publisher's Note: MDPI stays neutral with regard to jurisdictional claims in published maps and institutional affiliations.

© 2020 by the authors. Licensee MDPI, Basel, Switzerland. This article is an open access article distributed under the terms and conditions of the Creative Commons Attribution (CC BY) license (http://creativecommons.org/licenses/by/4.0/).

sustainability

MDPI

Article

Sustainability of Traditional Rice Cultivation in Kerala, India—A Socio-Economic Analysis

Jayasree Krishnankutty [1], Michael Blakeney [2,*], Rajesh K. Raju [1] and Kadambot H. M. Siddique [3]

[1] Communication Centre, Kerala Agricultural University, Mannuthy, Kerala 680656, India; jayasree.krishna@kau.in (J.K.); rajkr369@gmail.com (R.K.R.)

[2] Faculty of Law, The University of Western Australia, Crawley, WA 6009, Australia

[3] The UWA Institute of Agriculture, The University of Western Australia, Perth, WA 6001, Australia; kadambot.siddique@uwa.edu.au

* Correspondence: michael.blakeney@uwa.edu.au

Abstract: Traditional rice cultivars and cultivation are on the decline in most rice-growing areas, mainly as a result of their low productivity. Packed with nutritionally, environmentally and locally superior qualities, traditional cultivars hold the key for sustainability in rice cultivation. This study explored the dynamics of traditional rice cultivation in Kerala, India. It examined the economic, institutional and socio demographic factors involved in the production and marketing of traditional rice. We employed a multinomial logit model and discriminant function analysis to extract the key factors governing farmers' marketing behaviour, and various cost measures to study the economics of rice enterprises. The socio-demographic factors were analysed using descriptive statistical tools. Holding size and institutional support were the main factors governing the marketing behaviour of farmers. Even though traditional rice farming was not found to be cost-effective in implicit terms, it was remunerative when imputed personal labour and owned land costs were not considered. The study found that traditional farmers are ageing, have a lower education and use limited marketing channels. However, the majority of them were satisfied with their farm enterprise. By streamlining the market support mechanism and processing facilities, traditional rice would most likely gain momentum in key areas.

Keywords: traditional rice economics; institutional; socio-demographic factors; multinomial logit model; constraints

Citation: Krishnankutty, J.; Blakeney, M.; Raju, R.K.; Siddique, K.H.M. Sustainability of Traditional Rice Cultivation in Kerala, India—A Socio-Economic Analysis. *Sustainability* **2021**, *13*, 980. https://doi.org/10.3390/su13020980

Received: 12 December 2020
Accepted: 13 January 2021
Published: 19 January 2021

Publisher's Note: MDPI stays neutral with regard to jurisdictional claims in published maps and institutional affiliations.

Copyright: © 2021 by the authors. Licensee MDPI, Basel, Switzerland. This article is an open access article distributed under the terms and conditions of the Creative Commons Attribution (CC BY) license (https://creativecommons.org/licenses/by/4.0/).

1. Introduction

In September 2000, world leaders at the United Nations Millennium Summit recognized a collective responsibility to work toward "a more peaceful, prosperous and just world" [1]. Following this Summit, the nations of the world committed to achieving certain goals across eight priority areas of social and economic development by 2015. Goal 1 of these millennium development goals (MDGs) was the eradication of extreme poverty and hunger. A stocktaking on the achievement of these goals undertaken by the United Nations Conference on Sustainable Development in Rio de Janeiro, June 2012, precipitated a process to develop a new set of Sustainable Development Goals (SDGs) intended to carry on the momentum generated by the MDGs beyond 2015 [2]. The SDGs sought to continue the fight against extreme poverty, but added the challenges of ensuring more equitable development and environmental sustainability. The UN General Assembly on 25 September 2015 adopted the 2030 Agenda for Sustainable Development, which in SDG2 aimed to "End hunger, achieve food security and improved nutrition and promote sustainable agriculture", and recognized the inter-linkages among supporting sustainable agriculture, empowering small farmers, promoting gender equality, ending rural poverty, ensuring healthy lifestyles and tackling climate change, among other issues [3].

Although promoted by the 2030 Agenda, the concept of sustainability is not a new concept and can be traced back at least to the 1987 report of the Brundtland Commission on Environment and Development, which defined sustainable development as that "which meets the needs of the present without compromising the ability of future generations to meet their own needs" [4]. This vagueness of meaning became transposed into the definition of sustainable agriculture. This has been described as having been formulated in a contested discourse [5] and the admitting of more than 70 definitions [6,7] suggest that it is an exercise in "measuring the immeasurable". Its multifarious meanings are described as open to value-judgment and leading to many different interpretations [8,9].

For our purposes we accept Professor Swaminathan's judgement that sustainable agriculture must achieve productivity in perpetuity without accompanying ecological and social harm [10] discussed in [11].

The forecast increase in the world population of up to 9.7 billion by 2050 and 10.9 billion in 2100 [12] will require a major effort to increase food production for the additional 2.6 billion people compared to today. Disregarding problems of distribution and food waste and loss, this would require an increase in global food production by at least one third, with the obvious potentially adverse ecological impacts which that might have.

A vigorous contemporary debate has been conducted around the sustainable agricultural strategies which might be adopted to meet the challenge of feeding the world's increasing population [13]. These range from precision farming [14] and the use of genetically modified crops [15–17] through to "ecogariculture" [18–21].

Declaring 2004 as the International Year of Rice, the UN General Assembly noted that rice is the staple food of more than half the world's population, and affirmed the need to heighten the awareness of the role of rice in alleviating poverty and malnutrition and reaffirmed the need to focus world attention on the role rice can play in providing food security and eradicating poverty (See http://www.fao.org/3/Y5167E/y5167e02.htm, accessed 7 December 2020.) During the Green Revolution of the 1960s, high yielding varieties (HYV) of rice were developed which increased yield, reduced the cropping period and increased cropping intensity to allow the cultivation of 2–3 crops per year, but which required the use chemical fertilizers, pesticides, tractors, mechanical threshers and controlled water supply to crops [22]. One of the major ecological consequences of the Green Revolution was the significant depletion in the number of traditional rice varieties, as the HYV had a very narrow and unstable genetic base compared with traditional varieties [23–26]. Traditional varieties have gradually disappeared as farmers have abandoned them in favour of monohybrid crops [27] In recent years, and in the face of climate change, it has been realised that traditional rice varieties represent a valuable gene pool for traits which may underpin the capacity of modern varieties of rice to adapt to climate change [28–30].

This article considers the contribution which the cultivation of traditional rice varieties can make to sustainable agriculture, by an examination of the socio-economic situation of farmers in traditional rice cultivation in Kerala.

Kerala lies in the south-western corner of the Indian peninsula, in the southern part of the Western Ghats adjoining Tamil Nadu and Karnataka in the east and north-east, and bounded by the Arabian Sea in the west. Kerala is topographically and ecologically diverse, consisting of a mix of coastland, wetlands and plains to the west, and the foothills of the Western Ghats to the east. The ecological conditions in the state have resulted in a considerable diversity of germplasm in both wild and cultivated rice [31]. It has a rich culture of rice cultivation, where rice farming is considered a symbol of prosperity and traditional lineage [32]. The rice requirements in this state are estimated as 3.5–4.0 million t/year. However, Kerala produces only one-fifth of this amount. The deficit in rice production is mounting each year, owing to the decline in the area under rice cultivation. Large-scale conversion of paddy lands for other crops or for residential purposes has caused a serious problem in the age-old practice of rice farming in Kerala [33].

The main reasons given for the decline in traditional rice varieties is their low productivity, longish growing duration, lack of price premium for some varieties and comparatively longer cooking time [34,35]. For many farmers, these factors outweigh the nutritive and environmental advantages of traditional rice varieties [36].

A more recent factor contributing to the decline in global rice production is climate change. Rainfed cultivation is estimated to account for about 25 per cent of global rice production, which makes it particularly vulnerable to fluctuations in rainfall, as well as heat stress from high temperatures [37,38]. With the expected demand for rice to increase in the coming years, food security will be imperilled, unless this situation can be improved. Farmers will have to increase yields by adopting high yielding varieties, or by utilising those traditional varieties which are suitable for marginal lands. A recent study [39] found that the cost of cultivating high yielding hybrid varieties in Assam was on average 29.43 percent higher than traditional rice. The higher costs were for plant protection chemicals (85.13 percent), irrigation (63.21 percent) and seed (62.81 percent). Under traditional rice cultivation, farmers used little or no plant protection chemicals, which makes cultivation environmentally sustainable, as well as economical [40].

2. Materials and Methods

Sustainability of an agricultural system is a time- and space-specific concept. Its assessment is closely linked to the context in which the specific farming system thrives [7]. Due to variations in biophysical and socioeconomic conditions, indicators used in one country may not be suitable for other countries and can be of great subjectivity [41]. The sustainability of traditional rice system will be analysed based on three broad dimensions in this study, which are adopted and adapted from the following:

Economic factors, especially cost of production and profitability—[42–47].

Socio demographic factors [48,49]—area; [50] experience; [15] age; [48,49]—education.

Institutional factors—[7,51].

Apart from production levels and profitability, of late, several authors [52,53] have postulated about their well-being as perceived by farmers, which is a derivative of their satisfaction in life, as a substantial indicator of sustainability. Therefore, we looked at the satisfaction level of traditional rice farmers, which will be a component to the system's sustainability.

With such a backdrop, the study addressed the following research questions in an attempt to link the answers to sustainability of traditional rice farming. (a) What is the varietial diveristy of traditional rice in the study area? (b) How profitable is this rice system and how best the farmers market their produce? (c) What are the socio-demographic characteristics of the farmers cultivating traditional rice varieties and how do they tell on sustainability of the system? (d) How satisfied are the traditional rice farmers? (e) What are the constraints experienced by traditional rice farmers?

2.1. Study Area and Sampling

This study examined the cultivation of traditional rice varieties in the principal traditional rice-cultivating districts of Kerala. Three of the fourteen districts in Kerala state, namely Palakkad, Wayanad and Malappuram, had the largest area under traditional rice and were selected for the study. Palakkad is known as the rice bowl of Kerala, Wayanad is known for its hill area rice cultivation and Malappuram is rich with traditional, family rice farms. A random sample of 100 traditional rice farmers was chosen from each district, such that the total sample size for the study was 300. These farmers were cultivating traditional rice by choice, for consumption and marketing and were not government-supported farmers producing traditional varieties for conservation purposes.

2.2. Data Collection and Analysis

Farmer responses were collected through personal interviews using a semi-structured pre-tested interview schedule, focus group discussions and direct observation. The data generated were classified, coded, tabulated and analysed using a host of statistical tools. The socio-demographic characteristics of respondents were classified and interpreted using simple descriptive statistical tools, including percentages and frequency distributions. Linear discriminant function analysis was undertaken as a multivariate test of differences between groups to determine the minimum number of dimensions needed to describe these differences. The influence of the explanatory variables on the marketing channel choices of farmers was explored in a multinomial logit model. The odds ratio was used to quantify the influence. A satiety index was developed to understand each farmer's level of satisfaction with traditional rice farming. Garrett's ranking analysis was used to prioritise farmer constraints.

2.3. *Tools for Data Analysis*

Cost Concepts Used in the Study

The cost concepts for farm management studies, used by the Commission on Agricultural Costs and Prices (CACP) of the Government of India, were employed. Data were collected on select physical indicators; the value of seed (purchased or home-grown), insecticides and fungicides, manure (owned and purchased), fertilisers, irrigation, machinery (own or hired machinery), human, animal and machine labour (own or hired), land revenue, rent paid for leased land or rental value of own land, interest on working capital, land revenue, depreciation of machinery and miscellaneous expenses.

The structure of different costs and their components [54] were as follows:

(i) Cost A_1 includes value of human labour (casual and permanent), hired bullock power, owned bullock power, owned machine power, hired machine power, seeds (farm produced and purchased), manure (owned and purchased), fertilizer, plant protection chemicals, herbicides, irrigation charges, land tax (Landowners in India pay tax to the government; the levy is based on the area owned. Land tax is paid to the respective village office) and other taxes, depreciation on farm implements and buildings, interest on working capital and miscellaneous expenses;
(ii) Cost A_2 = Cost A_1 + Rent paid for leased land;
(iii) Cost B_1 = Cost A_1 + Interest on the value of owned fixed capital assets (excluding land);
(iv) Cost B_2 = Cost B_1 + Rental value of owned land (less land revenue) and rent paid for leased land;
(v) Cost C_1 = Cost B_1 + Imputed value of family labour;
(vi) Cost C_2 (Cost of cultivation) = Cost B_2 + Imputed value of family labour;
(vii) Cost C_3 = Cost C_2 + 10 percent of cost C_2 (to account for managerial input of the farmer).

2.4. *Choice of Marketing Channels*

A multinomial logit model (MNL) was used to quantify the predictors of variables that affect the marketing channels (0: consumption alone, 1: selling to friends and relatives; 2: selling to local markets; and 3: selling directly to Supplyco[2] at the farmgate) of traditional rice farmers of Palakkad, Malappuram and Wayanad. The MNL model assumes autonomy in the choice of conventional techniques for estimating multi-category dependent variables [55]. The chances of alternative marketing channel choices among farmers are shown below:

$$Prob = (Y_i = j) = \frac{\exp\left(\beta'_j x_i\right)}{\sum_{j=1}^{4} \exp\left(\beta'_j x_i\right)} \; for\; j = 0, 1, 2, 3 \quad (1)$$

where:

Y_i is the probability that farmers choose market j, pr(Y_i = j);
j = 0: consumption alone; 1: selling to friends and relatives; 2: selling to local markets; 3: selling directly to Supplyco at the farmgate;
x_i is the vector of households, production and marketing variables;
β_j is the vector of coefficients associated with market choice j.

The odds ratio was estimated from the significant variables to identify the probability of improving the marketing channel choice

$$\text{Odds ratio} = \frac{Exp(B)}{1 + Exp(B)} \times 100 \quad (2)$$

2.5. Farmer Satisfaction with Traditional Agriculture

Farmer satisfaction was measured by calculating the satiety index, which should act as an indicator of the sustainability of traditional rice (Supplyco is an institutional mechanism in Kerala operated by the government to procure produce from farmers at a predetermined price. Supplyco is an integral part of public distribution system in Kerala) cultivation. An arbitrary scale comprising questions on the probability of remaining in traditional farming, opinions about profitability, prices fetched by the produce, the procurement system was used, with the responses marked on a five-point continuum. The following formula was used for the satiety index:

$$Satiety\ index = \frac{S_i}{S_M} \times 100$$

where:

S_i = Score obtained by the *ith* individual;
S_M = Maximum possible score.

The satiety index categorised farmers as dissatisfied (<50), moderately satisfied (50–75) or highly satisfied (>75).

Furthermore, as with marketing channel choices, the MNL was used to measure the likelihood of improving farmer satisfaction with traditional agriculture.

2.6. Constraints in Traditional Rice Farming

Constraints faced by traditional rice farmers were analysed by using Garrett's ranking. Farmers were asked to express the constraints experienced in traditional farming via an open-ended question. The constraints expressed by the farmers were imputed into a frequency table. The rank of each constraint was converted to percent position by using the following equation:

$$\text{Per cent position} = \frac{100(R_{ij} - 0.5)}{N_j}$$

where:

R_{ij} is the rank for *ith* constraint faced by the *jth* individual;
N_j is the number of constraints ranked by the *jth* individual.

The rank obtained was an interval on a scale where its midpoint expressed the interval, such that 0.5 was subtracted from each rank. The percent position was converted into a score by using Garrett's table [56]. By using the score obtained from each constraint, the mean score was calculated and ranked according to the mean score.

3. Results and Discussion

3.1. Varieties under Cultivation

Despite having a government policy that does not support conversion of agricultural land for non-agricultural purposes, paddy lands are being used for other crops and non-agricultural purposes in Kerala state. As reported by the Kerala Bio Diversity Board, out of the nearly 160 rice varieties of Wayanad, 55 traditional varieties are now extinct [57,58]. A study conducted by Kerala Agricultural University and Dept of Agriculture and Farmers Welfare identified 63 cultivated traditional rice varieties in Wayanad district [59].

The major rice varieties grown in the study area are listed in Table 1. In Wayanad, the surveyed famers reported the highest productivity for variety "*Valichoori*" (>5000 kg/ha), which is protected under the Protection of Plant Varieties and Farmers' Rights (PPVRFA). Regardless of their unique characteristics, the low productivity reported for varieties *Jeerakasala*, *Gandhakasala* and *Navara* were considered less preferred for widespread cultivation. In Palakkad, the main traditional cultivated varieties were *Chitteni*, *Chettadi*, *Thavalakkannan* and *Chenkazhama*, which form part of the registration of the geographical indication (GI) *Palakkadan Matta* under the Geographical Indications of Goods (Registration and Protection) Act 1999. These varieties were also grown in the Malppuram district. Most farmers in each region were unaware of their legal status with regard to cultivating these varieties.

Table 1. Registration status and area under cultivation of traditional varieties identified in the study area.

Sl No	Variety	Registration Status	Area (ha)	Percentage of Total Area Surveyed	Year of Registration
1	Valichoori	Farmer Variety Reg No 221	24.93	20.87	2015
2	Gandhakasala	Farmer Variety Reg No 57 Geographical Indication Certificate no. 34	2.15	1.80	2013 2010
3	Jeerakasala	Farmer Variety Reg No 59 Geographical Indication Certificate no. 34	1.2	1.00	2013 2010
4	Adukkan	Farmer Variety Reg No 23	12.15	10.17	2016
5	Navara	Geographical Indication Certificate no. 17	3.31	2.77	2007
6	Mullankaima	Farmer Variety Reg No 220	0.20	0.17	2015
7	Chomala	Farmer Variety Reg No 59	0.20	0.17	2013
8	Chitteni *	Geographical Indication Certificate No 40	33.99	28.45	2013
9	Rakthasali		2.90	2.43	
10	Thavalakkannan *	Geographical Indicator Certificate No 40	2.67	2.24	2013
11	Thondi	Farmer Variety Reg No. 61	12.53	10.49	2013
12	Chettadi *	Geographical Indicator Certificate No 40	20.5	17.16	2013
13	Chenkazhama *	Geographical Indicator Certificate No 40	1.21	1.01	2013
14	Kattamodan		0.70	0.59	
15	Kochumannan		0.21	0.18	
16	Vella kayama		0.21	0.18	
17	Thekkancheera		0.40	0.33	
	Total		97.73	100	

Note: * The varieties under GI Certificate 40 are together labelled as Palakkadan matta and not as their individual names.

Traditional rice cultivation is declining in Kerala, as evidenced by the total area sown by the 300 surveyed farmers (97.73 ha). Legal and governmental measures exist to protect and support the traditional systems, but neither appear to be making a difference for farmers. In most cases, the farmers are unaware of the legal support mechanisms in place for their cultivated variety. For example, one reason for the low marketing efficiency of traditional rice is that farmers sell it as raw rather than de-husked grains because of the lack of suitable milling facilities for traditional rice.

The 300 surveyed respondents cultivated a total area of 97.93 ha of traditional rice (average 0.33 ha per farm), which is a true reflection of the very low area under traditional rice cultivation. This can be indicative of the substantial decline in varietal diversity in practice, because, by default, *padasekharams* (a collection of rice fields owned by different farmers) in Kerala have at most two or three varieties. Farmers keep this uniformity for ease and efficiency in management. Our survey results showed only 17 varieties being cultivated for consumption or marketing in the study area. This indicates a clear and ominous erosion of the gene pool of valuable traits.

3.2. Respondent Profile–Socio Demographic Characteristics

Table 2 shows that most of the surveyed respondents were 40–70-years-old (81%), male (87%), and had, at most, a high school education (77%), more than 30 years farming experience (63%) and cultivated less than 1 ha of land. Aging of traditional farmers has been reported in other rice-growing parts of the world. [60].

Table 2. Distribution of respondents according to socio-demographic characteristics.

		Palakkad (n = 100)	Malappuram (n = 100)	Wayanad (n = 100)	Total (N = 300)
			Age		
Frequency	<40	2	13	16	31 (10.33)
	40–54	34	33	38	105 (35.0)
	54–69	49	49	38	136 (45.33)
	>70	15	5	8	28 (9.33)
	Total	100	100	100	300 (100)
			Gender		
Category	Male	84	92	85	261 (87.00)
	Female	16	8	15	39 (13.00)
	Total	100	100	100	300 (100)
			Education		
Category	Primary	41	36	58	135 (45.00)
	High school	32	38	28	98 (32.67)
	SSLC and above	20	18	8	46 (15.33)
	College and above	7	8	6	21 (7.00)
	Total	100	100	100	300 (100)
			Experience		
Frequency	<15	5	14	4	23 (7.67)
	15—30	37	23	28	88 (29.33)
	30–45	36	44	38	118 (39.33)
	>45	22	19	30	71 (23.67)
	Total	100	100	100	300 (100)
			Area(ha)		
Category	<0.404	57	59	41	157 (52.33)
	0.404–0.809	31	33	39	103 (34.33)
	0.809–1.21	10	3	12	25 (8.33)
	>1.21	2	5	8	15 (5.00)
	Total	100	100	100	300 (100)

This picture of an ageing, less educated, experienced body of small holder farmers is similar in most of parts of the rice cultivating world [61] and these are the farmers who keep up traditional rice systems. This poses a serious question in terms of the sustainability of these farming systems in eras to come.

3.3. Economics of Traditional Rice Farming

Economic viability is a major consideration for sustainable farming. The total rice area under cultivation in 2017–18 was 75,415 ha, 7864 ha and 8026 ha in Palakkad, Malappuram and Wayanad districts, respectively, or 47 per cent of the total rice paddy area in Kerala [62]. These three districts contribute 46.8 per cent of the total rice production in Kerala, with production higher than the state average in Malappuram, lower than the state average in Palakkad and Wayanad. The costs of cultivation and production in 2016–2017 was Rs. 112,862/ha and Rs.31.40/kg, respectively (Table 3; [63].

Table 3. Area, production, and costs of cultivation and production of rice in Kerala.

Districts	Area (ha) (2017–18)	Production (tonnes) (2017–18)	Productivity (kg/ha) (2017–18)	Cost of Cultivation C) (Rs/ha) (2016–17)	Cost of Production (Rs/kg) (2016–17)
Palakkad	75,415	198,626	2633		
Malappuram	7864	23,571	2999		
Wayanad	8026	21,792	2715		
Kerala	194,235	521,310	2757	112,862	31.40

Source: Agricultural statistics 2017–18 Department of Economics and Statistics, GoK. Report on cost of cultivation of important crops in Kerala 2016–17, Department of Economics and Statistics, GoK.

The calculation for the cost of farming of traditional rice in Palakkad, Malappuram and Wayanad was based on three cost concepts detailed in Methods. Cost A in general covers the total paid up costs by the farmer, Cost B the interest and rental values, and Cost C the imputed values and imputed managerial costs. The calculation outcomes for each district are in Table 4. Since no land was rented in the study area, Cost A_1 and A_2 were the same; thus, Cost A_1 and B_1 were the same as the only fixed resources was land. The workers used their own equipment, and were paid accordingly in their wages. The Malappuram district had the highest values for Cost C_1, C_2, and C_3 and the Wayanad district had the lowest.

Table 4. Cost of cultivation and cost of production of traditional rice in the study area.

Cost of Cultivation (Rs./ha)				Cost of Production (Rs./kg)			
Districts/Cost	**Palakkad**	**Malappuram**	**Wayanad**	**Districts/Cost**	**Palakkad**	**Malappuram**	**Wayanad**
Cost A_1	50,806	52,181	51,063	Cost A_1	18.51	19.31	15.39
Cost A_2	50,806	52,181	51,063	Cost A_2	18.51	19.31	15.39
Cost B_1	50,806	52,181	51,063	Cost B_1	18.51	19.31	15.39
Cost B_2	68,306	74,181	68,563	Cost B_2	25.14	27.43	20.71
Cost C_1	57,374	56,525	54,622	Cost C_1	22.94	21.57	17.06
Cost C_2	74,874	78,525	72,122	Cost C_2	29.57	29.69	22.38
Cost C_3	82,361	86,378	79,334	Cost C_3	32.53	32.66	24.62
		Average yield/ha			2675	2877	3970

The cost of cultivation in all three districts was much lower than the state average for small-holding rice (Rs 112,862/ha, DoES, 2017) making traditional rice cultivation an attractive option in these areas. The rental value of owned land escalated Cost C_3, resulting in a slightly higher than average value (Rs. 31.40) except for the Wayanad district.

The Malappuram and Wayanad districts had higher productivity than the state average (2757 kg/ha), Wayanad had the highest productivity and Palakkad had the lowest (Table 4). The higher productivity in Wayanad can be attributed to the variety *Valichoori* which yields >5000 kg//ha, as reported by the farmers.

Different measures of income were used to identify the economic viability of traditional rice farming in the Palakkad, Malappuram and Wayanad districts (Table 5). According to the survey response, farmers in Wayanad had the highest average gross income (Rs. 85,281/ha), while those in Palakkad had the lowest (Rs. 71,578/ha). Similarly, Wayanad had the highest farm business income (Rs. 34,218/ha) and Palakkad had the lowest (Rs. 20,772/ha). Three speciality rices—*Navara* (medicinal variety) and *Gandhakasala* and *Jeerakasala* (aromatic varieties)—grow best in Wayanad, generating a higher price than other varieties, which would explain the higher farm business income in Wayanad.

Table 5. Estimates of different measures of income (Rs./ha).

Measures	Palakkad	Malappuram	Wayanad
Gross income (GI)	71,578	7936	85,281
Farm business income (GI-Cost A_1)	20,772	27,147	34,218
Family labour income (GI-Cost B_2)	3272	5144	16,718
Net Income (GI-Cost C_3)	−10,783	−7054	5947
BC (GI:C_3)	0.87	0.91	1.07
BC At Explicit (GI:A_1)	1.40	1.52	1.67

The net income and benefit:cost ratio indicated that farming was a loss-making business for respondents in Palakkad and Malappuram if family labour and land values were considered. In contrast, the positive net income and benefit:cost ratio in Wayanad was attributed to the high productivity of *Valichoori* and the higher price fetched by varieties

Jeerakasala and *Gandhakasala*. The benefit:cost ratio for explicit cost (i.e., Cost A_1) was >1 for all districts, with the highest in Wayanad (1.67) followed by Malappuram (1.52) and Palakkad (1.40).

This led us to consider the marketing behaviour of traditional rice farmers, with many rectifiable gaps identified.

3.4. Marketing Channels

Marketing is a concern for farmers, irrespective of the crops they cultivate. A common feature of grain marketing systems in developing countries is the co-existence of a government marketing agency (parastatal) and a parallel private marketing channel with many private intermediaries. These parastatals are assigned to control or regulate the system [64]. The survey revealed that most farmers sell their produce through the state-owned procurement and distribution agency, the parastatal in Kerala, named Supplyco, for a pre-determined price (Rs.25.3/kg at the time of the study).

In Wayand three marketing channels were identified (Figure 1A), of which Supplyco dominated with 52.46 per cent of the marketed volume. However, Supplyco, had limited penetration in some regions, such that farmers were forced to sell through other channels for a lower price (Table 6). Only varieties such as *Jeerakasala* and *Gandhakasala* obtained a higher price owing to their unique characteristics.

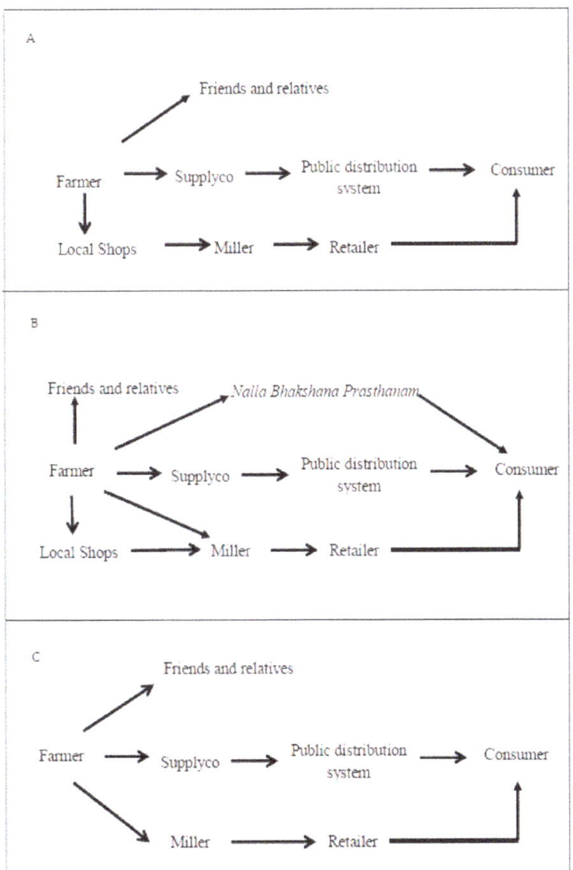

Figure 1. Marketing channels in Wayanad, Malappuram and Palakkad.

Table 6. The volume of rice marketed through each channel in the three districts.

Districts	Marketing Channels	Volume (%)
Wayanad	Channel 1	37.6
	Channel 2	52.46
	Channel 3	9.94
	TOTAL	100
Malappuram	Channel 1	4.04
	Channel 2	22.73
	Channel 3	59.44
	Channel 4	4.52
	Channel 5	9.27
	TOTAL	100
Palakkad	Channel 1	8.68
	Channel 2	79.37
	Channel 3	11.95
	TOTAL	100

In Malappuram, five marketing channels were identified (Figure 1B), with Supplyco having the highest share of marketed volume (Table 6). *Nalla Bhakshana Prasthanam*, an NGO work to provide "*safe to eat*" products, and is another rice marketing channel. Members of the society chose this avenue, as they received a higher price than at Supplyco. Farmers cultivating varieties such as *Navara* and *Rakthasali* were doing contract farming with a local miller who provided the seeds and procured the produce at a higher price than the prevailing local market price.

In Palakkad, Supplyco had the greatest marketing share of the marketed volume (almost 80%). The other avenues were selling to millers and friends and relatives (Figure 1C, Table 6).

The greatest quantity of rice is transacted through Supplyco; however, its performance had been criticised [65] since its establishment. Procurement lags and delays in payment often put farmers under stress [66]. The farmers only receive a standard price for traditional rice, as fixed by the agency, irrespective of the variety.

To fit the multinomial logit model, the farmers were classified into three categories according to their marketing behavior, such as, using their produce for consumption alone, selling in local markets and selling through Supplyco. These were considered as the first, second and third stages (Table 7). The significant variables at each stage indicate that they are the explanatory variables for farmers' progress from one stage to the other. In the first stage, age, education, area cultivated and yield were the decisive variables for their transition to the next. Interestingly, the awareness of marketing options can essentially promote their move from local markets to Supplyco, that is, from the second to the third stage. Therefore, enhancing the institutional support for marketing itself can help farmers improve their marketing. Education and production amount were identified as the main factors governing farmers' choice of marketing channels [67] attitude to risk, asset ownership, institutional variables, transaction costs and market attributes [68] and age of household, education status, credit access, off-farm income and total land-holding size [69].

To assess model fit in MLR, the most commonly used tool is the likelihood ratio test. Significance at less than 0.05 suggests a model fit [70]. To evaluate the goodness-of-fit of logistic models, several pseudo R-squared models have been developed. One method that is endorsed repeatedly [71,72] is the one proposed by [73]. According to McFadden, pseudo R^2 values from 0.2–0.4 indicate excellent model fit.

Marketing options will improve as institutional support improves and yields increase as well as the area cultivated by farmers (Table 8). In other words, these three factors differentiate the marketing behavior of farmers in the three districts, which were also significant in the case of potato farmers [74].

Table 7. Variables explaining farmers' choice of marketing channels—multinomial logit model.

Marketing Channels	Parameters	Odds Ratio	Chance of Improvement
0–1	Age	0.580	36.69
	Education	0.678	40.42
	Area	0.037	3.53
	Yield	0.999	49.99
1–2	Area	0.318	24.15
	Yield	0.999	49.98
2–3	Age	0.587	36.98
	Education	0.663	39.88
	Mobility	0.621	38.29
	Yield	0.999	49.98
	Awareness	2.263	69.36

Model Fitting Information

Model	Model Fitting Criteria	Likelihood Ratio Tests		
	−2 Log Likelihood	Chi-Square	df	Sig.
Intercept Only	708.760			
Final	546.513	162.247	27	0.000

Pseudo R-Square

McFadden	0.228

Table 8. Standardized Canonical Discriminant Function Coefficients for the choice of marketing channel.

Variable	Function
	1
Institutional support	0.897615
Yield	0.359001
Area	0.348232

3.5. *Are Farmers Satisfied?*

There are concerns about the economic viability and sustainability of traditional rice varieties. One study reported that the yields of traditional and modern rice varieties do not differ significantly [75], while another concluded that farmers rely on modern varieties because of their higher yields [76]. Traditional landraces are often more resilient to trying environmental conditions and produce more reliable yields across many situations than modern varieties [76]. We evaluated the farmers' satisfaction with probability of continuing with traditional rice and this was calculated as a 'satiety index' (Table 9).

Table 9. Percentage distribution of farmers according to satiety index.

Category	Per Cent (N = 300)
>50	10
50–75	66.67
75–100	26.33
Grand total	**100**

Of the 300 survey respondents, only 10 per cent were dissatisfied with traditional rice farming, while 26.33 were highly satisfied, which is a promising result for the sustainability of traditional rice cultivation. Furthermore, we evaluated the possibility of improving the satisfaction level using a multinomial logit model and calculated the odds ratio (Table 10).

The multinomial logit model showed that age, education, years of experience, mobility (access to the market), cultivated area and yield are significantly linked with farmer satisfaction in the first tier. Improvement in any of these variables would improve the satisfaction level of farmers in category I (dissatisfied) to category II (moderately satisfied).

Education, cultivated area [77], years of experience and mobility (access to transport) were the factors responsible for maximizing the farmers' satisfaction, i.e., in the second tier.

Table 10. Odds ratio of the satiety index of farmers in traditional rice cultivation with tables.

Satiety Index	Parameters	Odds Ratio	Chance of Improvement (%)	
1–2	Age	0.418	29.49	
	Education	0.591	37.16	
	Years of experience	1.793	64.20	
	Mobility	0.709	41.49	
	Area	1.423	58.73	
	Yield	1.000	50.01	
2–3	Age	0.476	32.25	
	Education	0.812	44.81	
	Year of experience	1.469	59.49	
	Mobility	0.649	39.35	
	Area	1.420	58.67	
Model Fitting Information				
Model	Model Fitting Criteria	Likelihood Ratio Tests		
	−2 Log Likelihood	Chi-Square	df	Sig.
Intercept Only	708.760			
Final	545.697	163.063	30	0.000
Pseudo R-Square				
McFadden			0.229	

3.6. Farmers' Constraints

Table 11 summerises the constraints expressed by the traditional rice farmers. Most of them (82%) do not rely on milling because they sell the rice as raw grains. Factors including grain shape (length:width ratio) and hardness affect milling efficiency. Round grains (raw) with low length:width ratios are difficult to break, while slender grains with higher ratios are easy to break. The surface hardness of the brown rice kernel is a varietal characteristic that determines the extent to which the grain can resist the forces applied during milling. Lower surface hardness facilitates breakage during milling, thus reducing milled rice recovery and quality [78]. High hulling percentages increase the recovery of rice [79]. Traditional rice varieties are more prone to breakage during milling and low hulling percentages than modern varieties, which reduces their commercial value and deters wholesalers from procuring traditional rice. Farmers who try to market traditional rice alone are aware of this drawback. Subsequently, traditional rice sold to wholesalers is mixed with other (modern) rice and sold under brand names that do not identify the rice varieties. Most urban consumers buy this type of rice.

Table 11. Constraints faced by the respondents.

Sl No.	Constraints	Mean Score	Rank
1	Shortage of skilled labour	64.64	1
2	Delay in payment	54.72	2
3	Shortage of water/rain	53.48	3
4	Lack of institutional support	53.09	4
5	Low productivity of labour	48.89	5
6	Transportation facility	47.53	6
7	Neighbourhood practices	46.09	7
8	Lack of milling facility	43.93	8
9	Animal attack	39.87	9

4. Conclusions

The drivers of agricultural sustainability in developing countries has been observed to encompass a range of demographic, natural, socio-economic, political, institutional and management factors [80]. Traditional rice cultivation could be a case study for sustainable agriculture. It has lower costs of cultivation than modern varieties because traditional varieties have evolved locally and have thrived for generations, resulting in fewer pest and disease issues and the ability to withstand climatic variations. Their medicinal, nutritive and safety values are considerable [23]. This study shows that in explicit terms traditional rice cultivation is less costly, which is a point of further thought for their promotion on larger scale in developing countries of a comparable situation.

The study supports the observation of [30] that farmers' decisions are influenced to a large extent by socioeconomic factors and that holding size, education status and yield influenced cultivation decisions. Because of the value of traditional varieties to sustainable agriculture and on-farm conservation, innovative government support policies are counselled to strengthen and sustain the traditional rice system. However, as [80] have observed, government policies to promote the cultivation of traditional rice varieties in the Western Ghats, by the promotion of on-farm conservation of crop genetic resources has been ineffective, because of an absence of financial incentives and education as to how these varieties satisfy their livelihood concerns, such as avoidance of risk, yield maximization, input suitability, yield stability and tolerance to environmental stress and marketability. The current study echoes those results in counselling educational extension activities by the agricultural authorities [81]. It highlights the need for consistent and timely institutional support in marketing.

Holding size and institutional support were the main factors governing the marketing behaviour of farmers. These promoted them to look for profitable marketing avenues. Even though traditional rice farming was not found to be cost-effective in implicit terms, it was remunerative when imputed personal labour and owned land costs were not considered. For the farmers involved in traditional rice cultivation, the strict economic validations of these factors did not matter so much, as they held their farming as more of a cultural heritage. The study also found that traditional farmers are ageing, have lower education and use limited marketing channels. But still, the majority of them were satisfied with their farm enterprise, which poses a positive note on sustainability. The reason for this might be that most traditional rice growers were also traditional in nature. The concept of the relative advantage in growing a potential high value crop seems to be lost on them and this is one area where concentrated awareness generation is necessary, as many of the respondents of the study said that they raised the traditional varieties only as a continuation of ancestral practice. This reality prevails in other Asian rice cultivating countries which see an erosion of young generation from rice farming, especially traditional rice farming. Thus, this study calls for a continuing dialog between the extension, research and policy systems, which should lead to the next generations being educated and supported in holding up the intrinsic value of these heritage crops and their sustainability at a time of climate change.

Author Contributions: J.K. wrote the report of the study. M.B. provided the mateial on the agricultural policy context. R.K.R. co-ordinated the surveys of farmers and relevant stakeholders. K.H.M.S. provided an over-view of all contributions. All authors have read and agreed to the published version of the manuscript.

Funding: This research was funded by the Australian Research Council, grant number DP170100747.

Institutional Review Board Statement: The study was conducted according to the guidelines of the Declaration of Helsinki, and approved by the Ethics Committee of the University of Newcastle (protocol code H-2017-0054 date of approval 1 May 2019).

Informed Consent Statement: Informed consent was obtained from all subjects involved in the study.

Data Availability Statement: The data presented in this study are available on request from the corresponding author.

Conflicts of Interest: The authors declare no conflict of interest.

References

1. United Nations (UN). United Nations Millennium Declaration, General Assembly Resolution 55/2 of 8 September 2000, UN doc. A/RES/55/2. 2000. Available online: https://www.un.org/en/development/desa/population/migration/generalassembly/docs/globalcompact/A_RES_55_2.pdf (accessed on 9 October 2019).
2. UN. Future We Want—Outcome Document, UN doc. A/RES/66/288. 2012. Available online: https://sustainabledevelopment.un.org/futurewewant.html (accessed on 9 October 2019).
3. UN. Resolution Adopted by the General Assembly on 25 September 2015, UN doc. A/RES/70/1. 2015. Available online: https://onlinelibrary.wiley.com/doi/10.1002/9781119541851.app1 (accessed on 9 October 2019).
4. WCED (World Commission on Environment and Development). *Our Common Future*; Oxford University Press: Oxford, UK, 1987.
5. McNeill, D. The Contested Discourse of Sustainable Agriculture. *Glob. Policy* **2019**, *10*, 16–27. [CrossRef]
6. Bell, S.; Morse, S. *Sustainability Indicators: Measuring the Immeasurable?* Earthscan Pubs: London, UK, 2008.
7. Zhen, L.; Routray, J.K. Operational Indicators for Measuring Agricultural Sustainability in Developing Countries. *Environ. Manag.* **2003**, *32*, 34–46. [CrossRef] [PubMed]
8. Pretty, J.N. Agricultural sustainability: Concepts, principles and evidence. *Phil. Trans. R. Soc. B* **2008**, *363*, 447–465. [CrossRef] [PubMed]
9. Walter, C.; Stützel, H. A new method for assessing the sustainability of land-use systems (I): Identifying the relevant issues. *Ecol. Econ.* **2009**, *68*, 1275–1287. [CrossRef]
10. Swaminathan, M.S. *Sustainable Agriculture: Towards Food Security*; Economics: Konark, Delhi, India, 1996.
11. Mohapatra, D.R. Sustainable Agriculture: A Key to Sustainable Development. *Int. J. Acad. Res.* **2017**, *4*, 39–45.
12. UN (Department of Economic and Social Affairs, Population Division). World Population Prospects 2019, UN doc. ST/ESA/SER.A/423. 2019. Available online: https://population.un.org/wpp/ (accessed on 9 October 2019).
13. Pham, L.V.; Smith, C. Drivers of agricultural sustainability in developing countries: A review. *Environ. Syst. Decis.* **2014**, *34*, 326–341. [CrossRef]
14. Srinivasan, A. *Handbook of Precision Agriculture: Principles and Applications*; Haworth Press: New York, NY, USA, 2006.
15. Huang, J.; Pray, C.; Rozelle, S. Enhancing the crops to feed the poor. *Nature* **2002**, *418*, 678–684. [CrossRef]
16. Phipps, R.H.; Park, J.R. Environmental Benefits of Genetically Modified Crops: Global and European Perspectives on Their Ability to Reduce Pesticide Use. *J. Anim. Feed Sci.* **2002**, *11*, 1–18. [CrossRef]
17. Prasifka, J.R.; Hellmich, R.L.; Weiss, M.J. Role of biotechnology in sustainable agriculture. In *Integrated Pest Management. Concepts, Tactics, Strategies and Case Studies*; Radcliffe, E.B., Hutchison, W.D., Cancelado, R.E., Eds.; Cambridge University Press: New York, NY, USA, 2009; pp. 260–272.
18. Altieri, M.A. *Agroecology: The Science of Sustainable Agriculture*; Westview Press: Boulder, CO, USA, 1995.
19. Gliessman, S.R. *Agroecology: Ecological Processes in Sustainable Agriculture*; CRC Press: Boca Raton, FL, USA, 1997.
20. McNeely, J.A.; Scherr, S.J. *Ecoagriculture. Strategies to Feed the World and Save Biodiversity*; Island Press: Washington, DC, USA, 2003.
21. Wezel, A.; Casagrande, M.; Celette, F.; Vian, J.-F.; Ferrer, A.; Peigné, J. Agroecological practices for sustainable agriculture. A review. *Agron. Sustain. Dev.* **2014**, *34*, 1–20. [CrossRef]
22. Janaiah, A.; Debdutt, B. The Rice Seed System in India: Structure, Performance, and Challenges, Intellectual Property Rights, Innovation and Rice Strategy for India. In *The Future Rice Strategy for India*; Mohanty, S., Ed.; Elsevier Science & Technology: London, UK, 2017; pp. 359–382.
23. Ashraf, A.M.; Lokanadan, S. A Review of Rice Landraces in India and its Inherent Medicinal Values—The Nutritive Food Values for Future. *Int. J. Curr. Microbiol. Appl. Sci.* **2017**, *6*, 348–354. [CrossRef]
24. Nelson, A.R.L.E.; Ravichandran, K.; Antony, U. The impact of the Green Revolution on indigenous crops of India. *J. Ethnic Foods* **2019**, *6*, 8. [CrossRef]
25. Parayil, G. The green revolution in India: A case study of technological change. *Technol. Cult.* **1992**, *33*, 737–756. [CrossRef]
26. Roy, K.; Mukherjee, A.; Maity, A.; Shubha, K.; Nag, A. Protecting Non-Basmati Indigenous Aromatic Rice Varieties of West Bengal, India Under Geographical Indication: A Critical Consideration. In *The Role of Intellectual Property Rights in Agriculture and Allied Sciences*; Roy, C., Ed.; Apple Academic Press: Oakville, ON, USA, 2019; pp. 273–295.
27. Sathiya, A. Are the Indian Rice Landraces a Heritage of Biodiversity to Reminisce their Past or to Reinvent for Future? *Asian Agrihist* **2013**, *17*, 221–232.
28. Rekha, T.; Martin, K.P.; Sreekumar, V.B.; Madassery, J. Genetic diversity assessment of rarely cultivated traditional *Indica* rice (*Oryza sativa* L.) varieties. *Biotechnol. Res. Int.* **2011**, 1–7. [CrossRef] [PubMed]
29. Thrupp, L.A. Linking Agricultural Biodiversity and Food Security: The Valuable Role of Agrobiodiversity for Sustainable Agriculture. *Int. Aff.* **2000**, *76*, 265–281. [CrossRef] [PubMed]
30. Zhu, Y.; Wang, Y.; Chen, H.; Lu, B.-R. Conserving Traditional Rice Varieties through Management for Crop Diversity. *Bioscience* **2003**, *53*, 158. [CrossRef]
31. Sharma, R.D. *The Story of Rice*; National Book Trust: New Delhi, India, 1991; p. 68.
32. Richharia, R.; Govindasamy, S. *Rices of India*; Academy of Development Science: Karjat, India, 1990.
33. Kumari, S.L. Status Paper on Rice. Rice Knowledge Management Portal (RKMP) Directorate of Rice Research, Rajendranagar, Hyderabad 500030. 2007. Available online: http://www.rkmp.co.in/sites/default/files/ris/rice-state-wise/Status%20Paper%20on%20Rice%20in%20Kerala.pdf (accessed on 15 January 2021).

34. NICRA [National Innovations on Climate Resilient Agriculture]. 2019. Available online: http://www.nicra-icar.in/nicrarevised/index.php?option=com_content&view=article&layout=edit&id=184 (accessed on 15 January 2021).
35. Xie, J.; Wu, X.; Tang, J.-J.; Zhang, J.-E.; Luo, S.-M.; Chen, X. Conservation of Traditional Rice Varieties in a Globally Important Agricultural Heritage System (GIAHS): Rice-Fish Co-Culture. *Agric. Sci. China* **2011**, *10*, 754–761. [CrossRef]
36. Sinha, H. Rediscovering the Traditional Paddy Varieties in Jharkhand: Conservation Priority in Hybrid Rice Era. *J. Rural Dev.* **2016**, *35*, 285–307.
37. Singh, K. Mapping regional risks from climate change for rainfed rice cultivation in India. *Agric. Syst.* **2017**, *156*, 76–84. [CrossRef]
38. Soora, N.K.; Aggarwal, P.K.; Saxena, R.; Rani, S.; Jain, S.; Chauhan, N. An assessment of regional vulnerability of rice to climate change in India. *Clim. Chang.* **2013**, *118*, 683–699. [CrossRef]
39. Gogoi, J.; Hzaraika, J.P.; Barman, U.; Deka, N. Comparative Study of Input Use, Productivity and Profitability of Hybrid and Traditional Rice Cultivation in Assam, India. *Econ. Aff.* **2020**, *65*, 389–394. [CrossRef]
40. Huang, J.; Qiao, F.; Zhang, L.; Rozelle, S. *Farm Pesticide, Rice Production, and Human Health*; Research Report (No. 2001-RR3); Economy and Environment Program for Southeast Asia: Singapore, 2001.
41. Heryadi, D.Y.; Noor, T.I.; Deliana, Y.; Hamdani, J.S. Assessing the performance and factors affecting organic rice agribusiness sustainability. *Int. J. Devel. Res.* **2018**, *8*, 24110–24115.
42. Brklacich, M.; Bryant, C.R.; Smith, B. Review and appraisal of concept of sustainable food production systems. *Environ. Manag.* **1991**, *15*, 1–14. [CrossRef]
43. Senanayake, R. Sustainable agriculture: Definition and parameters for measurement. *J. Sustain. Agric.* **1991**, *1*, 7–28. [CrossRef]
44. Simon, D. Sustainable development: Theoretical construct or attainable goal? *Environ. Conserv.* **1989**, *16*, 41–48. [CrossRef]
45. Smith, C.S.; McDonald, G.T. Assessing the sustainability of agriculture at the planning stage. *J. Environ. Manag.* **1998**, *52*, 15–37. [CrossRef]
46. Stockle, C.O.; Papendick, R.I.; Saxton, K.E.; Campbell, G.S.; Van Evert, F.K. A framework for evaluating the sustainability of agricultural production systems. *Am. J. Altern. Agric.* **1994**, *9*, 45–50. [CrossRef]
47. Tisdell, C. Economic indicators to assess the sustainability of conservation farming projects: An evaluation. *Agric. Ecosyst. Environ.* **1996**, *57*, 117–131. [CrossRef]
48. Ashfaq, M.; Hassan, S.; Naseer, M.Z.; Baig, I.A. Factors Affecting Farm Diversification in Rice-Wheat. *Pak. J. Agric. Sci.* **2008**, *45*, 91–94.
49. Koesling, M.F.; Lien, G. Factors Influencing the Conversion to Organic Farming in Norway (Spacial Issue: Continuity and Change in Organic Farming-Philosophy, Policy and Practice). *Int. J. Agric. Resour. Gov. Ecol.* **2008**, *7*, 365–374.
50. Chaharsughi Amin, H.; Mirdamadi, M. Identify and analyze the factors affecting sustainable agriculture among women farmers in Bandar Anzali city with emphasis on local knowledge of rice cultivation in the area. *J. Agric. Knowl. Iran* **2008**, *5*, 61–84.
51. Daryaei, N.; Moghaddam, K.R.; Salmanzadeh, C. Factors Affecting Sustainable Rice Farming: The Case of Rice Farmers in the Mazandaran Province, Iran. *Eur. J. Nat. Soc. Sci.* **2015**, *4*, 8–16.
52. Herrera, B.; Bentaya, M.G.; Knierim, A. Farm-level factors influencing farmers' satisfaction with their work. In Proceedings of the 30th International Conference of Agricultural Economists, Vancouver, BC, Canada, 28 July–2 August 2018.
53. Stiglitz, J.E.; Sen, A.; Fitoussi, J. *Mismeasuring Our Lives. Why GDP Doesn't Add Up*; The Report by the Commission on the Measurement of Economic Performance and Social Progress; New Press: New York, NY, USA, 2010.
54. CSO [Central Statistical Organization]. Manual on Cost of Cultivation Surveys. Ministry of Statistics and Programme Implementation. 2008. Available online: http://mospi.nic.in/sites/default/files/publication_reports/manual_cost_cultivation_surveys_23july08_0.pdf (accessed on 15 January 2021).
55. Wooldridge, J. *Introductory Econometrics: A Modern Approach*; Cengage LearningSouth-Western 5191 Natorp Boulevard Mason: Warren County, OH, USA, 2012.
56. Garrett, H.E.; Woodworth, R.S. *Statistics in Psychology and Education*; Vakils, Feffer and Simons Pvt. Ltd.: Bombay, India, 1969; p. 329.
57. Business Line. Traditional Rice Varieties Vanishing in Kerala: Study. 2012. Available online: https://www.thehindubusinessline.com/economy/agri-business/Traditional-rice-varieties-vanishing-in-Kerala-Study/article20476098.ece (accessed on 15 January 2021).
58. Lakshmikutty, D. Traditional Rice Varieties of India. 2018. Available online: https://www.esamskriti.com/e/Culture/Indian-Culture/Traditional-Rice-Varieties-of-India-1.aspx (accessed on 17 July 2019).
59. KAU [Kerala Agricultural University]. Wayanadan Nellinangal. In *Directory of Farmers Conserving Traditional Rice Varieties*; Government of Kerala Department of Agrl Development and Farmers Welfare-Kerala Agricultural University: Kerala, India, 2018.
60. Rambukwella, R.; Priyankara, E.A.C. *Production and Marketing of Traditional Rice Varieties in Selected Districts in Sri Lanka: Present Status and Future Prospects. Research Report 195*; Hector Kobbekaduwa Agrarian Research and Training Institute: Wijerama Mawatha Colombo, Sri Lanka, 2016.
61. Ginigaddara, G.A.S.; Disanayake, S.P. Farmers' Willingness to Cultivate Traditional Rice in Sri Lanka: A Case Study in Anuradhapura. In *Rice Crop—Current Developments*; IntechOpen: London, UK, 2018. [CrossRef]
62. DoES [Department of Economics and Statistics]. Agricultural Statistics 2017–2018. 2018. Available online: http://www.ecostat.kerala.gov.in/images/pdf/publications/Agriculture/data/2017-18/rep_agristat_1718.pdf (accessed on 10 July 2019).

63. DoES [Department of Economics and Statistics]. Report on Cost of Cultivation of Important Crops in Kerala 2016–2017. 2017. Available online: http://www.ecostat.kerala.gov.in/images/pdf/publications/Cost_of_Cultivation/data/cost1617.pdf (accessed on 10 July 2019).
64. Crawford, I.M. *Agricultural and Food Marketing Management*; FAO: Rome, Italy, 1997; p. 344.
65. Mathew, M.D.; Sany, N. *Relevance of Supplyco and Its Operations in Kerala's Current Scenario*; Centre for Public Policy Research: Kochi, Kerala, India, 2018.
66. Asianet News. 2017. Available online: https://www.youtube.com/watch?v=7VggF3iR9Pk (accessed on 9 October 2019).
67. Xaba, B.G.; Masuku, M.B. Factors Affecting the Choice of Marketing Channel by Vegetable Farmers in Swaziland. *Sustain. Agric. Res.* **2012**, *2*, 112–113. [CrossRef]
68. Mazibuko, B.P.D.; Ferrer, S.; Ortmann, G. Factors affecting the choice of marketing outlet selection strategies by smallholder farmers in Swaziland. *Afr. J. Sci. Technol. Innov. Devel.* **2019**, *11*, 569–577. [CrossRef]
69. Dessie, A.B.; Abate, T.M.; Mekie, T.M. Factors affecting market outlet choice of wheat producers in North Gondar Zone, Ethiopia. *Agric. Food Secur.* **2018**, *7*, 1–8. [CrossRef]
70. Petrucci, C.J. A Primer for Social Worker Researchers on How to Conduct a Multinomial Logistic Regression. *J. Soc. Serv. Res.* **2009**, *35*, 193–205. [CrossRef]
71. Menard, S. Coefficients of determination for multiple logistic regression analysis. *Am. Stat.* **2000**, *54*, 17–24.
72. Allison, D.P. Measures of Fit for Logistic Regression. Paper 1485–2014 SAS Global Forum. 2014. Available online: https://statisticalhorizons.com/wp-content/uploads/GOFForLogisticRegression-Paper.pdf (accessed on 15 January 2021).
73. McFadden, D. Conditional logit analysis of qualitative choice behavior. In *Frontiers in Econometrics*; Zarembka, P., Ed.; Academic Press: Cambridge, MA, USA, 1974; pp. 105–142.
74. Emana, B.; Ketema, M.; Mutimba, J.K.; Yousuf, J. Factors Affecting Market Outlet Choice of Potato Producers in Eastern Hararghe Zone, Ethiopia. *J. Econ. Sust. Devel.* **2015**, *6*, 159–173.
75. Hassen, B.N.; Monaco, F.; Facchi, A.; Romani, M.; Valè, G.; Sali, G. Economic Performance of Traditional and Modern Rice Varieties under Different Water Management Systems. *Sustainability* **2017**, *9*, 347. [CrossRef]
76. Ficiciyan, A.; Loos, J.; Glotzbach, S.S.; Tscharntke, T. More than Yield: Ecosystem Services of Traditional versus Modern Crop Varieties Revisited. *Sustainability* **2018**, *10*, 2834. [CrossRef]
77. Xiao, D.; Haipeng, N.; Fan, L.; Zhao, S.; Yan, H. Farmers' Satisfaction and its Influencing Factors in the Policy of Economic Compensation for Cultivated Land Protection: A Case Study in Chengdu, China. *Sustainability* **2019**, *11*, 5787. [CrossRef]
78. CIPHET [Central Institute of Post Harvest Engineering and Technology]. Rice Milling. 2019. Available online: https://vikaspedia.in/agriculture/post-harvest%20technologies/technologies-for-agri-horti-crops/rice-milling-plant-process (accessed on 15 January 2021).
79. Bhonsle, S.J.; Sellappan, K. Evaluation of Traditionally Cultivated Rice Varieties of Goa, India. *Recent Res. Sci. Technol.* **2010**, *2*, 88–97.
80. Prakash, T.N.; Mburu, J.; Chandrashekar, H.; Abebaw, D. Farmers' willingness to conserve crop genetic resources under different policy incentives. The case of traditional rice varieties in the Western Ghats of South India. *Outlook Agric.* **2007**, *36*, 137–143. [CrossRef]
81. The New Indian Express. Paddy Farmers in a Soup over Delay in Disbursal of Funds. 2019. Available online: http://www.newindianexpress.com/states/kerala/2019/may/16/paddy-farmers-in-a-soup-over-delay-in-disbursal-of-funds-1977399.html (accessed on 9 October 2019).

sustainability

MDPI

Article

Is a Training Program Sufficient to Improve the Smallholder Farmers' Productivity in Africa? Empirical Evidence from a Chinese Agricultural Technology Demonstration Center in Tanzania

George Mgendi [1], Shiping Mao [1,*] and Fangbin Qiao [2]

[1] Institute of Agricultural Economics and Development, Chinese Academy of Agriculture Sciences, Beijing 100081, China; gmgendi@gmail.com

[2] China Economics and Management Academy, Central University of Finance and Economics, Beijing 100081, China; fangbin.qiao@gmail.com

* Correspondence: maoshiping@caas.cn; Tel.: +86-10-8210-8388

Abstract: The article aims to analyze the effect of training programs on the yield of smallholder farmers. The empirical analysis employed a sample of data collected from a rice farming household in the Mvomero district of Tanzania. The results indicate that the yield outcome among trained and non-trained farmers with water access for irrigation was significantly more than double; however, the yield difference between trained and non-trained farmers was insignificant in non-irrigated plots. Our findings have policy implications for agricultural development in developing countries where training programs alone may not be a panacea for smallholder farmers' productivity improvement. Therefore, respective governments, policymakers, and other agricultural stakeholders, should consider both farm and non-farm factors altogether, which may increase agricultural training effectiveness to address the challenges of low yields.

Keywords: China-Africa cooperation; agricultural program; agricultural training; technology adoption

Citation: Mgendi, G.; Mao, S.; Qiao, F. Is a Training Program Sufficient to Improve the Smallholder Farmers' Productivity in Africa? Empirical Evidence from a Chinese Agricultural Technology Demonstration Center in Tanzania. *Sustainability* **2021**, *13*, 1527. https://doi.org/10.3390/su13031527

Academic Editor: Michael Blakeney

Received: 13 January 2021

Accepted: 27 January 2021

Published: 1 February 2021

Publisher's Note: MDPI stays neutral with regard to jurisdictional claims in published maps and institutional affiliations.

Copyright: © 2021 by the authors. Licensee MDPI, Basel, Switzerland. This article is an open access article distributed under the terms and conditions of the Creative Commons Attribution (CC BY) license (https://creativecommons.org/licenses/by/4.0/).

1. Introduction

The majority of the world's poor and undernourished people live in rural areas of developing countries. Often these people depend on agricultural activities for most of their lives. Agriculture plays an important role, not only as a source of food and household income, but also in contributing to economic development. In Tanzania, the agricultural sector contributes about 27% of Gross Domestic Product (GDP) and 67% of total employment [1]. However, despite an average growth rate of 4.4%, the country's agricultural sector is still characterized by traditional agriculture and its methods that cannot bring real change in contributing to more significant wealth creation and poverty reduction [2].

The low adoption of new agricultural technology may be linked to various factors, such as new technology not being profitable in the farmers' context, infrastructures that are not supportive, lack of enough money to purchase technology (even if it is profitable), or unwillingness of farmers to try due to risk it entails [3,4]. Still, the need for upscaling agricultural technology in Tanzania can never be undervalued because technological change is necessary for the development process [5]. The Tanzanian agricultural sector needs to grow at an average growth rate of around 6 to 8% to play a significant role in poverty reduction and increasing wealth [6,7].

The Tanzanian government's strategy to bring about greater efficiency in the agricultural sector includes strengthening various agricultural development strategies and technical cooperation, with local and international development partners to increase awareness and make technology easier to access. Tanzania is among the African countries that benefited from constructing the Agricultural Technology Demonstration Center (ATDC) between 2009 and 2013. The ATDC is a Chinese-funded program to promote agricultural technology and share Chinese farming experiences with Africa. Approximately USD $130 million was funded through a China-Africa cooperation to facilitate the construction of about 23 ATDCs across Africa to improve agricultural production and food security in rural areas. The program benefited Tanzanians with one ATDC located at Dakawa, which was constructed from 2009 to 2011, and cost about USD $6 million.

The demonstration, based on Chinese expert perspectives, implies doing on-station comparison studies and pilot research, generally within the compound of ATDC, to present how Chinese technology may perform in an African context [8]. In the initial trial of experimental plots in the station, the rice yields with Chinese hybrids were four times higher than that of local varieties [9]. However, the Chinese hybrid varieties were then neither locally reproduced nor disseminated to farmers due to intellectual property rights, seed regulation, and phytosanitary procedures [10]. Besides, experiment trials revealed that Chinese farm technology alone could contribute 20% to 30% of productivity improvement for the local rice varieties than traditional methods [11]. Thus, the actual rice technology received by farmers from ATDC was limited to their recommended rice management practices from seeding to harvesting, disseminated via training, and demonstration in helping farmers improving rice productivity in Tanzania

Nevertheless, Tanzania has continued to increase agricultural production and reduce food insecurity in recent years, but much more needs to be done. The country ranks second in rice production in Eastern and Southern Africa after Madagascar, and in the country, rice is the second staple food and commercial crop after maize. Nevertheless, rice productivity in Tanzania is still unsatisfactory. While Senthilkumar [12] reported the rice yield gap as high as 87%, the Food and Agriculture Organization of the united nations (FAO) statistic data indicate that the rice yield per hectare is lower (1.6–2.7 MT/ha) because most of the harvest is still dependent on rainfall during the growing season. In contrast to the few areas available for irrigation schemes, higher average yields (5–6 MT/ha) have been found [13]. Yet, nearly 71% of produced rice within the country relies on rainfall cultivation by smallholder farmers, constituting about 80% of the country's farmers. These small farmers usually cultivate an average of 0.2 to 3.0 hectares. However, they are experiencing weather variability while facing limited long-term strategies and investment in agricultural production due to the lack of financial capacity of the farming households. They therefore continue to experience limited technology that prevents them from achieving high yields.

Many farmers nowadays still relying on traditional methods that may be ineffective for rice production, and, thus, they continue to be small-scale producers, mostly to feed their families, and what remains is sold to generate income. However, improving rice production remains the government priority in the rice industry strategy, by which Tanzania is visioning to become a significant rice exporter in Africa. Achieving rice production growth is an essential requirement that necessitates the technology transfer process and smallholder farmers' adoption of rice yield increasing technology. However, researchers have shown that under the best condition, potential rice yields in Tanzania range from 4 to 6 MT/ha for uplands and 6 to 10 MT/ha for lowlands irrigated ecosystems of cultivation. Still, these levels of harvests always depend on the type of rice varieties and the level of management practices [13]. Therefore, agricultural training and advice are essential in enabling rice producers to realize potential productivity by giving them the proper tools, knowledge, and skills necessary.

To date, several national and international partners have been collaborating with the government to address this issue of low rice productivity of smallholder farmers in Tanzania. Agricultural training on rice cultivation practices has reached farmers through various extension services, such as village extension services under the district government office, the ministry of agriculture, agricultural research, training institutions, and Non-Governmental Organizations (NGOs). The field extension services are based on personal or group visits, scheduled farmers meetings, the use of key farmers (model farmer), farm demonstration plots, and lately, information and communications technology (ICT) application. It is argued that, in comparison to conventional farmers' practices, with a set of good agricultural practices (GAP) in a participatory farm trial in 2013 and 2014, concerning growing seasons in representative irrigated and rainfed fields of Kilombero valley in Tanzania, farmers achieved a yield increase of 1 MT/ha in 2013 and 2.7 MT/ha in 2014 when farmers followed these rice cultivation practices. However, under rainfed conditions, the GAP impact on yield was not significant [12].

A five-year household-level panel data from 2008 to 2012 in Tanzania's Ilonga irrigation scheme supports the finding that technology adoption increases immediately after training. Furthermore, non-trained farmers may catch up belatedly. In the case of technologies disseminated by Japan International Cooperation Agency (JICA) through project for supporting rice industry development in Tanzania (TANRICE) training, the paddy yield of the key farmers increased from 3.1 to 5.3 MT/ha, while ordinary farmers' yield increased from 2.6 to 3.7 MT/ha (Nakano et al., 2018). Such findings indicate the effectiveness of training programs to increase the productivity of smallholder farmers.

Nevertheless, while literature considered the low adoption rate of high-yielding technology, and inputs (such as fertilizer) as barriers to a "Green Revolution" in sub-Saharan Africa [14–16], the impact of ATDC on farmers' yield has not been evaluated empirically to bring insight into whether international agricultural technical assistance through the establishment of ATDCs may be an effective way to overcome these obstacles. At times, technology transfer is characterized by significant reticence and environmental sensitivity—this is worth considering for international agriculture technology transfer [17]. For instance, backward areas may require adaptive investments in the process of technology introduction and transfer to suit local agricultural production conditions. As such, this study is designed to understand the effectiveness of training, by the Chinese Agricultural Technology Demonstration Center, to improve yields of smallholder farmers in Tanzania.

Tanzania's farmers have benefited from training and demonstration program on rice cultivation technologies provided by the Chinese Agricultural Technology Demonstration Center (ATDC) since 2011. The ATDC program aimed to improve food security in recipient countries by transferring Chinese technology via demonstration in farms, extension, and training from Chinese experts [18]. However, little is known regarding the effect of ATDC training in addressing the productivity of communities that participated in the projects. Previous studies on ATDC have focused on the concept of establishment of ATDC, situation, performance, and challenges, but the farmers' side, as beneficiaries, required further investigation [8,10,19–24]. Thus, understand the impact of the ATDC program on beneficiaries yield improvement may assist with setting priorities, provide feedback to research programs, and guide policymakers and stakeholders involved in the process of technology transfer.

This study, therefore, utilizes representative household data collected in 2019 to contribute to the literature in two ways. First, it tests whether the farmer's access to agricultural training will enhance the agricultural productivity of small farmers. Second, this paper examines the main factors that can encourage participation in agricultural training and the improvement of farm yield among rice farmers.

The rest of this paper is organized as follows. Section 2 is about material and methods used, which introduce the Tanzania ATDC program in brief, methodology, and data used in this study. Section 3 presents estimation results on the participation of smallholder farmers in agricultural training programs, factors affecting yields among trained and non-trained farmers, training impact on technology adoption and rice yields of farmers, and the spillover effect of training programs. Section 4 concludes this study with some policy implications.

2. Materials and Methods

2.1. The Tanzania ATDC Program in Brief

The construction of Agricultural Technology Demonstration Center (ATDC) was initially announced in November 2006 at the Beijing Summit of China–Africa Cooperation, as a model for delivering agricultural aid. The criteria for establishing ATDC were based on the preliminary request from the government of a host country, comments from the Chinese embassy in a host country on the feasibility of the project, and tradition of co-operation between China and a host country. Tanzania followed the procedures and the protocols signed between China and Tanzania, one year after the summit, followed by the construction stage from 2009 to 2011. The ATDC was constructed in the Mvomero district in Dakawa, near the Tanzania Agriculture Research Institute (TARI–Cholima). The total area of ATDC is 62 ha, which includes three sections: office and training area (2 ha), experiment and display area (10 ha), and production and demonstration area (50 ha) with irrigation infrastructure. The overseer of the Chinese ATDC in Tanzania is a Chinese enterprise (Chongqing Sino-Tanzania Agricultural Development Co. Ltd., Chongqing, China) and the Chongqing Academy of Agricultural Sciences in China, with the cooperation of the Tanzanian Ministry of Agriculture, Food Security, and Cooperatives (MAFC). The primary function of ATDC includes the research experiment of improved cultivars (hybrid and conventional rice, maize, horticulture, and banana) and techniques, training, and demonstration of the technology. However, ATDC in Dakawa has focused on rice, with various trials and demonstrations of the ten different hybrid rice cultivars from the Chongqing Academy of Agricultural Sciences, but also promotes agro-mechanization tools from China. China dispatched the modern agricultural equipment, including tractors, hollows, plows, rice harvesters, to be used in training and demonstration activities, to stimulate the interests of farmers to purchase similar products in the future.

The establishment of the Chinese Agricultural Technology Demonstration Center program in Tanzania may be summarized in three stages: the construction stage, technical cooperation, and business operation stage. The construction stage refers to the establishment of ATDC, held for two years (2009–2011) in Tanzania. The technical cooperation stage followed from 2011, after the establishment of ATDC. In the technical cooperation phase, the center began to implement the primary function of the center, which took place in three years. The ATDC training involved a combination of classroom, on-site, and video teaching, which aimed to spread modern agricultural knowledge on the use of new varieties and apply new technologies. The skills covered in the rice-training program include sowing and paddy seed preparation, leveling and water lodging prevention, nursery care, transplanting techniques, management of the soil nutrients, and fertilizer application. In the business stage, which is also known as the sustainable development phase, ATDC is required to establish a market-oriented operation and integrated agribusiness value chain to support ATDC operations without Chinese government grants. Currently, the ATDC in Tanzania is in a business stage that may take up to 12 years. In the technical cooperation and sustainable development phase, local farmers living in nearby villages have been organized for training in the center, particularly in sowing and harvesting seasons.

2.2. Modeling the Yield Effect of ATDC Training Program

The primary purpose of this paper is to assess the rice yield effect of the ATDC training program in a quasi-experimental setting using a cross-sectional dataset collected in Tanzania. The main challenge of quantifying the impact of a program, such as ATDC, is that we observe only what happens to beneficiaries who receive the training program; we do not observe what would happen to the same farmers if they did not receive training from ATDC. Alternatively, the program would have been introduced in a randomized control design setting, where communities and households are randomly selected to receive the treatments (ATDC training program), where some would receive, and the rest would not. The households benefited from ATDC training were known as treated households; while the rest were known as control/untreated households. Apart from a randomized evaluation design, the study would apply a regression discontinuity design that requires a cut-line/threshold separating treated households and untreated households. However, a challenge that would arise is selection bias when treated households differ in some systematic way from untreated households, which would imply that those who choose to participate in ATDC training projects are already more likely to have better outcomes for reasons entirely other than their participation in the program. The randomization process would be suitable for internal validity as it could eliminate the validity threat of selection bias [25]. Thus, our ability to make a statement about the yield effect of the ATDC training program rests on how well we can address these main challenges of impact studies.

The method applied in this paper needs to address several issues given program intervention. First, we realize during the survey that the training program was disseminated to villages closer to ATDC without random distribution. Second, the ATDC model is based on farmers' invitation to the ATDC, to access training from Chinese experts, yet the farming households who participated in the Tanzania ATDC program were not randomly selected from the village population. Third, there were no precise quantitative or qualitative mechanisms used in the selection of farming households that participated in the training program. These factors may bring a challenge of disentangling the yield impact of the ATDC program using our observational data. Yet, they make it feasible not to apply a randomized evaluation design and a regression discontinuity design in our evaluation analysis. Following Shiferaw et al. [26], when treatment groups are not randomly assigned during an intervention, the yield impact is likely to be influenced both by unobservable and observable heterogeneity, which leads to invalid estimation. Thus, we adopted procedures of a matching method based on the inverse probability weighted regression adjusted (IPWRA) technique, as applied by various impact evaluation studies [26–29].

In IPWRA application, we understand that it can only address self-selection bias due to observable variables, but not estimation bias due to unobservable variables, such as land quality and farmer capacity, which are common omitted variables in agricultural productivity studies. Nevertheless, the instrumental variable method's implementation, such as endogenous switching regression as an alternative approach, faces minimal data problems to be feasible, while disentangling the ATDC program's effects against others remain important in this paper. The farmer's distance to the ATDC is an important factor determining whether the farmer participates in the training program or not, which means that the instrumental variables' correlation condition is satisfied. Since the location where the ATDC was constructed is not affected by the individual characteristics, the distance to the ATDC does not affect the farmer's productivity through any other channel than farmer training participation, satisfying the exogeneity condition of the instrumental variable. However, the distance to different trainings, particularly in farmers who participated in only other programs, is missing in our data. Thus, it may be feasible to use the distance to the ATDC from the farmer as an instrumental variable for whether or not to participate in the ATDC project, but not in other non-ATDC projects. As such, this analysis relies on the IPWRA matching method.

The treated and untreated samples are matched to estimate the average difference between outcome yields by observable characteristics in the matching approach. Here, the IPWRA implicitly compares every unit to every other, while placing higher weights on observations with a similar likelihood of being treated or untreated, and lower weights on dissimilar observations [28]. On the other hand, we have assumed that households who participated in the ATDC program faced the binary choices (1 = adopt improved rice cultivation practices that improve farm yield, 0 = otherwise). However, the decision an individual household takes depends on certain factor/characteristic/technology attributes. Thus, reconstructing the valid counterfactuals is essential for determining the unbiased causal effects of the training program. The selected households for comparison should have the same characteristics, apart from participation in the ATDC training program, so that observed rice yield differences may be estimated as the yield impact of ATDC, not otherwise.

The advantage of the IPWRA model facilitates the estimation of yield outcome for each farmer twice. First, from the perspective of the weighted probability of being treated, second, from the weighted probability of being in the untreated group. After that, the average yield for treated and untreated households is calculated as the difference between the two averages. On the other hand, IPWRA is chosen for this study because it estimates unbiased treatment effects, even when we experience confounding effects, but it also provides a consistent estimator in the presence of misspecification in the treatment or outcome model, but not both [29–31]. The double-robust property of the IPWRA estimator ensures consistent results as the method allows the outcome and the treatment model to account for misspecification in the estimation of average treatment effect for the treated (ATT).

In IPWRA, the ATT is estimated following the two steps: using linear regression and probit regression [29,32]. A probit model predicts the participation of households in the ATDC training program, and linear regression estimates the effects of covariates on yield per acre of the individual household. Under the assumption that the outcome model is in linear regression function: $Y_i = \beta_i + \alpha_i x_i + \varepsilon_i$ for $i = [0,1]$ and propensity score: $P(x;\gamma)$. The IPWRA necessitates all covariates that affect training participation to be specified in the treatment model, while covariates that affect the rice yield are also specified in the outcome model, for both treated households and untreated households. In our analysis, the households and farms covariates were used, but also, we added extension and technology adoption variables in outcome equations to estimates their effects between treated and control groups. Since beneficiaries were not randomly selected during ATDC intervention, IPWRA resolve the problem by weighting regressions where the weights are derived from inverse propensity score [28]. In the first step, the propensity score estimated as $P(x;\hat{\gamma})$ using probit regression, and followed by linear regression to estimate (β_0, α_0) and (β_1, α_1) using inverse probability weighted least square as given below:

$$\min_{\beta_0,\alpha_0} \sum_i^N Y_i - \beta_0 + \alpha_0 x_i / P(x,\hat{\gamma}) \; if \; T_i = 0 \text{ (Outcome model for untreated households)} \quad (1)$$

$$\min_{\beta_1,\alpha_1} \sum_i^N Y_i - \beta_1 + \alpha_1 x_i / P(x,\hat{\gamma}) \; if \; T_i = 1 \text{ (Outcome model for treated households)} \quad (2)$$

The estimator of ATT is the difference of Equations (1) and (2) as shown in Equation (3)

$$ATT = \frac{1}{N_w} \sum_i^{N_w} [(\hat{\beta}_i - \hat{\beta}_0) - (\hat{\alpha}_1 - \hat{\alpha}_0) x_i] \quad (3)$$

where: $(\hat{\beta}_1, \hat{\alpha}_1)$ and $(\hat{\beta}_0, \hat{\alpha}_0)$ are estimated inverse probability-weighted parameters for treated households and untreated households, respectively. (N_w) stands for the total number of treated households.

2.3. Study Data

The data used in this study were collected from October to December 2019 from the household survey in the Mvomero district of the Morogoro region of Tanzania. Rice and maize are the main crops grown in this area. The study took place in two stages. First, a preliminary survey was conducted to understanding the rice production characteristics and training conditions in the study area. Consultations were conducted with different stakeholders, including agricultural officials from Mvomero district, extension officers, and workers working directly with farmers from ATDC and TARI-Cholima.

We began by including all five villages that benefited from ATDC training programs from 2011 to 2019 to collect village preliminary information. The preliminary findings were used to shortlist 28 villages out of 125 villages. The nine untreated villages were randomly selected, by which four untreated villages are close to treated villages while the other five untreated villages are far away from the treated villages. Treated villages were not randomly selected; instead, we include all villages by which the ATDC rice-training program was implemented. This study classified treated villages as any village in the ATDC training program, while untreated villages are vice versa. In the treated village, about 40 treated rice farmers (participated in ATDC's program) and 20 untreated farmers (did not participate in ATDC's program) were randomly sampled from village farm household information, while in untreated villages, 30 rice farmers were randomly selected. Finally, we end up surveying 469 households from 14 villages using structured questionnaires.

The data were collected from rice farm households through face-to-face interviews by our trained enumerators, familiar with rice farming systems and who speak a local language. The study covered various information include household characteristics, rice cultivation practices, rice farm characteristics, and agricultural training information.

3. Empirical Results

3.1. Farmers Participation in Training Programs

From our data, it was revealed that rice farmers in the study area had undertaken various rice farming training programs between 2011 and 2019, as shown in Figure 1. However, in 2011, 2013, 2014, and 2016, farmers participated more in the ATDC training program, followed by other programs. In 2017–2019, most farmers were more involved in the Ripoma/Helvetas project, with participation in ATDC training decreasing in 2019 due to a lack of operating funds. Thus, our study has considered that various projects have been in place in the study area. Therefore, to fulfill the purpose of this study, we classified our analysis into four groups: (1) farmers who attended the ATDC training program only (here after refer as ATDC only/ATDC), (2) farmers who participated in all other training programs that excluded ATDC (here after refer as other only/non-ATDC/Other), (3) farmers who participated in both training programs, that is ATDC and non-ATDC (here after refer as ATDC and Other), and (4) farmers who did not attend any agricultural training programs (here after refer as non-trained/control). The distribution of the sample, according to these groups, indicates that 70% of the sample population (ATDC only (7%), ATDC and Other (21%), and Other only (42%)) had at least attended agricultural training, while 30% did not attend any agricultural training program. Information about the distribution of farmers in different training programs in treated and non-treated villages is available in Appendix A Table A1, which indicates that about 27.3% of farmers in our sample population attended the ATDC training program in treated villages. In comparison, 72.7% did not attend the ATDC program in treated villages, while 45.4% of farmers from our sample population were drawn from non-treated villages where the ATDC program was not implemented.

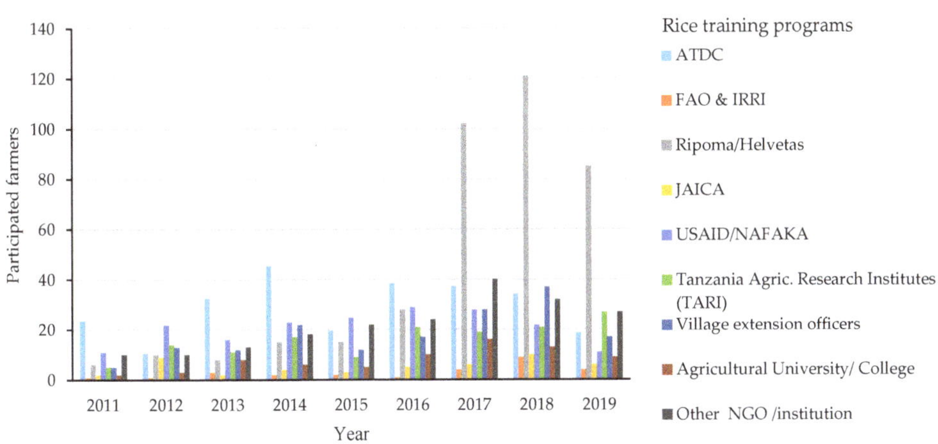

Figure 1. Participation of farmers in rice training programs between 2011 and 2019 in study area. Note: The abbreviations of training programs in Figure 1 are defined—ATDC (Agricultural Technology Demonstration Center); FAO (Food and Agriculture Organization of the united nations); IRRI (International Rice Research Institute); Ripoma/Helvetas (Rice Post-harvest Management and Marketing project implemented by Helvetas); JAICA (Japan International Cooperation Agency); USAID/NAFAKA (Feed the Future Tanzania-Nafaka program implemented by USAID); TARI (Tanzania Agricultural Research Institute); Other NGO/institution (government and non-government organizations and institutions not listed).

3.2. Rice Farming Households Characteristics

The characteristics of the surveyed households are presented in Table 1. Results of summary statistics were tested whether there was a significant difference between the pair of means between trained farmers and non-trained/control farmers. Data indicate that the farmers groups that participated in the training programs (ATDC only, ATDC and Other, Other only, all trained) are comparable against non-trained households, mainly in terms of household and farm characteristics. However, there are significant differences in the association memberships of farming households, access to credit, farm sizes, distance from farm to agro-dealers, distance from household to ATDC, and extension and adoption characteristics.

Table 1. Characteristics of surveyed households.

Characteristics	Mean Trained Farmers				Mean Non-Trained
	All Trained	ATDC	ATDC and Other	Other	
	Household characteristics				
Household size	4.810	5.313	5.122	4.574	4.711
Rice experience of household head (year)	16.067	19.094	16.980	15.122	15.585
Rice experience of spouse (year)	13.947	16.208	15.361	12.965	13.018
Spouse education (year)	7.076	6.875	7.143	7.076	6.754
Household head education (year)	7.410	6.688	7.347	7.558	7.465
Age of spouse (year)	40.099	39.200	43.016	38.979	39.698
Age of household head (year)	47.171	50.906	48.694	45.807	47.099
Association membership (1 = yes)	0.700 ***	0.594	0.745 ***	0.695 ***	0.401
Access to credit (1 = yes)	0.471 ***	0.281	0.510 ***	0.482 ***	0.275
From household to ATDC (distance in km)	19.559 ***	13.000 ***	13.311 ***	23.732	24.761
	Farm characteristics				
Average acres cultivated rice in 2019	3.295	4.992	3.525	2.905	3.889
Cultivated rice land (acre) in 2019 (direct seedling)	1.818 *	2.734	1.561	1.797	2.829
Cultivated rice land (acre) in 2019 (transplanting)	1.471 *	2.258 **	1.944 **	1.108	1.060
Share of transplanting (acre) in 2019	0.487 *	0.479	0.544 *	0.460	0.405
Access to water for irrigation (1 = yes)	0.609	0.594	0.561	0.635	0.500
Share of Irrigated rice farm (acres)	0.378	0.323	0.403	0.374	0.325
Irrigated rice farm (acres)	1.295	1.758	1.709 *	1.014	0.990
Non-irrigated rice farm (acres)	2.087 *	3.563	1.875	1.953	3.039
Total own land (acres)	6.152	10.916 *	6.148	5.380	6.029
Paddy field own (acres)	2.689	4.328	2.977	2.280	3.234
From farm to agro-dealer (distance in km)	3.595 *	4.344	3.064	3.737	4.432
	Extension and Adoption Characteristics				
Access to farm inputs (1 = yes)	0.884 ***	0.906 **	0.878 ***	0.883 ***	0.704
Agricultural advice in right time (1 = yes)	0.832 ***	0.844 ***	0.857 ***	0.817 ***	0.549
Access to any extension advice (1 = yes)	0.878 ***	0.813	0.888 ***	0.883 ***	0.641
Use water in seed selection (1 = yes)	0.676 ***	0.625	0.694 ***	0.675 ***	0.423
Seed pre-germination (1 = yes)	0.688 ***	0.625	0.684 ***	0.701 ***	0.465
Use of seedling bed (1 = yes)	0.740 ***	0.688	0.735 ***	0.751 ***	0.542
Land hallowing and levelling (1 = yes)	0.810 ***	0.875 **	0.816 ***	0.797 ***	0.613
Transplanting by standardized spacing (1 = yes)	0.654 ***	0.688	0.612	0.670 ***	0.465
Agrochemical usage (frequency)	1.454	1.393	1.337	1.526	1.323
Base fertilizer application (1 = yes)	0.364 ***	0.406 **	0.398 ***	0.340 ***	0.162
Topdressing fertilizer application (1 = yes)	0.657 ***	0.656	0.663 **	0.655 *	0.500
Rice yield (kg/acre)	1337.372 **	1405.382	1387.540	1301.368	1088.851
Participation (N = 469)	0.697	0.069	0.208	0.420	0.303

Note: Bonferroni multiple-comparison test *** $p < 0.01$, ** $p < 0.05$, * $p < 0.1$.

Household characteristics indicate no significant mean differences between farmers who attended ATDC only training and non-trained farmers. The results suggest that these two groups have similar characteristics in terms of household size, rice-farming experiences, years of education, age, association membership, and credit access. With the mean of all trained farmers, those who participated in both projects (ATDC and Other), and Other only, the association membership, and access to credit are significantly different from a non-trained group. It suggests that households who participated in non-ATDC programs (ATDC and Other, Other only) had much higher access to credit and affiliation membership in farming-based organization than those of non-trained households and ATDC only. Likewise, a mean distance from farmer households to ATDC was significantly smaller in those who participated in the ATDC program (ATDC only, and ATDC and Other), indicating they lived near ATDC.

In rice farming characteristics, there was no significant mean difference in acres of cultivated rice in 2019 among training groups. However, the share of transplanting acres was significantly higher in families participating in all training programs, particularly in ATDC and Other. The direct-seedling planted acres (broadcasting method) in the average of all trained households was significantly lower than non-trained families, with no significant difference within each trained group (ATDC only, ATDC and Other, and Other only). The mean transplanted acres were significantly high in all trained households, especially those who attended ATDC training programs (ATDC only, ATDC and Other), while the differences were not observed between other only and non-trained households. On the other hand, households who participated in ATDC and non-ATDC training (ATDC and Other) had a significantly larger size of irrigated farms than non-trained families, but the share differences of irrigated farm acres were not significant among training groups.

In contrast, Table 1 shows the mean differences of irrigation farms between the trained and non-trained households are insignificant with ATDC only, Other only, and the mean of all trained families. Among household groups, the access to water for irrigation is insignificantly different, implying that all farmers in irrigation areas have equal access to water for irrigation. However, characteristics of farming and training of surveyed households, based on access to water for irrigation (see Appendix A Table A2), suggest that farmers with access to water for irrigation have significantly high mean irrigation farms than others. These farmers with significant mean irrigation farms in Appendix A Table A2 probably imply their access to well-established irrigation schemes, in which water availability is guaranteed because irrigation facilities are well furnished. Nevertheless, Table 1 presents that the farm size of non-irrigated rice in all trained households is smaller than a non-trained group; however, the difference is not significant among groups (ATDC only, ATDC and Other, Other only). Moreover, there are no significant differences between trained and non-trained households in terms of their land allocated in paddy production rather than in terms of total farmland owned by a farmer, where the mean difference is more significant for households that attended only ATDC training. In the case of average distances from farm to agro-dealers, the trained farmers are closer to agro-dealers than non-trained farmers; however, those differences are not significant among trained groups.

Concerning extension and adoption behavior, our data indicate significant differences between the trained and non-trained groups except in the frequency of applying agro-chemicals. Access to farm inputs and agricultural advice at the right time of needs is more significant in trained households than non-trained households. The mean difference is not significant between farmers who participated only in ATDC training and non-trained groups in terms of accessing any extension advice, water practice in seed selection, seed pre-germination, use of seedling beds, and standardized transplanting. However, a considerable difference was observed in farmers who participated in both programs (ATDC and Other). On the other hand, the adoption of land hallowing and leveling, and base fertilizer application was more significant among trained households, including those who attended ATDC programs (ATDC only, ATDC and Other); topdressing fertilizer application was more substantial in those farmers who attended ATDC and Other. Moreover, the average yield of rice farmers who participated in the training (ATDC only, ATDC and Other, Other only) was higher than non-trained households, but not significantly different. Still, a significant mean yield difference is observed between trained farmers and non-trained farmers, suggesting that the trained households gained more yield than non-trained households.

In general, Table 1 indicates that training programs (includes ATDC) work because farmers use more technology/practices as suggested and obtained a high yield (check the third section of Table 1). On the other hand, the farmers who participated in training programs are like those who did not (Section 1), and their farms are not very different from those who did not participate in the training program (Section 2). Therefore, Table 1 may suggest that ATDC contributed to productivity growth, such as other training programs, and there are no vast differences between ATDC/other training programs.

3.3. Participation of Smallholder Farmers in Agricultural Training Programs

Table 2 presents farmers' participation results in agricultural training programs. The results show that the farmers' participation in the training program is likely to be influenced significantly by factors, such as education years of the household head, sex, size of farmland owned by farmers, distance from household to the training center (ATDC), association membership, access to extension advice, credit, and irrigation farm.

Table 2. Participation model results of inverse probability weighted regression adjusted (IPWRA) analysis.

Variables	ATDC	ATDC and Other	Other	All
Household size	0.0816	0.0636	−0.0257	0.0160
	(0.0661)	(0.0607)	(0.0453)	(0.0401)
Household head education (year)	−0.1360 **	−0.0090	0.0041	−0.0121
	(0.0565)	(0.0365)	(0.0311)	(0.0274)
Rice experience of household head (year)	0.0112	0.0006	−0.0030	−0.0032
	(0.0116)	(0.0101)	(0.0071)	(0.0064)
Sex of household head (1 = male)	−0.5130	−0.8290 ***	−0.5650 ***	−0.6470 ***
	(0.3460)	(0.2250)	(0.1870)	(0.1610)
From farm to agro-dealer (distance in km)	0.0063	−0.0051	−0.0150	−0.0133
	(0.0204)	(0.0241)	(0.0128)	(0.0123)
From household to ATDC (distance in km)	−0.0355 ***	−0.0409 ***		
	(0.0137)	(0.0076)		
Access to farm inputs (1 = yes)	0.2490	0.1240	0.3050	0.2860
	(0.4100)	(0.2770)	(0.2160)	(0.1840)
Access to any extension advice (1 = yes)	0.7070 **	0.8520 ***	0.5930 ***	0.6050 ***
	(0.3140)	(0.2870)	(0.1900)	(0.1660)
Agricultural advice in right time (1 = yes)	0.9010 ***	0.7630 ***	0.3850 *	0.4840 ***
	(0.3460)	(0.2610)	(0.1970)	(0.1680)
Association membership (1 = yes)	0.1380	0.6260 **	0.5050 ***	0.5020 ***
	(0.3650)	(0.2630)	(0.1870)	(0.1630)
Access to credit (1 = yes)	−0.5980 *	0.1060	0.1940	0.0654
	(0.3550)	(0.2500)	(0.1840)	(0.1620)
Access to water for irrigation (1 = yes)	0.0950	−0.3190	0.2920	0.0299
	(0.3330)	(0.2350)	(0.1920)	(0.1620)
Irrigated rice farm (acres)	0.0004	−0.0330	−0.1210 **	−0.0158
	(0.0867)	(0.0513)	(0.0611)	(0.0391)
Non-irrigated rice farm (acres)	0.0056	−0.1140 *	−0.0090	−0.0215
	(0.0439)	(0.0627)	(0.0318)	(0.0296)
Total own land(acres)	0.0263 **	0.0254	0.0264 *	0.0202
	(0.0113)	(0.0280)	(0.0138)	(0.0130)
Paddy field own (acres)	−0.0343	0.0189	−0.0423	−0.0147
	(0.0479)	(0.0439)	(0.0312)	(0.0310)
Constant	−0.9170	−0.5870	−0.5540	−0.2630
	(0.6830)	(0.5620)	(0.4120)	(0.3640)
Observations	174	238	341	469

Note: *** $p < 0.01$, ** $p < 0.05$, * $p < 0.1$, Standard errors in parentheses.

The results indicate that access to any extension advice and timely access to agricultural advice would increase participation by more than half in the ATDC program, which suggests that a closer relationship between farmers and local extension is essential for the success of the agricultural program. The access to any extension advice would increase farmers' participation in ATDC only program by 70.7% and 85.2% in the ATDC and Other program, while other only programs would rise to 59.3%, which would make the average participation in all training programs increased by 60.5%. Likewise, the access to extension advice in the right time of needs would increase farmers' participation by 90.1% in ATDC only program, 76.3% in ATDC and Other program, and 38.5% for the non-ATDC programs (Other only), which would count the average participation of 48.4% in all training programs implemented in the local area. On the other hand, farmers' farms would positively influence participation in training programs, indicating that a one-acre increase in total owned land is associated with participation increase by 2.6% and 2.6% in the ATDC only program and other only programs, respectively, implying that farmers with their lands participate more in agricultural training programs than otherwise.

Moreover, the membership of farmers in an association/group would increase participation by 62.6% in the ATDC and Other training program, while non-ATDC training programs (other only) increases by 50.5%. The result may also indicate that most non-ATDC training programs are disseminated via farmers' association/groups; therefore, being a member would increase the chance of benefiting from these training programs. This is different from the ATDC program, where the membership factor does not significantly influence farmer participation, which means being an association member is not a matter for ATDC participation. This finding may suggest that both farmers, members, and non-members in the group/association participated in ATDC training programs; however, the membership of farmers in association significantly affects the rice yield of trained farmers in the ATDC only program positively.

Factors, such as distance from household to the training center (ATDC), the sex and education years of the household head, credit, and whether the farmer has access to irrigation farms, would significantly decrease farmers' probability of participating in the training program. Results indicate that an increase of one kilometer from household to training center (ATDC) would reduce the likelihood of farmers participating in the ATDC only program by 3.6%, and 4.1% for those who participated in ATDC and Other, which implies that transportation to help bring farmers to the training center during training seasons would increase the farmers' participation in the ATDC training program. On the other hand, the increase in the academic year of the household head would decrease the likelihood of training participation by 13.6% in the ATDC program, only because most farmers in the study area have relative basic education and, therefore, those with higher education are likely to engage in non-farm activities. While the sex of the household head (i.e., whether the head is male) would reduce the probability of participating in the ATDC and Other program by 82.9%, other only programs by 56.5%, and all training programs by 64.7%; it suggests that training participation favored women than men. Therefore, women have a higher likelihood of participating in agricultural training programs than men, because women comprise the majority of the labor force in the Tanzania agricultural sector [33].

The access to credit would reduce the probability of participating in ATDC only program by 59.8%, which suggests that most farmers who attended ATDC only program had limited access to credit, which negatively affected the yield of trained farmers in the ATDC only program. The remedy of access to credit would improve farmers' participation in agricultural training programs and yield. Then again, an increase in an acre of irrigated rice farms would reduce farmers' participation in other only programs by 12.1%, while an increase in non-irrigated rice farms would reduce farmers' participation in ATDC and Other programs by 11.4%. These negative results have an important implication that participation in the non-ATDC program (other only) favors more farmers with non-irrigation farms because they have little access to irrigation farms; however, participation in ATDC and non-ATDC programs (ATDC and Other) favors more farmers with irrigation farms. Nevertheless, Table 2 reveals that those who participated in ATDC were not far from ATDC, while those who participated in other training programs were association members. Yet, farmers with more land, access to any extension advice, and agricultural advice at the right time were more likely to participate in ATDC and other programs.

3.4. Factors Affecting Yields among Trained and Non-Trained Smallholder Farmers

The IPWRA results in Table 3 indicate significant factors that negatively or positively influenced the mean yield outcome among trained (OME1) and non-trained smallholder farmers (OME0). Factors that negatively affected rice yield include household head education and rice farming experiences, distance from farm to agro-dealers, and access to farm input. Likewise, the membership of farmers in farming associations and farms owned by farmers influence positively the rice yield of trained farmers participating in ATDC only programs. However, farm size allocated to rice farming influences yields negatively, pointing out that land resources were not optimally utilized in Mvomero, Tanzania.

Table 3. Outcome model results of IPWRA Analysis.

Variables	ATDC		ATDC and Other		Other		All	
	OME0	OME1	OME0	OME1	OME0	OME1	OME0	OME1
Household size	24.04	76.62	44.47	−12.80	11.81	−45.70	11.17	−20.89
	(39.51)	(94.06)	(42.46)	(41.03)	(39.62)	(38.22)	(43.05)	(28.80)
Household head education (year)	−17.04	−2.24	−9.79	−13.30	−16.61	−44.96 *	−14.89	−17.51
	(23.32)	(60.82)	(24.50)	(29.78)	(20.45)	(24.16)	(21.56)	(20.18)
Rice experience of household head (year)	−12.43 *	−7.05	−13.32 **	−8.91	−13.25 **	0.85	−15.03 **	0.57
	(6.53)	(25.87)	(6.42)	(7.16)	(5.97)	(5.75)	(6.32)	(5.07)
Sex of household head (1 = male)	196.00	−732.50 *	203.90	360.20 ***	176.90	179.00	234.00	114.30
	(157.70)	(390.70)	(163.80)	(130.60)	(156.60)	(129.10)	(157.80)	(96.41)
From farm to agro-dealer (distance in km)	−11.13	150.50 ***	−9.47	−21.29	−11.59	−40.85 ***	−8.02	−22.36 **
	(9.24)	(36.28)	(10.56)	(26.18)	(10.37)	(9.42)	(11.46)	(10.08)
Access to farm inputs (1 = yes)	−209.30	−1323.00 **	−305.50 *	−292.30	−247.20	−127.60	−263.80	−56.86
	(178.80)	(587.30)	(181.00)	(252.30)	(159.80)	(167.00)	(171.20)	(136.10)
Access to any extension advice (1 = yes)	−8.93	988.70 ***	−83.50	−886.40 ***	−75.39	−70.55	−131.40	−243.10 *
	(138.80)	(298.90)	(144.70)	(183.40)	(152.10)	(167.50)	(160.60)	(143.40)
Agricultural advice at the right time (1 = yes)	103.10	−241.10	227.80	−202.40	227.10	180.40	268.90	164.20
	(167.00)	(319.30)	(184.20)	(187.70)	(167.30)	(143.70)	(179.50)	(118.20)
Association membership (1= yes)	−177.70	1386.00 ***	−278.30	−291.50	−219.00	31.29	−228.40	31.62
	(179.60)	(298.10)	(188.50)	(210.50)	(182.40)	(143.00)	(183.70)	(119.50)
Access to credit (1 = yes)	126.00	−660.50 **	190.50	202.20	94.08	88.74	127.70	−5.09
	(172.20)	(298.80)	(170.30)	(163.90)	(158.00)	(133.90)	(155.60)	(101.80)
Access to water for irrigation (1 = yes)	390.00	−1421.00 **	385.70	780.00 ***	374.20	258.20	347.00	158.90
	(262.70)	(599.90)	(289.00)	(251.10)	(244.20)	(212.70)	(266.10)	(193.90)
Irrigated rice farm (acres)	4.39	260.30	69.86	−129.10	14.38	143.10 *	0.24	137.30 *
	(128.70)	(283.20)	(147.80)	(114.40)	(134.60)	(73.58)	(134.90)	(78.15)

Table 3. *Cont.*

Variables	ATDC		ATDC and Other		Other		All	
	OME0	OME1	OME0	OME1	OME0	OME1	OME0	OME1
Non-irrigated rice farm (acres)	−199.20 ***	−138.90	−198.80 ***	−32.77	−195.30 ***	22.75	−190.70 ***	−36.65
	(32.66)	(200.60)	(38.54)	(75.87)	(34.08)	(132.10)	(32.11)	(78.71)
Total own land(acres)	−0.65	19.91	−0.46	−5.83	4.74	13.89	1.89	11.03 *
	(12.44)	(16.62)	(17.67)	(28.80)	(14.92)	(14.15)	(17.25)	(6.25)
Paddy field own (acres)	86.59 ***	−289.70 ***	93.40 ***	14.78	89.01 ***	−33.64	98.37 ***	−15.16
	(26.94)	(65.30)	(32.40)	(32.95)	(28.05)	(33.39)	(25.83)	(20.72)
Cultivated rice land (acre) in 2019(direct seedling)	107.50 ***	182.50	88.97 ***	−147.20	91.54 ***	−55.24	80.40 ***	−16.61
	(21.46)	(237.90)	(30.73)	(89.88)	(23.75)	(133.10)	(22.13)	(81.38)
Cultivated rice land (acre) in 2019 (transplanting)	−79.00	234.50	−175.20	177.20	−88.21	39.38	−94.26	−24.30
	(131.80)	(222.80)	(151.90)	(117.90)	(140.60)	(78.72)	(139.70)	(79.06)
Use water in seed selection (1 = yes)	−109.70	−186.80	−305.70	381.40 **	−198.20	50.94	−196.50	111.10
	(241.70)	(415.40)	(232.00)	(161.20)	(249.00)	(216.20)	(265.00)	(130.90)
Seed pre-germination (1 = yes)	277.60	956.60 **	450.20	233.10	151.00	232.20	270.70	246.30
	(314.40)	(452.80)	(325.70)	(193.20)	(354.70)	(213.90)	(352.00)	(149.80)
Use of seedling bed (1 = yes)	347.50	388.10	385.20	−406.30 *	367.30	124.70	283.40	195.20
	(287.50)	(520.60)	(317.80)	(216.30)	(302.70)	(292.70)	(307.90)	(193.30)
Land hallow and leveling (1 = yes)	−65.13	−80.44	−17.79	103.40	−32.23	421.70 ***	−17.35	283.20 **
	(138.20)	(327.00)	(138.60)	(168.50)	(139.60)	(133.90)	(146.60)	(121.30)
Transplanting in standardized spacing (1 = yes)	82.70	−134.00	200.70	320.50 *	90.77	−252.30	85.91	−170.80
	(237.40)	(310.80)	(248.90)	(180.50)	(254.50)	(218.40)	(266.50)	(155.50)
Agrochemical usage (frequency)	40.01	12.91	76.94	−34.01	110.10	60.03	100.40	11.62
	(77.15)	(137.80)	(77.26)	(86.26)	(73.56)	(65.31)	(76.78)	(53.98)
Base fertilizer application (1 = yes)	182.40	793.00 ***	308.30	397.60 **	202.10	284.10 *	273.20	299.40 **
	(240.30)	(254.90)	(250.00)	(159.60)	(217.10)	(151.40)	(239.30)	(131.10)
Topdressing fertilizer application (1 = yes)	134.60	187.70	65.10	211.40	167.80	4.01	197.70	182.30
	(242.20)	(463.50)	(290.10)	(245.80)	(235.20)	(164.40)	(264.80)	(156.10)
Constant	812.20 **	814.20	676.30 *	1988.00 ***	900.80 ***	884.00 **	879.10 ***	764.30 **
	(343.30)	(652.80)	(361.20)	(389.20)	(312.60)	(428.70)	(335.10)	(305.70)
Program (ATE)—trained vs. non-trained (Kg/acre)		−150.60		254.70 **		−40.15		39.41
		(290.00)		(115.80)		(96.95)		(92.67)
Potential household mean yield (Kg/acre)	1077.00 ***		1157.00 ***		1204.00 ***		1203.00 ***	
	(77.24)		(85.40)		(74.59)		(80.71)	
Observations	174	174	238	238	341	341	469	469

Note: *** $p < 0.01$, ** $p < 0.05$, * $p < 0.1$, Standard errors in parentheses.

The increase of household head education years seems to affect the rice yield of trained farmers from other programs negatively. This result suggest that household head engagement in rice farming jobs would shift to non-farm because farmers with high education are more likely to engage in non-farm activities [34].

The sex of the household head, if male, affects negative the yield of participated farmers in the ATDC only program, while favoring the yield of farmers participating in both ATDC and Other program. It indicates that the rice yield of female-headed households who participated in the ATDC training program was slightly higher than in male-headed households, suggesting that women can perform better when accessing the training program. However, access to productive resources is more critical, especially in the Tanzania context, where women are more constrained in adopting and benefiting from irrigation technologies compared to men [35]. This supports the significant coefficient of "sex of the household head" in the ATDC and Other, as participants in this program have a substantial share of irrigation farm (see Table 1). The FAO suggests that women are just as efficient as men and would achieve the same yields if they had equal access to productive resources and services; however, the limited access to resources may decrease women productivity, up to 30% [36].

Rice farming experience of the household head significantly affected non-trained farmers in either training program, which indicates that a one-year increase would reduce productivity. This finding is inconsistent with other studies [37,38], which noted that farming experience increases farmer technical efficiency and adoption of technology and, therefore, improves farm yield. However, another study found that the awareness of rice farming management practices was low, even among farmers with more farming experiences; thus, they continued with conventional rice production that had lower yield return [39]. In this case, our study's findings may indicate that farming experience alone is not enough to guarantee smallholder farmers' production efficiency that can increase significant productivity. Therefore, administering more agricultural training programs to instill the right skills and knowledge are crucial for enhancing farming productivity.

The distance from farms to agro-dealers negatively influenced the rice yield of farmers who participated in non-ATDC (other only) and all projects in general, which suggests an increase in distance from farm to the agro-dealer would challenge trained farmers from purchasing recommended farm inputs from agro-suppliers. However, distance from farm to agro-dealers would not averse farmers participated in ATDC only program; still, access to farm input would reduce the yield of trained farmers, suggesting that farm inputs may not be accessible when needed at the right time of production, even if farmers are willing to purchase. Thus, improving the accessibility of farm inputs at the right time and the distribution of agro-dealers close to farms may play a significant role in facilitating the use of improved farm inputs that may improve farm yields.

While supply chains can affect agricultural inputs accessibility, which may decrease adoption of technology and rice yields, improving the availability of farm inputs, in terms of distribution, transaction cost, and credit, may increase the possibility of attaining potential yield. On the other hand, access to farm input may still affect yield negatively if farmers face difficult access in acquiring inputs or inputs that are not delivered at the right time due to poor infrastructure. This may complement the access to extension advice that encourages farmers to acquire the right farm inputs, especially when the required inputs in farm production are not clear to farmers. Following our results, we noted a positive relation of access to any extension advice on rice yield of households in ATDC only programs; however, the negative results were also noted to those who participated in both ATDC and Other programs. The finding may likely suggest that extension advice from many sources may likely mislead farmers and ultimately reduce their productivity. Thus, extension advice and training should be more specific to farmers in order to guide them to the best decision in inputs allocation and utilization. In contrast, other researchers [40–43] have also noted the benefit of extension advice and training on adopting improved technology, which may increase farmers' productivity.

Access to irrigated fields has significantly increased the rice yields of farmers trained in non-ATDC (other only) programs and all training programs in general. Our results suggest that there is a positive relationship between irrigated fields and rice yields. However, rice cultivation in non-irrigated areas has significantly reduced the potential for high yields for trained and untrained farmers in ATDC and non-ATDC programs. Farming in non-irrigation areas means rainfed farming, whereby unpredictable rainfall usually affects yield outcome. Similarly, limited access to water for irrigation has reduced yields for some farmers who participated in ATDC programs only, while access to irrigation water increases the rice yields of farmers trained at ATDC and Other. These results may indicate that rice farming in non-irrigated farms, where water is not assured to approve better rice farming practices, places farmers at risk of adopting the technology.

When the farming environment does not guarantee better results, if the farmer invests in the best farming techniques, ultimately, trained and untrained farmers will continue to use traditional practices, even after training programs, as evidenced in our results. A positive relationship between yield and field size cultivated through direct seed, which is a traditional practice, has been revealed among participants of the non-ATDC programs. It shows that some farmers cultivate in non-irrigated areas (so not-transplanted), which may suggest that technology adoption is sometimes constrained by factors beyond the farmers' control, which needs further improvements for training to be practical.

Moreover, You [44] has noted the technology bias in agricultural research and development (R&D) towards irrigated rice because most modern rice varieties are bred under the irrigated conditions so that the rice seed would perform better with irrigation. Nevertheless, the adoption of improved seed, fertilizer, and other improved technology in irrigation farms will not work well independently. These technologies need to complement each other because their combined effects are more than the sum of their individual effects. Likewise, the benefit of access to an irrigation farm is like drought insurance, because with irrigation, the farmer will be willing to invest in fertilizer and seed, and other technologies, without fear of losing these pre-harvest investments in case of a drought. The argument is also supported by our data, which indicate that the adoption of agricultural technologies and rice yields are more significant among farmers with access to water for irrigation than otherwise (see Appendix A Table A2). On the other hand, among farmers with limited access to water for irrigation, there is no significant difference in yield between trained and non-trained farmers, even if technology adoption is evidenced among trained farmers. These results can provide micro-empirical evidence with important literature, such as the theory of induced innovation and adaptive investment in technology transfer.

3.5. Impact of Training Programs on Technology Adoption and Yield of Smallholder Farmer

Results in Table 3 show that trained farmers groups adopted some technology, which indicates that agricultural training programs may influence farm technology adoption and improve smallholder farmers' yield. For instance, the practice of seed selection by using water significantly increased adopters' rice yield by 381 kg/acre than those who did not adopt among trained farmers in ATDC and Other programs, though it is insignificant for ATDC only and Other only programs. This practice is the first crucial stage determining the seed quality and so germination of the plant and productivity. It ensures a better seed, which may result in a healthier seedling. Thus, the adoption of a seed selection procedure by using water facilitate farmers to plant only the highest quality rice seed, which leads to rice yield improvement. The finding is in line with International Rice Research Institute (IRRI) [45], which noted that the selection of good quality seeds might improve germination by more than 80% and increase the rice yield by 5–20%.

Seed pre-germination practice that improves seed establishment before sowing increased the yield of trained farmers in the ATDC only program. It was revealed that the yield of adopters significantly increased by 957 Kg/acre than non-adopters. This finding supports other studies that have revealed the positive impact of pre germination practice on yield, such as Tilahun-Tadesse et al. [46], who found planting pre-germinated rice seeds may increase yield advantage up to 58%, while Farooq et al. [47] noted 11–24% yield advantage.

On the other hand, training programs also impacted the adoption of basal fertilizer. Our research findings revealed that adopters of basal fertilizer among trainees groups improved the yield significantly by 793 kg/acre (ATDC only), 398 kg/acre (ATDC and Other), 284 kg/acre (other only), and 299 kg/acre (all training programs in general), which support the study of [12], who noted yield increases of 563 kg/acre when farmers applied basal fertilizer.

In comparison with farmers who participated in the ATDC and Other program, the non-adopters of seedling bed practices among non-participants lose significantly about 406 kg/acre, while those transplanting following specified spacing and row gained a yield increase by 321 kg/acre. The land hallowing and leveling practices improved the yield by 422 kg/acre and 283 kg/acre, respectively, among adopters in non-ATDC projects and all training in general.

Above all, this study revealed that both the ATDC training program and non-ATDC programs facilitated the adoption of improved rice practices, and, therefore, improved the yield of smallholder farmers in the study area. The result might provide a clue to the inconsistent finding of the impact of agricultural training programs in Africa. However, the average treatment effect on treated, which suggests the average yield impact of the program on beneficiaries, is neither significant for ATDC only beneficiaries nor non-ATDC only beneficiaries. Nevertheless, the impact training is significant on farmers who participated in both programs (ATDC and Other), suggesting that training has significantly improved yield by 255 kg/acre.

This study indicates that training increases productivity, especially in the context of irrigation, because farmers who participated in both ATDC and Other had the advantage of access to irrigation farms and water for irrigation than other farmers groups (see Table 1). Our further analysis revealed that trained and non-trained farmers with access to water for irrigation adopted technology significantly, and yield was more than double than their counterparts. Their yield effect estimation (average treatment effects—ATE) is significantly higher (1131 kg/acre and 940 kg/acre, respectively) than the comparison group (non-trained farmers without access to water) (see Appendix A Table A2). On the other hand, the yield difference between trained and non-trained farmers is insignificant in non-irrigated plots. In contrast, the ATE estimation on yield is a significant 862 kg/acre between "trained and irrigated vs. trained and non-irrigated", while in "non-trained and irrigated vs. trained and non-irrigated" it is 751 kg/acre.

Moreover, results of ATDC only and non-ATDC only beneficiaries suggest that a participation training program does not always improve the yield of smallholder farmers in developing countries, where farming infrastructure are neither perfect nor the accessibility of water for irrigation. Therefore, agricultural training programs should not be considered the only influencing factor for accelerating smallholder farmers' productivity. The influence of supportive factors for enhancing the adoption of agricultural technology and yield improvements (such as irrigation facilities, agricultural credit, farm inputs availability and accessibility, and access to agricultural advice at the right time) should be critically analyzed as it will ensure the best results of agricultural training programs. On the other hand, our analysis revealed a positive impact of agricultural training programs on irrigated plots, but not rain-fed plots. This finding might provide an answer to why a positive impact had been found in some previous studies [5,12,48] and not in other previous studies [49–51].

3.6. The Spillover Effect of Training Programs

The agricultural training programs may benefit non-training participants through the spillover effect of the training program, resulting from social interaction with treated farmers. The flow of technology information from person to person within the community networks (such as farmers' groups/associations) may improve the adoption of agricultural technology practices and yield for non-trained participants. Therefore, failure to recognize the spillover effect in our analysis may underestimate the effectiveness of a training program on technology adoption and yield because its effect on untreated will go unmeasured [52]. In this case, we decided to extrapolate whether non-participants in the agricultural training programs were indirectly affected by training intervention through farmers' interaction with treated farmers. Thus, we tested the role of farmers' social networks (farmers with membership in any group/association vs. those without membership) in facilitating the spread of technology adoption from trained to non-trained farmers.

Our data suggest that being a member of any farmers' group increases the possibility of access to agricultural training opportunities up to 49%, while not being a member reduces the accessibility by 28%. The mean adoption of technology among training participants and non-participants is indicated in Appendix A Table A3 based on farmers' group membership status. Our results show no significant differences in technology adoption between trained and non-trained farmers who are members of groups, except in base fertilizer application (trained vs. non-trained group member), which suggests the evidence of spillover effect of a training program on technology disseminated to trained farmers. The mean yields of trained farmers are higher than non-trained, although the difference is not significant. This suggests non-trained farmers also adopted technology through social interactions by observing and imitating trained farmers. In contrast, the significant mean difference in technology adoption and yield between trained in farmers groups against non-trained in non-farmers groups may suggest a lack of spillover effect between them (trained group member vs. non-trained who are not group member). The low adoption of technology is noted significantly among non-trained farmers who are not group members, which suggests that farmers belong to farmers groups, adopting more technologies that result in higher yields than other farmers.

Regarding the impact of training programs, by considering farmers' group membership, the ATEs are insignificant. However, it indicates the training programs would positively affect the yield of non-trained farmers who are not members of farmers groups by 144 kg/acre, than trained farmers from farming groups. The ATE analysis between trained and non-trained with the membership of farming groups indicates the impact of the training program, on non-trained in farmers groups, is 252 kg/acre, which is insignificant. These insignificant ATE results may suggest that the impact of agricultural training programs was not significantly reflected in the yields among those who did not benefit from such programs. However, the indirect effects of training programs may exist on technology adoption. Nevertheless, the ATE of non-trained who are in farming groups (252 kg/acre) is higher than trained who are not in farming groups (104 kg/acre), suggesting that the training programs have more effective results when disseminated via farming group members, because even indirect beneficiaries within the farmers groups can benefit more than direct beneficiaries outside farmers group. Inline, group membership can reduce the perceived risks to invest in improved technology because, within the group, knowledge and support are shared when applying new technology, which may indirectly benefit other farmers within the social network. Therefore, consideration of farmers' group members' participation during training dissemination is important. However, the ATE of group membership (group member vs. non-group member) is not significant, but its positive coefficient may indicate the positive synergy effect of the group membership.

4. Conclusions and Policy Implication

Understanding properly the impact of agricultural training programs on smallholder farmers' productivity is crucial for ensuring food security and poverty reduction in Africa. The Agricultural Technology Demonstration Center (ATDC) program is a Chinese technology transfer initiative established through a China–Africa agricultural cooperation (2009–2013) to support agricultural growth and food security in Africa through training and demonstration programs. With massive investments in ATDC through the China–Africa cooperation, all stakeholders had high hopes for its success. Nevertheless, our results in Tanzania case study indicates that the yield improvement of training participants with no access to water for irrigation was not significant. However, the yields among trained and non-trained farmers with access to water for irrigation were significantly more than double than the comparison group (non-trained farmers without access to water), while the yield difference between trained and non-trained farmers was insignificant in non-irrigated plots. The results might provide a clue to the inconsistent findings of previous studies on the impact of agricultural training programs in Africa.

The study may hold a lesson for the architects of agricultural development programs—that training programs should not be considered as the only influencing factors for accelerating smallholder farmers' productivity. The farming environment should be considered necessary. Rice technologies/practices that focus on non-irrigated fields should be addressed. Training that introduces new technologies/practices, such as seed varieties with higher yield, should consider the influence of supportive factors (such as irrigation facilities, agricultural credit, farm inputs availability and accessibility, and access to agricultural advice at the right time) to ensure the best results of agricultural training programs.

Additionally, the examination of social networks revealed no huge differences in technology adoption between trained and non-trained farmers who are members of groups, indicating evidence of a training program's spillover effect. However, the impact of agricultural training programs was not significantly reflected in yields. However, the ATE coefficient in non-trained in farmers groups was still higher than trained who are not in farming groups. This suggests that training programs have more effective results when disseminated via farming group representative members, because even indirect beneficiaries within farmers groups might benefit more than direct beneficiaries who are outside the farming groups.

Above all, referring to the implementers of agricultural development programs, such as ATDC, some recommendations might apply to make intervention more effective:

- First, the training program should be disseminated in the correct production calendar to enhance practicability and usefulness (in the ATDC program, some farmers participated during rice harvesting festivals, which questions the extent of their participation in the training program and the level of empowerment) [11,22]).
- Second, the close linkage of the agricultural training program to local extension services may be essential for farmers' participation and program effectiveness.
- Third, any agricultural program design should consider accountability and impact assessment within the program to prepare beneficiaries to become proficient and use proficiency in their field.

Our investigation noted that farmers experienced little consultation with Chinese experts—there is no formal feedback system and communication language is a challenge. We realized that there had been no rigorous monitoring and follow-up in ATDC activities to empower farmers and trace the theory of change, which may affect the program's effectiveness. Moreover, there were no clear guidelines on how much effort should be dedicated to training or commercial activities, which might incline ATDC to focus more on business rather than capacitating farmers to improve yields. Therefore, agricultural program architects should consider improving the program design, to enhance the farming training program's effectiveness and positively impact farmers' yields.

Moreover, further research may be required for identifying key factors that affect technology adoption and measures to improve agricultural technology program efficiency. The cross-section data employed in this study face limitations of capturing information in time and space, which cannot completely explain the incidences throughout the program intervention. Thus, the panel data may be recommended to study the impact of agricultural training programs (e.g., ATDC). The advantage of panel data is to address the dynamic behavior response of training program interventions that allow researchers to trace the impact change over time.

Author Contributions: S.M.: conceptualization, funding acquisition, and supervision of the study, G.M.: supervision, investigation, data collection, methodology, and writing of the manuscript. F.Q.: assisted with conceptualization and review. All authors have read and agreed to the published version of the manuscript.

Funding: This research received financial support from the Natural Science Foundation of China (NSFC), Projects of International Cooperation, and Exchanges (71761147005), and National Natural Sciences Foundation in China (71773150).

Data Availability Statement: The data underlying the results presented in the study are available from Institute of Agricultural Economics and Development, Chinese Academy of Agriculture Sciences.

Acknowledgments: We thank the cooperation from team members in Tanzania ATDC, the Tanzania Agricultural Research institute–Cholima, and the Mvomero District Agricultural Office, for their roles in data collection. We would also like to thank Cheng Xiang, Alex Ngereza, and Foya Hozeniel for their assistance during survey designing. Our appreciation also goes to Liangzhi You, Binlei Gong, and the anonymous reviewers for their constructive comments.

Conflicts of Interest: The authors declare no conflict of interest. The funders had no role in the design of the study; in the collection, analyses, or interpretation of data; in the writing of the manuscript, or in the decision to publish the results.

Appendix A

Table A1. A number of farmers participated in different training programs in treated and untreated villages.

	In Treated Villages		In Non-Treated Villages
	Treated Farmers	Non-Treated Farmers	
ATDC	128	128	213
FAO and IRRI	3	4	5
Ripoma/Helvetas	44	36	95
JAICA	9	7	6
USAID/NAFAKA	30	12	21
Tanzania Agric. Research Institutes (TARI)	25	7	27
Village extension officers	19	14	21
Agricultural University/College	7	1	16
Other NGO/institution	29	21	26
Total sample (N = 469)	128	128	213

Table A2. Characteristics of irrigation farming and training of the surveyed households.

Characteristics	Mean (Access to Water for Irrigation = 1)		Mean (Access to Water for Irrigation = 0)	
	Trained Farmers	Non-Trained Farmers	Trained Farmers	Non-Trained Farmers/Control
	Household characteristics			
Household size	4.714	4.690	4.961	4.732
Rice experience of household head (year)	16.090	15.746	16.031	15.423
Rice experience of spouse (year)	13.400	12.439	14.904	13.630
Spouse education (year)	7.126	6.915	7.000	6.592
Household head education (year)	7.442	7.817	7.359	7.113
Age of spouse (year)	39.354	37.712	41.388	41.754
Age of household head (year)	47.362	45.338	46.875	48.859
Association membership (1 = yes)	0.769 ***	0.535 ***	0.594 ***	0.268
Access to credit (1 = yes)	0.487 ***	0.324	0.445 **	0.225

Table A2. *Cont.*

Characteristics	Mean (Access to Water for Irrigation = 1)		Mean (Access to Water for Irrigation = 0)	
	Trained Farmers	Non-Trained Farmers	Trained Farmers	Non-Trained Farmers/Control
Farm characteristics				
Average acres cultivated rice land in 2019	4.041	3.359	2.135 **	4.419
Cultivated rice land (acre) in 2019 (direct seedling)	1.883 ***	1.489 **	1.740 ***	4.169
Cultivated rice land (acre) in 2019 (transplanting)	2.158 ***	1.870 ***	0.402	0.250
Share of transplanting (acre) in 2019	0.621 ***	0.645 ***	0.278	0.165
Share of irrigation acres	0.603 ***	0.597 ***	0.028	0.053
Irrigated rice farm (acres)	2.108 ***	1.835 ***	0.031	0.144
Non-irrigated rice farm (acres)	2.078 **	1.845 **	2.102 **	4.232
Total own land(acres)	6.619	4.937	5.426	7.120
Paddy field own (acres)	3.318	2.546	1.713 **	3.923
From farm to agro-dealer (distance in km)	3.545 ***	3.042 ***	3.672 **	5.821
Extension and Adoption Characteristics				
Access to farm inputs (1 = yes)	0.925 ***	0.817 ***	0.820 ***	0.592
Agricultural advice in right time (1 = yes)	0.874 ***	0.648 **	0.766 ***	0.451
Access to any extension advice (1 = yes)	0.894 ***	0.746 ***	0.852 ***	0.535
Use water in seed selection (1 = yes)	0.869 ***	0.718 ***	0.375 ***	0.127
Seed pre-germination (1 = yes)	0.905 ***	0.803 ***	0.352 ***	0.127
Use of seedling bed (1 = yes)	0.980 ***	0.901 ***	0.367 ***	0.183
Transplanting by standardized spacing (1 = yes)	0.894 ***	0.789 ***	0.281 *	0.141
Agrochemical usage (frequency)	1.618 ***	1.577 ***	0.813	0.732
Base fertilizer application (1 = yes)	0.528 ***	0.296 ***	0.109	0.028
Topdressing fertilizer application (1 = yes)	0.910 ***	0.831 ***	0.266	0.169
Rice yield (kg/acre)	1718.784 ***	1517.547 ***	744.395	660.154
Observation	199	71	128	71
Average treatment effects (ATE) estimation on rice yield (kg/acre): nearest-neighbor matching (Mahalanobis)				
Trained and irrigated vs. non-trained and non-irrigated	1131.000 *** (150.800)			
Non-trained and irrigated vs. non-trained and non-irrigated		940.000 *** (146.200)		
Trained and non-irrigated vs. non-trained and non-irrigated			102.400 (152.100)	
Trained and irrigated vs. non-trained irrigated	224.100 (137.800)			
Trained and irrigated vs. trained and non-irrigated				862.400 *** (159.100)
Non-trained and irrigated vs. trained and non-irrigated			750.800 *** (150.800)	
Observations	270	142	199	327

Note: Mean comparison is the Bonferroni multiple test, ATE's standard errors in parentheses, *** $p < 0.01$, ** $p < 0.05$, * $p < 0.1$.

Table A3. Adoption of technology and yield impact of training program among farmers groups.

Mean Adoption of Technology Among Farmers	Trained Farmer and Group Member	Non-Trained Farmer and Group Member	Trained Farmer and Not a Group Member	Non-Trained Farmer and Not a Group Member
	T = 1, E = 1	T = 0, E = 1	T = 1, E = 0	T = 0, E = 0
		Spillover Effect		No Spillover
Use water in seed selection (1 = yes)	0.712	0.544	0.592	0.341 ***
Seed pre-germination (1 = yes)	0.729	0.596	0.592 *	0.376 ***
Use of seedling bed (1 = yes)	0.795	0.667	0.612 ***	0.459 ***
Land hallowing and levelling (1 = yes)	0.812	0.684	0.806	0.565 ***
Transplanting by standardized spacing (1 = yes)	0.703	0.579	0.541 **	0.388 ***
Agrochemical usage(frequency)	1.389	1.439	1.102	0.965 ***
Base fertilizer application (1 = yes)	0.428	0.193 ***	0.214 ***	0.141 ***
Topdressing fertilizer application (1 = yes)	0.716	0.596	0.520 ***	0.435 ***
Rice yield (kg/acre)	1432.047	1164.396	1116.141 *	1038.191 **
Training participation (1 = yes)	0.488		0.209	
Sample (N = 469)	229	57	98	85
Average treatment effects (ATE) estimation on rice yield (kg/acre): nearest-neighbor matching (Mahalanobis)				
Trained vs. non-trained group member		252.300 (131.700)		
Trained group member vs. trained who are not group member			104.100 (120.800)	
Trained group member vs. non-trained who are not group member				143.900 (189.300)
Group member vs. non-group member	3.700 (96.600)			
Observations	469	286	327	314

Note: Mean comparison is the Bonferroni multiple test, ATE's standard errors in parentheses, *** $p < 0.01$, ** $p < 0.05$, * $p < 0.1$. T = Training intervention (1 = received, 0 = otherwise), E = Member of farmers group (1 = have membership, 0 = otherwise).

References

1. World Bank. *Tanzania Economic Update: Transforming Agriculture-Realizing the Potential of Agriculture for Inclusive Growth and Poverty Reduction*; Macroeconomics, Trade & Investment Global Practice. Africa Region; World Bank Group: Washington, DC, USA, 2019; Volume 84.
2. Kinuthia, B.K.; Mabaya, E. *The Impact of Agriculture Technology Adoption on Farmers' Welfare in Uganda and Tanzania*; Partnership for Economic Policy (PEP): Nairobi, Kenya, 2017; pp. 1–22.
3. Yigezu, Y.A.; Mugera, A.; El-Shater, T.; Aw-Hassan, A.; Piggin, C.; Haddad, A.; Khalil, Y.; Loss, S. Enhancing adoption of agricultural technologies requiring high initial investment among smallholders. *Technol. Forecast. Soc. Chang.* **2018**, *134*, 199–206. [CrossRef]
4. Mahajan, A.; Corral, C.; Seira, E.; Giné, X. *Basis Brief Identifying the Root Causes of Low Technology Adoption amongst Maize Growers in Mexico*; University of California: Davis, CA, USA, 2014.
5. Nakano, Y.; Tsusaka, T.W.; Aida, T.; Pede, V.O. Is farmer-to-farmer extension effective? The impact of training on technology adoption and rice farming productivity in Tanzania. *World Dev.* **2018**, *105*, 336–351. [CrossRef]
6. Kavia, F.; Loconto, A.; Simbua, E. Institutional collaboration for sustainable agriculture: Learning from the tea sector in the Southern Highlands of the United Republic of Tanzania. In *Innovative Markets for Sustainable Agriculture: How Innovations in Market Institutions Encourage Sustainable Agriculture in Developing Countries*; FAO—Food and Agriculture Organization of the United Nations: Rome, Italy, 2016; pp. 303–326.
7. Mirza, H.; Speller, W.; Dixie, G.; Goodman, Z. *The Practice of Responsible Investment Principles in Larger-Scale Agricultural Investments: Implications for Corporate Performance and Impact on Local Communities*; World Bank Group: Washington, DC, USA, 2014.
8. Xu, X.; Li, X.; Qi, G.; Tang, L.; Mukwereza, L. Science, technology, and the politics of knowledge: The case of China's agricultural technology demonstration centers in Africa. *World Dev.* **2016**, *81*, 82–91. [CrossRef]
9. Bräutigam, D.; Tang, X. *An Overview of Chinese Agricultural and Rural Engagement in Tanzania*; IFPRI Discussion Paper 01214; International Food Policy Research Institute: Washington, DC, USA, 2012.
10. Mgendi, G.; Shiping, M.; Xiang, C. A review of agricultural technology transfer in Africa: Lessons from Japan and China case projects in Tanzania and Kenya. *Sustainability* **2019**, *11*, 6598. [CrossRef]
11. Makundi, H. *Diffusing Chinese Rice Technology in rural Tanzania: Lessons from the Dakawa Agro-Technology Demonstration Center*; Johns Hopkins School of Advanced International Studies: Washington, DC, USA, 2017.
12. Senthilkumar, K.; Tesha, B.J.; Mghase, J.; Rodenburg, J. Increasing paddy yields and improving farm management: Results from participatory experiments with good agricultural practices (GAP) in Tanzania. *Paddy Water Environ.* **2018**, *16*, 749–766. [CrossRef]
13. Ngailo, J.A.; Mwakasendo, J.A.; Kisandu, D.B.; Tippe, D.E. Rice farming in the southern highlands of Tanzania: Management practices, socio-economic roles and production. *Eur. J. Res. Soc. Sci.* **2016**, *4*, 18–39.
14. World Bank. *World Development Report 2008: Agriculture for Development*; The World Bank: Washington, DC, USA, 2007.

15. Pingali, P.L. Green Revolution: Impacts, limits, and the path ahead. *Proc. Natl. Acad. Sci. USA* **2012**, *109*, 12302–12308. [CrossRef]
16. Duflo, E.; Kremer, M.; Robinson, J. Nudging farmers to use fertilizer: Theory and experimental evidence from Kenya. *Am. Econ. Rev.* **2011**, *101*, 2350–2390. [CrossRef]
17. Hayami, Y.; Ruttan, V.W. *Agricultural Development: An International Perspective*; The Johns Hopkins Press: Baltimore, MD, USA; London, UK, 1971; ISBN 0801812593.
18. Jiang, L.; Harding, A.; Anseeuw, W.; Alden, C. Chinese agriculture technology demonstration centres in Southern Africa: The new business of development. Available online: https://agritrop.cirad.fr/582983/ (accessed on 1 June 2020).
19. Chen, K.; Badiane, O.; Liwen, Z.; Collins, J.; Yunyi, Z. *China-Africa Agricultural Modernization Cooperation: Situation, Challenges and the Path Ahead*; International Food Policy Research Institute: Washington, DC, USA, 2018.
20. Grimm, S. China-Africa cooperation: Promises, practice and prospects. *J. Contemp. China* **2014**, *23*, 993–1011. [CrossRef]
21. Huang, M.; Xu, X.; Mao, X. *South-South Cooperation and Chinese Foreign Aid*; Springer: Singapore, 2018; ISBN 9789811320019.
22. Nalwimba, N.; Qi, G.; Mudimu, G.T. Towards an understanding of the Chinese agriculture technology demonstration centre(s) in Africa. *Eur. J. Soc. Sci. Stud.* **2019**, *4*, 140–158.
23. Tugendhat, H.; Alemu, D. Chinese agricultural training courses for African officials: Between power and partnerships. *World Dev.* **2016**, *81*, 71–81. [CrossRef]
24. Zhang, Y. Technology and knowledge transfer: A case study of China's agricultural technology demonstration center in the United Republic of Tanzania. In *South-South Cooperation and Chinese Foreign Aid*; Springer: Singapore, 2019; pp. 113–125.
25. Ragasa, C.; Mazunda, J. The impact of agricultural extension services in the context of a heavily subsidized input system: The case of Malawi. *World Dev.* **2018**, *105*, 25–47. [CrossRef]
26. Shiferaw, B.; Kassie, M.; Jaleta, M.; Yirga, C. Adoption of improved wheat varieties and impacts on household food security in Ethiopia. *Food Policy* **2014**, *44*, 272–284. [CrossRef]
27. Asfaw, S.; Shiferaw, B.; Simtowe, F.; Lipper, L. Impact of modern agricultural technologies on smallholder welfare: Evidence from Tanzania and Ethiopia. *Food Policy* **2012**, *37*, 283–295. [CrossRef]
28. Berhane, G.; Hoddinott, J.F.; Kumar, N. *The impact of Ethiopia's Productive Safety Net Programme on the Nutritional Status of Children: 2008–2012*; International Food Policy Research Institute: Addis Ababa, Ethiopia, 2017.
29. Wossen, T.; Abdoulaye, T.; Alene, A.; Haile, M.G.; Feleke, S.; Olanrewaju, A.; Manyong, V. Impacts of extension access and cooperative membership on technology adoption and household welfare. *J. Rural Stud.* **2017**, *54*, 223–233. [CrossRef]
30. Słoczyński, T.; Wooldridge, J.M. A general double robustness result for estimating average treatment effects. *Econom. Theory* **2018**, *34*, 112–133. [CrossRef]
31. Rahman, A. *Statistics for Data Science and Policy Analysis*; Springer: Singapore, 2020; ISBN 9789811517358.
32. Imbens, G.W.; Wooldridge, J.M. Recent developments in the econometrics of program evaluation. *J. Econ. Lit.* **2009**, *47*, 5–86. [CrossRef]
33. Leavens, M.K.; Anderson, C.L. *Gender and Agriculture in Tanzania*; Evans School Policy Analysis and Research (EPAR): Washington, DC, USA, 2011.
34. Tegegne, G. *Nonfarm Activities and Production Decision of Farmers: The Cases of Damotgale and Kachabira Woredas in Southern Region of Ethiopia*; Social Science Research Report Series, No. 15; OSSREA—Organization for Social Science Research in Eastern and Southern Africa: Addis Ababa, Ethiopia, 2000.
35. Theis, S.; Lefore, N.; Bryan, E.; Ringler, C.; Meinzen-Dick, R.S. *Integrating Gender Into Small-Scale Irrigation*; Feed the Future Innovation Lab for Small Scale Irrigation (FTF-ILSSI) Project Notes; International Food Policy Research Institute (IFPRI): Washington, DC, USA; International Water Management Institute (IWMI): Washington, DC, USA, 2017; Volume 2.
36. FAO. *Women in Agriculture: Closing the Gender Gap for Development*; FAO—Food and Agriculture Organization of the United Nations: Rome, Italy, 2011.
37. Tan, S.; Heerink, N.; Kuyvenhoven, A.; Qu, F. Impact of land fragmentation on rice producers' technical efficiency in South-East China. *NJAS Wagening. J. Life Sci.* **2010**, *57*, 117–123. [CrossRef]
38. Tsinigo, E.; Behrman, J.R. Technological priorities in rice production among smallholder farmers in Ghana. *NJAS Wagening. J. Life Sci.* **2017**, *83*, 47–56. [CrossRef]
39. Oo, S.P. Farmers' awareness of the low yield of conventional rice production in Ayeyarwady region, Myanmar: A case study of Myaungmya district. *Agriculture* **2020**, *10*, 26. [CrossRef]
40. Baba, S.; Dagong, M.I.A.; Sohrah, S.; Utamy, R.F. Factors affecting the adoption of agricultural by-products as feed by beef cattle farmers in Maros regency of South Sulawesi, Indonesia. *Trop. Anim. Sci. J.* **2019**, *42*, 76–80. [CrossRef]
41. Ghimire, R.; Huang, W.C.; Shrestha, R.B. Factors affecting adoption of improved rice varieties among rural farm households in Central Nepal. *Rice Sci.* **2015**, *22*, 35–43. [CrossRef]
42. Admassie, A.; Ayele, G. Adoption of improved technology in Ethiopia. *Ethiop. J. Econ.* **2010**, *19*, 155–179. [CrossRef]
43. Donkoh, S.A.; Azumah, S.B.; Awuni, J.A. Adoption of improved agricultural technologies among rice farmers in Ghana: A multivariate probit approach. *Ghana J. Dev. Stud.* **2019**, *16*, 46. [CrossRef]
44. You, L. A tale of two countries: Spatial and temporal patterns of rice productivity in China and Brazil. *China Econ. Rev.* **2012**, *23*, 690–703. [CrossRef]
45. IRRI. How to ensure seed quality. In *Seed Quality*; IRRI Rice Knowledge Bank: Los Baños, Philippines, 2013.

46. Tilahun-Tadesse, F.; Nigussie-Dechassa, R.; Bayu, W.; Gebeyehu, S. Effect of hydro-priming and pre-germinating rice seed on the yield and terminal moisture stress mitigation of rain-fed lowland rice. *Agric. For. Fish.* **2014**, *2*, 89.
47. Farooq, M.; Siddique, K.H.M.; Rehman, H.; Aziz, T.; Lee, D.-J.; Wahid, A. Rice direct seeding: Experiences, challenges and opportunities. *Soil Tillage Res.* **2011**, *111*, 87–98. [CrossRef]
48. Bravo-Ureta, B.E.; Higgins, D.; Arslan, A. Irrigation infrastructure and farm productivity in the Philippines: A stochastic meta-frontier analysis. *World Dev.* **2020**, *135*. [CrossRef]
49. Feder, G.; Murgai, R.; Quizon, J.B. Sending farmers back to school: The impact of farmer field schools in Indonesia. *Rev. Agric. Econ.* **2004**, *26*, 45–62. [CrossRef]
50. Lemba, J.; D'Haese, M.; D'Haese, L.; Frija, A.; Speelman, S. Comparing the technical efficiency of farms benefiting from different agricultural interventions in Kenya's drylands. *Dev. S. Afr.* **2012**, *29*, 287–301. [CrossRef]
51. Mdemu, M.V.; Mziray, N.; Bjornlund, H.; Kashaigili, J.J. Barriers to and opportunities for improving productivity and profitability of the Kiwere and Magozi irrigation schemes in Tanzania. *Int. J. Water Resour. Dev.* **2017**, *33*, 725–739. [CrossRef]
52. Angelucci, M.; Di Maro, V. Program evaluation and spillover effects. *J. Dev. Eff.* **2015**, *8*, 22–43. [CrossRef]

sustainability MDPI

Communication

Agricultural Innovations to Reduce the Health Impacts of Dams

Andrea J. Lund [1,*], David Lopez-Carr [2], Susanne H. Sokolow [3,4], Jason R. Rohr [5] and Giulio A. De Leo [6]

1 Emmett Interdisciplinary Program in Environment and Resources, Stanford University, Stanford, CA 94305, USA
2 Department of Geography, University of California, Santa Barbara, CA 93106, USA; davidlopezcarr@ucsb.edu
3 Woods Institute for the Environment, Stanford University, Stanford, CA 94305, USA; shsokolow@gmail.com
4 Marine Science Institute, University of California, Santa Barbara, CA 93106, USA
5 Department of Biological Sciences, University of Notre Dame, Notre Dame, IN 46556, USA; jrohr2@nd.edu
6 Hopkins Marine Station, Stanford University, Pacific Grove, CA 93950, USA; deleo@stanford.edu
* Correspondence: andrea.janelle.lund@gmail.com

Abstract: Dams enable the production of food and renewable energy, making them a crucial tool for both economic development and climate change adaptation in low- and middle-income countries. However, dams may also disrupt traditional livelihood systems and increase the transmission of vector- and water-borne pathogens. These livelihood and health impacts diminish the benefits of dams to rural populations dependent on rivers, as hydrological and ecological alterations change flood regimes, reduce nutrient transport and lead to the loss of biodiversity. We propose four agricultural innovations for promoting equity, health, sustainable development, and climate resilience in dammed watersheds: (1) restoring migratory aquatic species, (2) removing submerged vegetation and transforming it into an agricultural resource, (3) restoring environmental flows and (4) integrating agriculture and aquaculture. As investment in dams accelerates in low- and middle-income countries, appropriately addressing their livelihood and health impacts can improve the sustainability of modern agriculture and economic development in a changing climate.

Keywords: dams; agriculture; livelihoods; health; schistosomiasis; restoration; sustainable development; climate adaptation

Citation: Lund, A.J.; Lopez-Carr, D.; Sokolow, S.H.; Rohr, J.R.; De Leo, G.A. Agricultural Innovations to Reduce the Health Impacts of Dams. *Sustainability* **2021**, *13*, 1869. https://doi.org/10.3390/su13041869

Academic Editor: Michael Blakeney
Received: 16 January 2021
Accepted: 6 February 2021
Published: 9 February 2021

Publisher's Note: MDPI stays neutral with regard to jurisdictional claims in published maps and institutional affiliations.

Copyright: © 2021 by the authors. Licensee MDPI, Basel, Switzerland. This article is an open access article distributed under the terms and conditions of the Creative Commons Attribution (CC BY) license (https://creativecommons.org/licenses/by/4.0/).

1. Introduction

As infrastructure that enables production of both food and energy, dams are a key component of climate change adaptation [1]. They are meant to improve agricultural productivity and reduce vulnerability to droughts [2]. Especially in semi-arid regions experiencing desertification, water management infrastructure for food and energy production is necessary to sustain a growing human population and promote economic development in a changing climate. Yet, dams face many criticisms, ranging from cost overruns [3] and population displacement [4] to ecosystem disruption [5] and increased transmission of infectious disease [6]. Addressing these shortcomings—namely, the uneven distribution of dams' costs and benefits and the health impacts of dams driven by their environmental disruption—is crucial to maximize the social return on infrastructure investment and ensure the sustainability of modern agriculture in dammed landscapes.

Dams have displaced an estimated 40–80 million people globally in the last 50 years [4]. Dams designed to generate hydropower often transmit electricity to urban areas without extending the same benefits to rural areas [7,8]. While in some places dams supporting irrigation increase agricultural production and reduce vulnerability to droughts, in other places, they can increase poverty and variability in crop yields [9,10]. These disparities and uncertainties may be driven in part by the investment required to undertake irrigated agriculture compared to traditional practices. Compared to traditional cultivation of recession crops in floodplains, for example, irrigated agriculture requires greater investment in and use of agrochemical inputs such as fertilizers and pesticides, up-front costs that may be difficult for subsistence farmers to afford [2]. Furthermore, by altering natural flood regimes, dams have also made it difficult or impossible to cultivate recession crops in floodplains [11–14], forcing farmers to invest in inputs for irrigated crops or turn to another economic activity entirely.

A number of infectious diseases are also associated with both dams [6,15] and agricultural activity [16]. In particular, the transmission of schistosome parasites increases with proximity to dams and irrigation infrastructure [17]. Changes in hydrological regimes favor the aquatic snail populations that transmit the parasites, while the promotion of agricultural livelihoods keeps people in contact with fresh surface water [18], the route of exposure to infection. The convergence of these processes increases the likelihood of infection with schistosome parasites in endemic areas. In addition, repeated reinfection after treatment threatens the success of ongoing efforts to control the disease [19,20]. The inflammatory pathology of prolonged and repeated infection can lead to organ failure, cancer, and infertility [21] and disrupt cognitive development in children [22] and labor productivity in adults [23,24]. By putting people in contact with a favorable environment for parasite transmission, the negative health impacts of dams likely diminish the intended benefits of increased agricultural activity [25].

The livelihood and health impacts of dams have their roots in a combination of poverty, land-use change, and the associated environmental impacts. Dams compromise both terrestrial and aquatic biodiversity [26,27]. By stemming natural flood regimes, dams disrupt the reproduction of many fish species, affecting inland fisheries [11]. Disruptions to fisheries can, in turn, affect fishing livelihoods and increase the number of people at risk for nutritional deficiencies [28]. Environmental disruption can also facilitate the transmission of infectious diseases. In addition to the hydrological changes that favor populations of medically important snails, biodiversity loss—particularly the loss of aquatic invertebrates that prey on the snails that transmit schistosome parasites [29]—can lead to unchecked growth in snail populations and corresponding increases in parasite transmission. The transition from traditional to modern agricultural regimes also increases reliance on agrochemicals, as the sediment that provides nutrients for recession agriculture remains impounded behind dam barriers along with flood waters. The use of agrochemicals can exacerbate disease dynamics, because insecticides can cause greater harm to snail predators than to snails and fertilizers can promote the growth of both snail food and vegetative habitat [30,31].

While some of these impacts prompted funders to withdraw from dam projects in previous decades [32], the pace of investment in dam infrastructure is growing again [33]. The increasingly tangible impacts of climate change as well as the development trajectories of low- and middle-income countries make dams a key, but controversial, tool in the toolbox of sustainable development. Given this, addressing the livelihood, health and environmental impacts of dams will become central to improving the sustainability of modern agriculture and economic development in a changing climate. We outline agricultural innovations that can mitigate negative impacts of dams and promote sustainable development.

2. Agricultural Innovations to Reduce the Negative Impacts of Dams

2.1. Restoring Natural Predators of Parasite-Bearing Snails in Dammed Watersheds

Almost half of the 800 million people living in schistosomiasis endemic regions are estimated to be at risk of schistosomiasis because of the loss of snail predators following dam construction [34]. The disappearance of snail predators upstream of dams is thought to have contributed to the burden of schistosomiasis [29]. Aquatic species, such as the African river prawn (*Macrobrachium vollenhovenii*), are known to consume snails that transmit schistosomes [35], but dams interrupt their ability to migrate downriver to reproduce in brackish, estuarine water. A field experiment in northern Senegal shows that re-introducing *M. vollenhovenii* to river access sites in villages reduced the abundance of infected snails as well as the prevalence of schistosome infections in people [36]. While these data were limited in their geographic extent, the native range of *M. vollenhovenii* and closely related species extends across much of coastal Africa, where 90% of cases of schistosomiasis occur.

This impact extends hundreds of kilometers upstream of the dams themselves, such that restoring the ecological connectivity of African rivers could have a far-reaching impact on disease transmission, aquatic biodiversity, and inland fisheries [34].

Investment in aquaculture of *M. vollenhovenii* on the African continent could play an important role in both disease control and food security [37]. However, with dam barriers in place, aquatic migratory species like *M. vollenhovenii* would need to be continuously introduced upstream of dams. Sustained ecological restoration of these species will require modification to dam infrastructure to accommodate their aquatic migration. Designing prawn and/or fish ladders for existing and future dams could enable the restoration and preservation, respectively, of native aquatic species and the ecosystem services they provide for both disease control and food production [38]. Restoring prawns in the wild could support both the biological control of schistosome transmission as well as food production through the re-establishment of an inland fishery that has been lost to dam construction [39]. The establishment of catch guidelines that allow for the harvest of older, larger prawns for human consumption could ensure that younger, smaller prawns remain in the ecosystem for the sake of biological control [37,40]. Such a management system would allow dams (and their benefits) to remain in place while mitigating their negative impacts on livelihoods, food security and health.

2.2. *Removing Submerged Aquatic Vegetation and Transforming It into an Agricultural Resource*

By retaining water and slowing flows, dams encourage the growth of often-invasive aquatic vegetation [41,42]. Additionally, the agricultural expansion and intensification facilitated by dams likely increases fertilizer use in the landscape, which runs off into water sources and further promotes the growth of aquatic vegetation. Invasive aquatic species threaten biodiversity and productive use of river environments [43] while supporting the snails that transmit schistosomes [44,45]. Aquatic vegetation has been associated with both increased shedding of schistosome parasites from snails [46] and increased prevalence of infection in humans [47]. There is also evidence for a mutualistic relationship between snails that transmit schistosomes and submerged vegetation from the genus *Ceratophyllum* [46]. The removal of aquatic vegetation from natural water bodies [48], reservoirs [49], and irrigation canals [50], thus, might be an important intervention for reducing snail abundance and disease transmission while maintaining agricultural productivity.

A potential challenge with sustaining vegetation removal, however, is that it is a resource- and labor-intensive process that likely suffers from the tragedy of the commons. Without private incentives to remove, process and/or transport vegetation, the common good of disease control through vegetation removal may not be realized [51]. One agricultural innovation that might incentivize sustained vegetation removal is to transform removed vegetation into an agricultural resource. For example, submerged aquatic vegetation could be turned into compost to facilitate crop production or used as livestock feed to increase agricultural profitability [51–53]. In essence, these approaches would help to close the nutrient loop, returning nutrient-rich runoff from agricultural fields that is captured in aquatic plants back to food production [51]. Preliminary findings suggest that transforming submerged aquatic vegetation to compost can increase crop production in the Sahel region of Africa (unpublished data). Additionally, when we have harvested submerged vegetation, we have observed cattle and donkeys readily consuming it.

Another agricultural use of vegetation is as fuel for biodigesters. The breakdown of vegetation, produces both a nutrient-rich amendment for crops while simultaneously offering natural gas that can be used for cooking or for powering generators. By reducing reliance on fertilizers, these approaches might, indirectly, further reduce schistosome transmission [30,31,54,55]. These innovations have great potential for using aquatic vegetation to improve agricultural production and profitability. In turn, they help sustain the public health benefits of vegetation removal. Removal of vegetation should be undertaken carefully, however, as the opening of shorelines might attract human activity and inadvertently increase transmission [44,45,56]. We encourage researchers to quantify both the public health and private agricultural benefits of harvesting snail-inhabited vegetation and to test whether privatization helps to sustain the public health benefits.

2.3. Restoring Environmental Flows

In addition to the 40–80 million people directly displaced by dams, almost 500 million people globally live downstream of dams and have been affected by hydrologic alteration [8]. Flow regulation leading to the loss of natural seasonal flooding in dammed watersheds is associated with corresponding losses in inland fisheries, flood recession agriculture and dry season grazing [11]. Such traditional practices form the basis of subsistence livelihoods and food security among river-dependent populations [14,57], losses of which have been estimated to outweigh the benefits of dam construction in some settings [42,58]. The loss of seasonal floods also likely contributes to the increased risk of schistosomiasis in dammed landscapes. The destabilizing effects of seasonal flooding can stem the invasion of aquatic vegetation and keep snail populations from becoming perennially established [6].

Controlled dam releases can be used to restore and maintain ecological processes and natural resources [7]. Dam operations that restore environmental flows can accommodate traditional fishing and cultivation practices that rely on natural flood regimes without imposing substantial trade-offs on the conventional agricultural and energy production purposes of dams [13,59–61]. Particularly for recession agriculture, controlled releases can reduce sediment retained by dam barriers, which provides nutrients for cultivation in the floodplain [62]. These controlled releases, in turn, reduce the need for chemical fertilizers, whose use can support medically important snail populations and, by extension, schistosome transmission [31]. Some studies indicate that modifying dam operations in these ways could reduce schistosome transmission [63,64]. Similar studies of dam management for malaria control suggest that this can be accomplished with few trade-offs in irrigation or electricity generation [65,66].

Investigating how environmental and social benefits can be recovered by adjusting dam management could help diversify agricultural practices and improve agricultural outcomes, while mitigating the negative impacts of dams on livelihoods and health. These factors are seldom explicitly accounted for in decision-making about dam construction and operations [8,67], but they represent a significant opportunity for improving the well-being of the hundreds of millions of river-dependent people affected by dams. The estimated economic benefits of traditional uses of natural floodplains are substantial, even exceeding those of irrigated agriculture [42]. Reductions in the often-unquantified negative health impacts could add to this margin.

2.4. Integrating Agriculture and Aquaculture

The benefits of *Macrobrachium* spp. and other aquatic species can be cultivated in aquaculture settings and potentially integrated with irrigation systems [68]. These species can control snail populations within irrigation infrastructure and reduce occupational exposure to schistosomes. Modeling analyses indicate that prawn aquaculture is a profitable undertaking for smallholders that also sustainably controls medically important snails [37]. Especially in the water-intensive cultivation of rice, integrating prawn farming with crop cultivation may be particularly advantageous, efficient, and beneficial [69]. This practice is already common in South Asia [70,71], and could be applied to the rapidly growing area under rice cultivation on the African continent [69]. Identifying and using insecticides that minimize harm to aquaculture species would be necessary for this strategy to succeed [55].

3. Discussion

The agricultural innovations we propose represent an opportunity to reconcile the apparent trade-offs between the economic development that motivates dam investment and their unintended consequences. The agricultural landscapes created by dams have been demonized for displacing people, disrupting traditional livelihoods, and making people sick. Especially in areas endemic for environmentally-mediated infectious diseases like schistosomiasis, poverty, food insecurity, malnutrition, and infection can all reinforce each other in disease-driven poverty traps [72]. These dynamics likely diminish the intended benefits of dams. As a result, finding ways to restore dammed watersheds' ecological integrity and making ecologically sound design and management choices in the construction of new dams are important in alleviating poverty, food insecurity, and disease transmission for rural populations that depend on rivers and whose livelihoods and health are negatively impacted by dams. Such reconciliation of trade-offs could therefore promote the resilience of coupled human–natural systems in the face of climate change.

None of the four proposed agricultural innovations requires sophisticated technology, but all will require careful planning and management in addition to the technical capacity and political will to assess and address the hydrologic, environmental and social impacts of dams. The goals and implementation of new management schemes are likely to vary across contexts. The ecological and geographical characteristics of a watershed, the human needs, water uses, structural characteristics and limitations of dam infrastructure will dictate what management strategies are feasible and potentially beneficial for livelihoods, food security and health [7,73]. The Nature Conservancy has set an example in its Sustainable Rivers program [74], undertaking flow restoration projects in the United States in cooperation with the Army Corps of Engineers and documenting 850 projects worldwide that demonstrate how flow restoration can be undertaken at many existing dams as well as designed into new dam construction.

Health and disease control could become an objective in existing and future environmental management and agricultural innovation programs. Design or retrofitting of fish and prawn passages could be considered for all new and existing dams, respectively, within 500 km of coasts. The production and distribution of soil amendments and livestock feeds derived from aquatic vegetation holds promise as a social enterprise, while disease control could become a new criterion in the multi-objective frameworks used to manage dams and their reservoirs. Finally, the common practice of integrating agriculture and aquaculture could be extended to new geographic settings, where they may yield health as well as food security benefits.

By design, these innovations will also require cooperation between disciplines and sectors as well as attention to the needs of communities and the capacity and political will of governments. These requirements are in keeping with the focus of the Sustainable Development Goals (SDGs) on interlinkages between disciplines and sectors, bridging between economic development, health and the environment. The agricultural innovations we have proposed have the potential to improve food security (SDG2), health (SDG3) and equity (SDG10) in sustainable economic development in dammed watersheds. Cross-sectoral cooperation has the potential to yield cross-sectoral benefits.

4. Conclusions

As the pace of dam development quickens in low- and middle-income countries, managing the health and livelihood impacts of new and existing infrastructure is crucial to ensuring the sustainability of modern agriculture and economic development in a changing climate. Particularly in schistosomiasis-endemic regions, agricultural innovations involving the restoration of aquatic migratory species, the harvest and transformation of invasive aquatic vegetation, the restoration of environmental flows, and the integration of agriculture and aquaculture hold promise for reducing disease transmission while also improving economic development and food security. These innovations will require careful planning and management as well as cooperation between sectors, but strategically implementing such agricultural innovations can start to turn vicious cycles of poverty and disease into virtuous cycles of health and development.

Author Contributions: Conceptualization, A.J.L., D.L.-C. and G.A.D.L.; writing—original draft preparation, A.J.L.; writing—review and editing, A.J.L., D.L.-C., S.H.S., J.R.R. and G.A.D.L.; funding acquisition, D.L.-C., S.H.S., J.R.R. and G.A.D.L. All authors have read and agreed to the published version of the manuscript.

Funding: A.J.L. was supported by a James and Nancy Kelso Fellowship through the Stanford Interdisciplinary Graduate Fellowship program. D.L.-C. was partially supported by the National Science Foundation (BCS-1414102). G.A.D.L. and S.H.S. were partially supported by grants from the National Science Foundation (ICER-202438 and DEB-2011179) as well as a grant from the Bill & Melinda Gates Foundation (#OPP1114050). J.R.R. was supported by grants from the National Institutes of Health (R01TW010286-01), the National Science Foundation (DEB-2017785, IOS-1754868) and the Indiana Clinical and Translational Sciences Institute.

Institutional Review Board Statement: Not applicable.

Informed Consent Statement: Not applicable.

Data Availability Statement: Not applicable.

Conflicts of Interest: The authors declare no conflict of interest.

References

1. IPCC Global Warming of 1.5 °C. Available online: https://www.ipcc.ch/sr15/ (accessed on 8 January 2021).
2. Dillon, A.; Fishman, R. Dams: Effects of Hydrological Infrastructure on Development. *Annu. Rev. Resour. Econ.* **2019**, *11*, 125–148. [CrossRef]
3. Ansar, A.; Flyvbjerg, B.; Budzier, A.; Lunn, D. Should We Build More Large Dams? The Actual Costs of Hydropower Megaproject Development. *Energy Policy* **2014**, *69*, 43–56. [CrossRef]
4. World Commission on Dams. *Dams and Development: A New Framework for Decision-Making—The Report of the World Commission on Dams*; Earthscan: Sterling, VA, USA, 2000; ISBN 1-134-89805-3.
5. Nilsson, C.; Berggren, K. Alterations of Riparian Ecosystems Caused by River Regulation: Dam Operations Have Caused Global-Scale Ecological Changes in Riparian Ecosystems. How to Protect River Environments and Human Needs of Rivers Remains One of the Most Important Questions of Our Time. *BioScience* **2000**, *50*, 783–792. [CrossRef]
6. Jobin, W. *Dams and Disease: Ecological Design and Health Impacts of Large Dams, Canals and Irrigation Systems*; Routledge: New York, NY, USA, 1999.
7. Acreman, M. *Managed Flood Releases from Reservoirs: Issues and Guidance*; Centre for Ecology and Hydrology: Wallingford, UK, 2000; p. 96.
8. Richter, B.D.; Postel, S.; Revenga, C.; Scudder, T.; Lehner, B.; Churchill, A.; Chow, M. Lost in Development's Shadow: The Downstream Human Consequences of Dams. *Water Altern.* **2010**, *3*, 29.
9. Duflo, E.; Pande, R. Dams. *Q. J. Econ.* **2007**, *122*, 601–646. [CrossRef]
10. Strobl, E.; Strobl, R.O. The Distributional Impact of Dams: Evidence from Cropland Productivity in Africa. *J. Dev. Econ.* **2011**, *96*, 432–450. [CrossRef]
11. Adams, W.M. *Wasting the Rain: Rivers, People and Planning in Africa*; Earthscan: London, UK, 1992; ISBN 978-1-85383-089-1.
12. Saarnak, N.L. Flood Recession Agriculture in the Senegal River Valley. *Geogr. Tidsskr. Dan. J. Geogr.* **2003**, *103*, 99–113. [CrossRef]
13. Owusu, K.; Obour, P.B.; Nkansah, M.A. Downstream Effects of Dams on Livelihoods of River-Dependent Communities: The Case of Ghana's Kpong Dam. *Geogr. Tidsskr. Dan. J. Geogr.* **2017**, *117*, 1–10. [CrossRef]
14. Maingi, J.K.; Marsh, S.E. Quantifying Hydrologic Impacts Following Dam Construction along the Tana River, Kenya. *J. Arid Environ.* **2002**, *50*, 53–79. [CrossRef]
15. Lerer, L.B.; Scudder, T. Health Impacts of Large Dams. *Environ. Impact Assess. Rev.* **1999**, *19*, 113–123. [CrossRef]
16. Rohr, J.R.; Barrett, C.B.; Civitello, D.J.; Craft, M.E.; Delius, B.; DeLeo, G.A.; Hudson, P.J.; Jouanard, N.; Nguyen, K.H.; Ostfeld, R.S.; et al. Emerging Human Infectious Diseases and the Links to Global Food Production. *Nat. Sustain.* **2019**, *2*, 445–456. [CrossRef]
17. Steinmann, P.; Keiser, J.; Bos, R.; Tanner, M.; Utzinger, J. Schistosomiasis and Water Resources Development: Systematic Review, Meta-Analysis, and Estimates of People at Risk. *Lancet Infect. Dis.* **2006**, *6*, 411–425. [CrossRef]
18. Lund, A.J.; Sam, M.M.; Sy, A.B.; Sow, O.W.; Ali, S.; Sokolow, S.H.; Merrell, S.B.; Bruce, J.; Jouanard, N.; Senghor, S.; et al. Unavoidable Risks: Local Perspectives on Water Contact Behavior and Implications for Schistosomiasis Control in an Agricultural Region of Northern Senegal. *Am. J. Trop. Med. Hyg.* **2019**, *101*, 837–847. [CrossRef]
19. Ross, A.G.P.; Olveda, R.M.; Chy, D.; Olveda, D.U.; Li, Y.; Harn, D.A.; Gray, D.J.; McManus, D.P.; Tallo, V.; Chau, T.N.P.; et al. Can Mass Drug Administration Lead to the Sustainable Control of Schistosomiasis? *J. Infect. Dis.* **2015**, *211*, 283–289. [CrossRef]
20. Kittur, N.; King, C.H.; Campbell, C.H., Jr.; Kinung'hi, S.; Mwinzi, P.N.M.; Karanja, D.M.S.; N'Goran, E.K.; Phillips, A.E.; Gazzinelli-Guimaraes, P.H.; Olsen, A.; et al. Persistent Hotspots in Schistosomiasis Consortium for Operational Research and Evaluation Studies for Gaining and Sustaining Control of Schistosomiasis after Four Years of Mass Drug Administration of Praziquantel. *Am. J. Trop. Med. Hyg.* **2019**, *101*, 617–627. [CrossRef]
21. Colley, D.G.; Bustinduy, A.L.; Secor, W.E.; King, C.H. Human Schistosomiasis. *Lancet* **2014**, *383*, 2253–2264. [CrossRef]
22. Ezeamama, A.E.; Bustinduy, A.L.; Nkwata, A.K.; Martinez, L.; Pabalan, N.; Boivin, M.J.; King, C.H. Cognitive Deficits and Educational Loss in Children with Schistosome Infection—A Systematic Review and Meta-Analysis. *PLoS Negl. Trop. Dis.* **2018**, *12*, e0005524. [CrossRef]

23. Fenwick, A.; Figenschou, B.H. The Effect of Schistosoma Mansoni Infection on the Productivity of Cane Cutters on a Sugar Estate in Tanzania. *Bull. World Health Org.* **1972**, *47*, 567–572.
24. Barbosa, F.; Pereira Da Costa, D. Incapacitating Effects of Schistosomiasis Mansoni on the Productivity of Sugar-Cane Cutters in Northeastern Brazil. *Am. J. Epidemiol.* **1981**, *114*, 102–111. [CrossRef]
25. Ersado, L. *Small-Scale Irrigation Dams, Agricultural Production, and Health: Theory and Evidence from Ethiopia*; World Bank Policy Research; The World Bank: Washington, DC, USA, 2005.
26. Castello, L.; Macedo, M.N. Large-Scale Degradation of Amazonian Freshwater Ecosystems. *Glob. Chang. Biol.* **2016**, *22*, 990–1007. [CrossRef]
27. Johnson, P.T.; Olden, J.D.; Vander Zanden, M.J. Dam Invaders: Impoundments Facilitate Biological Invasions into Freshwaters. *Front. Ecol. Environ.* **2008**, *6*, 357–363. [CrossRef]
28. Golden, C.D.; Shapero, A.; Vaitla, B.; Smith, M.R.; Myers, S.S.; Stebbins, E.; Gephart, J.A. Impacts of Mainstream Hydropower Development on Fisheries and Human Nutrition in the Lower Mekong. *Front. Sustain. Food Syst.* **2019**, *3*, 93. [CrossRef]
29. Alkalay, A.S.; Rosen, O.; Sokolow, S.H.; Faye, Y.P.W.; Faye, D.S.; Aflalo, E.D.; Jouanard, N.; Zilberg, D.; Huttinger, E.; Sagi, A. The Prawn Macrobrachium Vollenhovenii in the Senegal River Basin: Towards Sustainable Restocking of All-Male Populations for Biological Control of Schistosomiasis. *PLoS Negl. Trop. Dis.* **2014**, *8*, e3060. [CrossRef]
30. Halstead, N.T.; Hoover, C.M.; Arakala, A.; Civitello, D.J.; De Leo, G.A.; Gambhir, M.; Johnson, S.A.; Jouanard, N.; Loerns, K.A.; McMahon, T.A.; et al. Agrochemicals Increase Risk of Human Schistosomiasis by Supporting Higher Densities of Intermediate Hosts. *Nat. Commun.* **2018**, *9*, 837. [CrossRef]
31. Hoover, C.M.; Rumschlag, S.L.; Strgar, L.; Arakala, A.; Gambhir, M.; de Leo, G.A.; Sokolow, S.H.; Rohr, J.R.; Remais, J.V. Effects of Agrochemical Pollution on Schistosomiasis Transmission: A Systematic Review and Modelling Analysis. *Lancet Planet. Health* **2020**, *4*, e280–e291. [CrossRef]
32. Khagram, S. *Dams and Development: Transnational Struggles for Water and Power*; Cornell University Press: Ithaca, NY, USA, 2004; ISBN 978-0-8014-4228-5.
33. Zarfl, C.; Lumsdon, A.E.; Berlekamp, J.; Tydecks, L.; Tockner, K. A Global Boom in Hydropower Dam Construction. *Aquat. Sci.* **2015**, *77*, 161–170. [CrossRef]
34. Sokolow, S.H.; Jones, I.; Jocque, M.; La, D.; Cords, O.; Knight, A.; Lund, A.; Wood, C.L.; Lafferty, K.D.; Kuris, A.M.; et al. Nearly 400 Million People Are at Higher Risk of Schistosomiasis Because Dams Block the Migration of Snail-Eating River Prawns. *Philos. Trans. R. Soc. Lond. B* **2017**, *372*, 20160127. [CrossRef]
35. Swartz, S.J.; De Leo, G.A.; Wood, C.L.; Sokolow, S.H. Infection with Schistosome Parasites in Snails Leads to Increased Predation by Prawns: Implications for Human Schistosomiasis Control. *J. Exp. Biol.* **2015**, *218*, 3962–3967. [CrossRef]
36. Sokolow, S.H.; Huttinger, E.; Jouanard, N.; Hsieh, M.H.; Lafferty, K.D.; Kuris, A.M.; Riveau, G.; Senghor, S.; Thiam, C.; N'Diaye, A.; et al. Reduced Transmission of Human Schistosomiasis after Restoration of a Native River Prawn That Preys on the Snail Intermediate Host. *Proc. Natl. Acad. Sci. USA* **2015**, *112*, 9650–9655. [CrossRef]
37. Hoover, C.M.; Sokolow, S.H.; Kemp, J.; Sanchirico, J.N.; Lund, A.J.; Jones, I.J.; Higginson, T.; Riveau, G.; Savaya, A.; Coyle, S.; et al. Modelled Effects of Prawn Aquaculture on Poverty Alleviation and Schistosomiasis Control. *Nat. Sustain.* **2019**, *2*, 611–620. [CrossRef]
38. Brauman, K.A.; Daily, G.C.; Duarte, T.K.; Mooney, H.A. The Nature and Value of Ecosystem Services: An Overview Highlighting Hydrologic Services. *Annu. Rev. Environ. Resour.* **2007**, *32*, 67–98. [CrossRef]
39. Hammerschlag, N.; Schmitz, O.J.; Flecker, A.S.; Lafferty, K.D.; Sih, A.; Atwood, T.B.; Gallagher, A.J.; Irschick, D.J.; Skubel, R.; Cooke, S.J. Ecosystem Function and Services of Aquatic Predators in the Anthropocene. *Trends Ecol. Evol.* **2019**. [CrossRef]
40. Sokolow, S.H.; Lafferty, K.D.; Kuris, A.M. Regulation of Laboratory Populations of Snails (Biomphalaria and Bulinus Spp.) by River Prawns, Macrobrachium Spp. (Decapoda, Palaemonidae): Implications for Control of Schistosomiasis. *Acta Trop.* **2014**, *132*, 64–74. [CrossRef]
41. Cogels, F.X.; Coly, A.; Niang, A. Impact of Dam Construction on the Hydrological Regime and Quality of a Sahelian Lake in the River Senegal Basin. *Regul. Rivers Res. Manag.* **1997**, *13*, 27–41. [CrossRef]
42. Manikowski, S.; Strapasson, A. Sustainability Assessment of Large Irrigation Dams in Senegal: A Cost-Benefit Analysis for the Senegal River Valley. *Front. Environ. Sci.* **2016**, 4. [CrossRef]
43. OMVS/GEF Transboundary Diagnostic Environmental Analysis of the Senegal River Basin. Available online: https://iwlearn.net/documents/3196 (accessed on 8 February 2021).
44. Klumpp, R.K.; Chu, K.Y. Importance of the Aquatic Weed Ceratophyllum to Transmission of Schistosoma Haematobium in the Volta Lake, Ghana. *Bull. World Health Org.* **1980**, *58*, 791–798.
45. Thomas, J.D.; Tait, A.I. Control of the Snail Hosts of Schistosomiasis by Environmental Manipulation: A Field and Laboratory Appraisal in the Ibadan Area, Nigeria. *Philos. Trans. R. Soc. Lond. B* **1984**, *305*, 201–253. [CrossRef]
46. Haggerty, C.J.E.; Bakhoum, S.; Civitello, D.J.; Leo, G.A.D.; Jouanard, N.; Ndione, R.A.; Remais, J.V.; Riveau, G.; Senghor, S.; Sokolow, S.H.; et al. Aquatic Macrophytes and Macroinvertebrate Predators Affect Densities of Snail Hosts and Local Production of Schistosome Cercariae That Cause Human Schistosomiasis. *Plos Negl. Trop. Dis.* **2020**, *14*, e0008417. [CrossRef]
47. Wood, C.L.; Sokolow, S.H.; Jones, I.J.; Chamberline, A.J.; Lafferty, K.D.; Kuris, A.M.; Jocque, M.; Hopkins, S.H.; Adams, G.; Buck, J.C.; et al. Precision Mapping of Snail Habitats Provides a Powerful Indicator of Human Schistosomiasis Transmission. *Proc. Natl. Acad. Sci. USA* **2019**, *116*, 23182–23191. [CrossRef]

48. World Health Organization. *Field Use of Molluscicides in Schistosomiasis Control Programmes: An Operational Manual for Programme Managers*; World Health Organization (WHO): Geneva, Switzerland, 2017; ISBN 978 92 4 151199 5.
49. Chu, K.Y. Trials of Ecological and Chemical Measures for the Control of Schistosoma Haematobium Transmission in a Volta Lake Village. *Bull. World Health Organ.* **1978**, *56*, 313.
50. Boelee, E.; Laamrani, H. Environmental Control of Schistosomiasis through Community Participation in a Moroccan Oasis. *Trop. Med. Int. Health* **2004**, *9*, 997–1004. [CrossRef]
51. Quilliam, R.S.; van Niekerk, M.A.; Chadwick, D.R.; Cross, P.; Hanley, N.; Jones, D.L.; Vinten, A.J.A.; Willby, N.; Oliver, D.M. Can Macrophyte Harvesting from Eutrophic Water Close the Loop on Nutrient Loss from Agricultural Land? *J. Environ. Manag.* **2015**, *152*, 210–217. [CrossRef]
52. Traore, O.; Traore, K.; Yaye, A. Characterization of Three Invading Aquatic Plants in Burkina Faso and Their Possible Use for Crop Production. *Int. J. Biol. Chem. Sci.* **2009**, *3*. [CrossRef]
53. Reddy, K.R.; Busk, W.F.D. Nutrient Removal Potential of Selected Aquatic Macrophytes. *J. Environ. Qual.* **1985**, *14*, 459–462. [CrossRef]
54. Becker, J.; Ganatra, A.A.; Kandie, F.; Muhlbauer, L.; Ahlheim, J.; Brack, W.; Torto, B.; Agola, E.L.; McOdimba, F.; Hollert, H.; et al. Pesticide Pollution in Freshwater Paves the Way for Schistosomiasis Transmission. *Sci. Rep.* **2020**, *10*, 3650. [CrossRef]
55. Haggerty, C.J.E.; Delius, B.; Jouanard, N.; Ndao, P.D.; Leo, G.A.D.; Lund, A.J.; Lopez-Carr, D.; Remais, J.V.; Riveau, G.; Sokolow, S.H.; et al. Identifying Low Risk Insecticides to Address Both Food Shortages and the Biocontrol of Human Schistosomiasis. *bioRxiv* **2021**. [CrossRef]
56. Waddy, B.B. Research into the Health Problems of Man Made Lakes, with Special Reference to Africa. *Trans. R. Soc. Trop. Med. Hyg.* **1975**, *69*, 39–50. [CrossRef]
57. DeGeorges, A.; Reilly, B.K. Dams and Large Scale Irrigation on the Senegal River. Impacts on Man and the Environment. Available online: http://hdr.undp.org/sites/default/files/degeorges_andre.pdf (accessed on 19 February 2018).
58. Lautze, J.; Kirshen, P. Dams, Health, and Livelihoods: Lessons from the Senegal, Suggestions for Africa. *Int. J. River Basin Manag.* **2007**, *5*, 199–206. [CrossRef]
59. Sabo, J.L.; Ruhi, A.; Holtgrieve, G.W.; Elliott, V.; Arias, M.E.; Ngor, P.B.; Räsänen, T.A.; Nam, S. Designing River Flows to Improve Food Security Futures in the Lower Mekong Basin. *Science* **2017**, *358*, eaao1053. [CrossRef]
60. Chen, W.; Olden, J.D. Designing Flows to Resolve Human and Environmental Water Needs in a Dam-Regulated River. *Nat. Commun.* **2017**, *8*, 2158. [CrossRef]
61. Raso, L.; Bader, J.-C.; Weijs, S. Reservoir Operation Optimized for Hydropower Production Reduces Conflict with Traditional Water Uses in the Senegal River. *J. Water Resour. Plan. Manag.* **2020**, *146*, 05020003. [CrossRef]
62. Richter, B.D.; Thomas, G.A. Restoring Environmental Flows by Modifying Dam Operations. *Ecol. Soc.* **2007**, *12*, 12. [CrossRef]
63. Jobin, W. *Rapid Health Impact Assessment of Manantali Dam OMVS Energy Project West Africa*; World Bank (Africa Technical Department): Washington, DC, USA, 1993; p. 57.
64. Ofoezie, I.E.; Asaolu, S.O. Water Level Regulation and Control of Schistosomiasis Transmission: A Case Study in Oyan Reservoir, Ogun State, Nigeria. *Bull. World Health Organ.* **1997**, *75*, 435–441.
65. Bianchi, P.R.; Gianelli, L. *Water-Related Disease Control via Dam Operation: Balancing Hydropower Production and Malaria Spreading on Kariba Reservoir*; Politecnico Milano: Milan, Italy, 2017.
66. Kibret, S.; Wilson, G.G.; Ryder, D.; Tekie, H.; Petros, B. Can Water-Level Management Reduce Malaria Mosquito Abundance around Large Dams in Sub-Saharan Africa? *PLoS ONE* **2018**, *13*, e0196064. [CrossRef]
67. Dunham, J.B.; Angermeier, P.L.; Crausbay, S.D.; Cravens, A.E.; Gosnell, H.; McEvoy, J.; Moritz, M.A.; Raheem, N.; Sanford, T. Rivers Are Social–Ecological Systems: Time to Integrate Human Dimensions into Riverscape Ecology and Management. *Wires Water* **2018**, *5*, e1291. [CrossRef]
68. Marques, H.L.A.; New, M.B.; Boock, M.V.; Barros, H.P.; Mallasen, M.; Valenti, W.C. Integrated Freshwater Prawn Farming: State-of-the-Art and Future Potential. *Rev. Fish. Sci. Aquac.* **2016**, *24*, 264–293. [CrossRef]
69. Balasubramanian, V.; Sie, M.; Hijmans, R.; Otsuka, K. Increasing Rice Production in Sub-Saharan Africa: Challenges and Opportunities. *Adv. Agron.* **2007**, *94*, 55–133. [CrossRef]
70. Ahmed, N.; Garnett, S.T. Sustainability of Freshwater Prawn Farming in Rice Fields in Southwest Bangladesh. *J. Sustain. Agric.* **2010**, *34*, 659–679. [CrossRef]
71. Kurup, B.M.; Ranjeet, K. Integration of Freshwater Prawn Culture with Rice Farming in Kuttanad, India. *Naga WorldFish Cent. Q.* **2002**, *25*, 16–19.
72. Garchitorena, A.; Sokolow, S.H.; Roche, B.; Ngonghala, C.N.; Jocque, M.; Lund, A.; Barry, M.; Mordecai, E.A.; Daily, G.C.; Jones, J.H.; et al. Disease Ecology, Health and the Environment: A Framework to Account for Ecological and Socio-Economic Drivers in the Control of Neglected Tropical Diseases. *Philos. Trans. R. Soc. Lond. B* **2017**, *372*, 20160128. [CrossRef]
73. Matthews, N.; McCartney, M. Opportunities for Building Resilience and Lessons for Navigating Risks: Dams and the Water Energy Food Nexus. *Environ. Prog. Sustain. Energy* **2018**, *37*, 56–61. [CrossRef]
74. The Nature Conservancy Sustainable Rivers Project. Available online: https://www.nature.org/en-us/what-we-do/our-priorities/protect-water-and-land/land-and-water-stories/sustainable-rivers-project/ (accessed on 15 January 2021).

sustainability

Review

Agricultural Innovation and Sustainable Development: A Case Study of Rice–Wheat Cropping Systems in South Asia

Aman Ullah [1], Ahmad Nawaz [2], Muhammad Farooq [1,3,4] and Kadambot H. M. Siddique [4,*]

1 Department of Plant Sciences, College of Agricultural and Marine Sciences, Sultan Qaboos University, Al-Khoud 123, Oman; amanullahmalik12@gmail.com (A.U.); farooqcp@squ.edu.om (M.F.)
2 Centre for Agriculture and Biosciences International (CABI), Central and West Asia (CWA), Opposite 1-A, Data Gunj Baksh Road, Satellite Town, Rawalpindi 46300, Pakistan; ahmadnawaz2006@gmail.com
3 Department of Agronomy, University of Agriculture, Faisalabad 38040, Pakistan
4 The UWA Institute of Agriculture, The University of Western Australia, Perth WA 6001, Australia
* Correspondence: kadambot.siddique@uwa.edu.au

Abstract: The rice–wheat cropping system is the main food bowl in Asia, feeding billions across the globe. However, the productivity and long-term sustainability of this system are threatened by stagnant crop yields and greenhouse gas emissions from flooded rice production. The negative environmental consequences of excessive nitrogen fertilizer use are further exacerbating the situation, along with the high labor and water requirements of transplanted rice. Residue burning in rice has also severe environmental concerns. Under these circumstances, many farmers in South Asia have shifted from transplanted rice to direct-seeded rice and reported water and labor savings and reduced methane emissions. There is a need for opting the precision agriculture techniques for the sustainable management of nutrients. Allelopathic crops could be useful in the rotation for weed management, the major yield-reducing factor in direct-seeded rice. Legume incorporation might be a viable option for improving soil health. As governments in South Asia have imposed a strict ban on the burning of rice residues, the use of rice-specific harvesters might be a pragmatic option to manage rice residues with yield and premium advantage. However, the soil/climatic conditions and farmer socio-economic conditions must be considered while promoting these technologies in rice-wheat system in South Asia.

Keywords: rice–wheat cropping system; South Asia; water requirements; nitrogen; direct seeding

Citation: Ullah, A.; Nawaz, A.; Farooq, M.; Siddique, K.H.M. Agricultural Innovation and Sustainable Development: A Case Study of Rice–Wheat Cropping Systems in South Asia. *Sustainability* **2021**, *13*, 1965. https://doi.org/10.3390/su13041965

Academic Editor: Roberto Mancinelli
Received: 19 January 2021
Accepted: 6 February 2021
Published: 11 February 2021

Publisher's Note: MDPI stays neutral with regard to jurisdictional claims in published maps and institutional affiliations.

Copyright: © 2021 by the authors. Licensee MDPI, Basel, Switzerland. This article is an open access article distributed under the terms and conditions of the Creative Commons Attribution (CC BY) license (https://creativecommons.org/licenses/by/4.0/).

1. Introduction

Rice–wheat cropping systems (RWCS) provide staple food to 15% of the world's population [1]. The major issue for the sustainability of conventional RWCS in South Asia is soil quality degradation associated with resource scarcity [2]. Other factors include water scarcity, low soil organic matter, nutrient imbalances, labor/energy crises, complex insect and weed flora, herbicide-resistant weeds, and greenhouse gas (GHG) emissions [3]. Moreover, conventional puddled transplanted rice (PTR) cultivation has over-exploited the groundwater leading to an alarming fall in the water table in South Asia [4].

The conventional rice production systems are no longer suitable as they require large amounts of water (3000–5000 L of water to produce one kg of rice) [5,6]. It has been reported that 15–20 Mha of conventional rice production systems will face water shortages by 2025 [7]. In some parts of Pakistan and India, groundwater tables are declining by 1.0–3.5 m and 6 m year^{-1}, respectively [8].

Puddling, as practices in conventional rice-wheat system, increases the soil bulk density, which causes soil compaction [9] and affects root development in post-rice crops [3]. Nitrogen uptake in puddled rice fields declines by 12–35% in the following wheat crop due to subsoil compaction [10]. The evolution of herbicide resistant weeds and shift in weed flora (a mixture of broadleaf and grassy weeds) have further exacerbated the scenario in

RWCS to harvest optimum crop yields [11]. Little seed canary grass (*Phalaris minor* Retz.) has been reported to decrease wheat yields by 10–65% with occasional crop failure [12], while smartweed (*Polygonum hydropiper* L.) can reduce the rice and wheat yields by 15–25% and 15–30%, respectively [13]. Furthermore, the rice and wheat monocultures in RCWS have increased disease and pest problems [14] and has caused macro- and micro-nutrient deficiencies [3,15,16].

In this scenario, resource conservation technologies, such as direct-seeded rice (DSR), no-till wheat, and laser-assisted land leveling, can be used to improve the sustainability of yields in RWCS [3]. Several studies reported that residue retention and no-tillage enhance the nitrogen and carbon pools in soil [17,18].

This case study focuses on the problems of conventional RWCS (i.e., nutrient mining, GHG emissions, and reduced profits) and alternative options such as DSR, use of advanced rice harvesters for harvest, no-till wheat, precision agriculture, and crop rotation to improve the yields, sustainability, and the conservation of scarce natural resources.

2. Review Methodology

We searched more than 180 articles, including 10 review and 170 research articles, using four databases: Scopus, Web of Science, Google Scholar, and Center for Agriculture and Bioscience International (CABI). These databases are large collections of mainstream articles and are widely used for searching. The different keywords as (rice–wheat cropping system, greenhouse gases emission, direct-seeded rice, zero tillage wheat, agricultural innovation systems, profit margin in the conventional rice-wheat cropping system, crop rotation, precision agriculture, nutrient mining, agricultural sustainability, and rice-specific harvesters) were used to search the articles from these databases. To take additional information from these articles, we used references from these articles as well. The articles other than South Asia and published before the year 2000 were not included in this review.

3. Problems in Conventional Rice–Wheat Systems

3.1. Greenhouse Gas Emissions

In the Indo-Gangetic Plains (IGP), conventional RWCS is the major source of atmospheric nitrous oxide (N_2O), carbon dioxide (CO_2), and methane (CH_4) emissions due to the use of intensive agricultural inputs [19], particularly the injudicious use of nitrogen fertilizers Table 1 [20,21], aerobic and anaerobic soil cycling [22], and residue burning. In northwest India, 2.5 M farmers burn 23 MMT of rice stubble each year (October to November) to prepare field for wheat crop, causing massive air pollution affecting millions of people across the IGP [23,24]. Annual residue burning emits GHGs, including CO_2 (379 Tg), carbon monoxide (CO; 23 Tg), CH_4 (0.68 Tg), NOx (0.96 Tg), and sulfur dioxide (SO_2) (0.10 Tg) [25]. The RWCS supplied with 75 kg N ha^{-1} had mean annual emissions of N_2O of 1.49 kg N ha^{-1}, or 2.97–3.04 kg N ha^{-1} when supplied with >150 kg N ha^{-1} [26].

Table 1. Greenhouse gas emissions from different rice production systems.

Greenhouse Gas	Quantity Emitted from DSR	Quantity Emitted from Transplanted Rice	Reference
Methane (CH_4)	0.49 mg m^{-2} day^{-1}	3.10 mg m^{-2} day^{-1}	[27]
Nitrous oxide (N_2O)	0.97 mg m^{-2} day^{-1}	1.03 mg m^{-2} day^{-1}	
Carbon dioxide (CO_2)	600 mg m^{-2} day^{-1}	1800 mg m^{-2} day^{-1}	
Nitrous oxide (N_2O)	0.90 kg ha^{-1}	0.56 kg ha^{-1}	[28]
Methane (CH_4)	23.3 kg ha^{-1}	32.8 kg ha^{-1}	
Methane (CH_4)	18.9 kg ha^{-1}	28.4 kg ha^{-1}	[29]
Nitrous oxide (N_2O)	0.95 kg ha^{-1}	0.65 kg ha^{-1}	
Nitrous oxide (N_2O)	25 kg ha^{-1}	48 kg ha^{-1}	[30]
Nitrous oxide (N_2O)	0.12 kg ha^{-1}	0.11 kg ha^{-1}	
Methane (CH_4)	0.2 kg ha^{-1}	1.1 kg ha^{-1}	[31]
Carbon dioxide (CO_2)	1.2 kg ha^{-1}	1.3 kg ha^{-1}	
Nitrous oxide (N_2O)	0.6 kg ha^{-1}	0.4 kg ha^{-1}	

Table 1. *Cont.*

Greenhouse Gas	Quantity Emitted from DSR	Quantity Emitted from Transplanted Rice	Reference
Methane (CH_4)	6.98 kg ha^{-1}	18.49 kg ha^{-1}	[32]
Nitrous oxide (N_2O)	3.35 kg ha^{-1}	3.71 kg ha^{-1}	
Methane (CH_4)	25 kg ha^{-1}	60 kg ha^{-1}	[33]
Methane (CH_4)	220 kg ha^{-1}	315 kg ha^{-1}	[34]
Methane (CH_4)	25 kg ha^{-1}	60 kg ha^{-1}	[35]
Nitrous oxide (N_2O)	0.12 kg ha^{-1}	0.10 kg ha^{-1}	
Methane (CH_4)	129 kg ha^{-1}	271 kg ha^{-1}	[36]
Methane (CH_4)	269 kg ha^{-1}	229 kg ha^{-1}	[37]
Methane (CH_4)	75 kg ha^{-1}	89 kg ha^{-1}	[38]

The flooding conditions in rice cause the anaerobic decomposition of organic matter, which produces methane (CH_4) in the soil [39]. Globally, rice contributes ~20% of the total CH_4 emissions [40]. The warming potential of CH_4 is 25–30 times greater than CO_2 [40,41]. In 2005, the concentration of atmospheric CH_4 reached 1774 ppb [40]. Several studies reported that PTR produces more CH_4 emissions than DSR, while DSR produces more N_2O than PTR [3]. In one study, DSR and PTR produced N_2O emissions of 1.2 t CO_2eq ha^{-1} and 0.4 t CO_2eq ha^{-1}, respectively and CH_4 emissions of 0.1 t CO_2eq ha^{-1} and 0.6 t CO_2eq ha^{-1}, respectively [42]. In conclusion, the adoption of monocultures in RWCS contributes to global GHG emissions due to intensive agricultural inputs use and residue burning.

3.2. Nutrient Mining and Unwise Nutrient Use

Continuous monoculture cropping has threatened the long-term sustainability and has caused macro- and micro-nutrient imbalances in RWCS [3]. In the IGP, the mining of major nutrients, including nitrogen (N), phosphorus (P), potassium (K), and sulfur (S), has created a major nutrient imbalance in RWCS. The production of 1 t of rice/wheat depletes 20.1/24.5 kg N, 4.9/3.8 kg P, and 25.0/27.3 kg K, respectively, from the soil [43], which decreases the soil productivity [44] if these nutrients are not replenished. In the IGP, the removal of crop residues removes five times more K than that supplied through fertilizers [45].

Among the micro-nutrients, Zn deficiency is more common in rice, while manganese (Mn) deficiency is more prevalent in wheat [46]. In India, 49% of soil samples were Zn deficient, followed by 33% deficient in B, 12% in Fe, 5% in Mn, and 3% in Cu [47]. In the IGP, most rice and wheat farmers apply N fertilizers following blanket recommendations based on crop response data, leading to under- or over-fertilization as there is wide spatial variability in the indigenous nutrient supply capacity of soils in different agro-ecologies [26]. Diagnostic surveys of the IGP showed that farmers apply more N and P fertilizers than recommended while under/overlooking the supply of K, other secondary macronutrients, and micro-nutrients [48]. The inadequate and imbalanced use of nutrients reduces nutrient use efficiencies and profitability and increases environmental hazards [49]. In conclusion, the continuous growing of rice and wheat has resulted in the mining of major (N, P, K, and S) and trace (Zn, B, and Fe) nutrients due to over- or under-fertilization.

3.3. Reduced Profit Margins

The PTR has smaller profit margins than DSR due to the high labor costs Table 2 [50]. With industrialization, the migration of people to cities reduced labor availability for agricultural activities, which increased labor costs. Labor shortages delay the transplantation of rice seedlings into puddled fields [51], delaying maturation, and decrease yields [3].

Table 2. Profit margins in different rice production systems.

Name of Input	Type of Soil	Unit Cost in DSR ha^{-1} ($)	Unit Cost in Transplanted Rice ha^{-1} ($)	Reference
Farmyard manure	Sandy loam clay	18.40	13.26	[52]
Fertilizer	Sandy loam clay	97.56	80.88	
Plant protection measures (weeds, insect pests and disease control)	Sandy loam clay	54.63	42.09	
Land preparation	Sandy loam clay	59.01	69.49	
Human labor charges	Reclaimed alkali soils	163.01	174.56	[53]
Machine use charges	Reclaimed alkali soils	60.62	103.34	
Cost of seeds	Reclaimed alkali soils	15.86	7.49	
Cost of plant protection chemicals	Reclaimed alkali soils	31.03	38.21	
Irrigation charges	Reclaimed alkali soils	36.57	47.15	
Micronutrients	Sandy loam clay	14.25	12.49	[52]
Irrigation	Sandy loam clay	84.07	152.13	
Nursery and transplanting/seed and sowing	Sandy loam clay	21.53	65.11	
Cost of weedicides	Reclaimed alkali soils	33.61	26.78	[54]
Preparatory Tillage	Reclaimed alkali soils	61.74	97.21	
Pre-Sowing Irrigation	Reclaimed alkali soils	12.80	15.70	
Harvesting/threshing	Reclaimed alkali soils	49.06	49.06	
Plant protection	Reclaimed alkali soils	76.52	80.23	
Hoeing and weeding	Reclaimed alkali soils	37.04	18.92	
Irrigation	Reclaimed alkali soils	76.50	125.82	
Fertilizer application	Reclaimed alkali soils	6.12	6.60	
Nitrogen	Reclaimed alkali soils	17.21	19.48	
Phosphate	Reclaimed alkali soils	15.20	20.04	
Zinc sulphate	Reclaimed alkali soils	7.96	8.47	
TYM	Reclaimed alkali soils	56.30	56.30	
Seed	Reclaimed alkali soils	13.76	7.26	
Cost of fertilizers	Reclaimed alkali soils	49.41	48.50	[53]

All the values in $ are converted according to rate of 10 January 2021; 1 Pakistani rupee = 0.0062$; 1 Indian Rupee = 0.014$.

Late transplantation of rice due to labor shortage causes heat stress during the reproductive stage; temperatures >33.7 °C at anthesis causes panicle sterility due to poor anther dehiscence [55] and >34 °C during grain formation substantially reduce grain yield [56]. Temperatures >35 °C (above optimal) during reproductive development affect flowering and grain formation in rice [57].

4. Agricultural Innovations for Sustainable Development of Rice–Wheat Systems

Adaptation of innovative agricultural practices, such as conservation agriculture (CA), improves and sustains the productivity of RWCS and preserves scarce natural resources, such as water, energy, environmental quality, time, and labor [58]. The adaptation of CA-based systems is most beneficial in extreme climatic conditions, mitigating the negative impact of climatic stresses, such as water and heat stress, and increasing crop yields (0.4–0.8 t ha^{-1} per season), when compared with the conventional system [59].

The CA improves energy efficiency and carbon sequestration and reduces GHG emissions [2,60–62]. The incorporation of crop residues favors N immobilization (biotic and abiotic), which conserves active soil N by, (i) decomposing crop residues for a source of C for microorganisms and as an energy source to strengthen their metabolism which results in N immobilization in biomass, and (ii) incorporating N into the soil organic matter through ammonium fixation by clay minerals, nitrosation of nitrite with phenolic compounds, and condensation of ammonia with phenol [63].

Immobilized N can serve as temporary N sink [63]. Residue retention increases total organic C and available nutrients, mainly available P (16%), available K (12%), available sulfur (6%), and DTPA-extractable Zn (11%), relative to no-residue retention [64]. The adoption of resource-conserving technologies, such as DSR, harvesting rice with advanced rice harvesters, no-till wheat, crop rotation, and precision agriculture for better nutrient management, can mitigate climate change, reduce environmental pollution, and conserve natural resources.

4.1. Direct-Seeded Rice

In the IGP, increasing shortages of energy, water, and labor force farmers to switch from conventional PTR to a smart seeding system, i.e., DSR. In many studies, DSR produced higher yields, maximum profitability, and water-saving (25%) than PTR [62,65,66] with improved soil health (Table 3). DSR is an economically feasible alternative as it reduces production costs by 11–17% (with 25–30% irrigation water saving) and saves INR 5000 (on fuel and labor) [67] for the same yields as PTR [62]. In a study, DSR used 7–13.9% less water than the conventional PTR system [68]. Other studies in South Asia have reported that DSR uses 20–57% less water than PTR [69,70]. Rice produced through DSR also matures earlier than PTR, requires less water, and enables the timely sowing of following wheat and other crops [51].

Table 3. Soil quality in different rice production systems.

Soil Property	Unit	Soil Type	Value in DSR	Value in Transplanted Rice	Reference
Total organic carbon	g kg^{-1}	Silt clay	7.24	7.25	[64]
Aggregate associated carbon	g kg^{-1}	Silt clay	12.56	11.94	
Aggregate size class (0.25–2 mm)	%	Silt clay	48	48.9	
Mean weight diameter	mm	Silt clay	1.61	1.61	
Aggregate ratio		Silt clay	5.06	5.58	
Water stable macro-aggregates		Silt clay	83.2	83.8	
Water-holding capacity		Loam	0.346	0.331	[71]
Available water	cm^3 cm^{-3}	Loam	0.170	0.164	
Geometric mean diameter	mm	Loam	0.86	0.80	
Soil moisture potential (75 kPa)		Loam	0.166	0.170	
Crack depth (60 kPa)	cm	Loam	13	23	
Bulk density (6–10 cm)	Mg m^{-3}	Clay, silt, sand	1.60	1.61	[72]
WSA (>0.25 mm)		Clay, silt, sand	67.24	64.44	
Steady-state infiltration rate		Clay, silt, sand	0.33	0.29	
Water stable micro-aggregates		Silt clay	16.8	16.2	[64]
pH		Silt clay	7.39	7.41	
Electrical conductivity	dS m^{-1}	Silt clay	0.79	0.75	
Available N	kg ha^{-1}	Silt clay	195.5	185.0	
Available P	kg ha^{-1}	Silt clay	28.4	27.5	
Available K	kg ha^{-1}	Silt clay	264.3	222.4	
Crack width (60 kPa)	cm	Loam	3	7	[71]
Crack length (60 kPa)	cm	Loam	300	420	[71]
Total nitrogen	g kg^{-1}	Sandy loam	0.29	0.27	[73]
Total soil organic carbon	g kg^{-1}	Sandy loam	3.40	3.14	
Soil microbial biomass carbon	μg g^{-1}	Sandy loam	155.6	150.28	
Soil microbial biomass nitrogen	μg g^{-1}	Sandy loam	586.3	551.78	
Soil aggregates (>0.25 mm)		Silt loam	60	51	[69]
MWD of soil aggregates	mm	Silt loam	1.56	1.33	
Bulk density (0–7 cm)	Mg m^{-3}	Silt loam	1.60	1.50	
Penetration resistance (5–10 cm)	MPa	Silt loam	1.2	0.75	

WSA, water stable aggregates; MWD, mean weight diameter.

In DSR, the crop is directly sown into the field, avoiding transplantation injuries, thus reducing exposure to terminal drought due to timely stand establishment [74]. Moreover, DSR improves soil health for post-rice winter cereals [3] by enhancing total porosity and decreasing soil bulk density [9], enabling deeper root penetration and facilitating nutrient and water uptake [3]. In RWCS, DSR has been reported to reduce methane emissions and production costs, with increased profitability (Table 1; [51]).

Weeds are a major challenge in DSR; however, the application of weedicides can control the issue. For example, pre-emergence application of pendimethalin (1.5 kg ha^{-1}) followed by bispyribac-Na (25 g ha^{-1}) at post-emergence and hand weeding 35 days after sowing provided better weed control and higher rice yields (123–130%), net returns (327–806%) and net benefit: cost ratios than PTR [75]. However, diversification of weed flora has been reported in DSR in Pakistan which are very difficult to control and many farmers are afraid to plant rice in the DSR system. This needs the immediate attention of the government agencies in the region.

In conclusion, switching from PTR to DSR in RWCS increases profitability reduces production costs and GHG emissions, and is environmentally friendly, apart from the weed management issue during early growth.

4.2. Zero-Tillage Wheat

Using zero tillage (ZT) wheat in RWCS benefits the timeliness of wheat sowing and economics when compared with conventional tillage [59,76]. Zero tillage improves soil health and enhances nutrient concentrations at the soil surface Table 4 [77,78].

Table 4. Soil quality in different wheat production systems.

Soil Property	Units	Soil Type	Value in ZT	Value in PT	Reference
Bulk density	Mg m^{-3}	Siltic soils (Haplic Solonetz)	1.63	1.67	[2]
Soil pH		Siltic soils (Haplic Solonetz)	7.84	8.06	
EC	dS m^{-1}	Siltic soils (Haplic Solonetz)	0.25	0.21	
Total N	%	Silty soils (Haplic Solonetz)	0.19	0.14	
Bulk density	Mg m^{-3}	Sandy loam	1.54	1.50	[79]
Infiltration rate	mm h^{-1}	Sandy loam	1.5	0.3	
MWD	mm	Sandy loam	1.9	1.7	
WSA (>0.25 mm)	%	Sandy loam	73	57	
Bulk density	Mg m^{-3}	Sandy loam	1.24	1.38	[80]
Soil temperature	°C	Sandy loam	33.15	35.29	
PAWC (0–15 cm)	mm	Sandy loam	16.70	14.7	
Infiltration rate	mm h^{-1}	Sandy loam	9.58	11.40	
Bulk density	Mg m^{-3}	Sandy loam	1.52	1.48	[81]
Infiltration rate	mm h^{-1}	Sandy loam to loam	18.0	42.0	[66]
β-Glucosidase (p-NP)	μg g^{-1} h^{-1}	Loam	51.24	36.23	[82]
Bulk density	Mg m^{-3}	Sandy loam	1.44	1.46	[83]
Earthworm count	ha^{-1}	Sandy loam	380,000	60,000	[84]
Dehydrogenase activity	μg g^{-1} d^{-1}	Sandy loam	166.6	29.5	
SOC	g kg^{-1}	Sandy loam	2.51	1.47	
Bulk density	Mg m^{-3}	Sandy loam	1.60	1.56	[68]
SOC stock	kg m^{-3}	Sandy loam	6.88	5.91	
Oxidizable organic C	g kg^{-1}	Fine loam (Typic Natrustalf)	8.1	4.9	[85]
WSA (>0.25 mm)	%	Sandy loam	70	59	[69]
MWD	mm	Sandy loam	2.68	1.62	
Infiltration rate	mm h^{-1}	Sandy loam	5.0	4.7	
Bulk density	Mg m^{-3}	Sandy loam	1.52	1.57	
Crack width	cm	Sandy loam (typic ustrochrept)	0.53	2.68	[86]
Least limiting water range	%	Sandy loam	6.2	3.3	[72]
WSA (>0.25 mm)	%	Sandy loam	67.24	52.66	
Bulk density	Mg m^{-3}	Sandy loam	1.55	1.48	
Infiltration rate	mm h^{-1}	Sandy loam	3.3	1.8	
Penetration resistance	MPa	Sandy loam	1.4	1.0	
Volume of crack ($\times 10^{-4}$)	m^3 m^{-2}	Clayey	77.21	55.57	[87]
Bulk density	Mg m^{-3}	Clay	1.5	1.5	[88]

Table 4. *Cont.*

Soil Property	Units	Soil Type	Value in ZT	Value in PT	Reference
Infiltration rate	mm h^{-1}	Clay	17.30	15.55	
PAWC (0–15 cm)	mm	Clay	40	36	
Bulk density	Mg m^{-3}	Clay	1.24	1.28	[89]
WSA (>0.25 mm)	%	Clay	60.47	51.36	
Alkaline phosphatase (p-nitrophenol) (0–10 cm)	µg g^{-1} h^{-1}	Silty clay	287.7	269.8	[90]
Carbon build up	%	Silty clay	14.56	5.44	
Fluorescein diacetate activity	mg kg^{-1} h^{-1}	Silty clay	49.54	43.54	
Bulk density	Mg m^{-3}	Illitic, Ustic Typic Calciorthent	1.46	1.55	[91]
Carbon input addition	Mg h^{-1}	Illitic, Ustic Typic Calciorthent	14.64	3.10	
Fluorescein diacetate activity	µg g^{-1} h^{-1}	Mixed loamy sand	27.9	13.3	[92]
Total C	g kg^{-1}	Sandy clay loam	7.25	6.95	[93]
$KMnO_4$ C	g kg^{-1}	Sandy clay loam	0.43	0.39	
Soil water retention	mm	sandy clay loam	4.6	4.2	[94]
WSA (>0.25 mm)	%	Sandy loam (Typic Ustochrept)	77.3	68.4	[95]
MWD	mm	Sandy loam	0.74	0.71	[96]
Effective porosity	%	Sandy loam	18.7	17.4	
Bulk density	Mg m^{-3}	Sandy loam	1.43	1.39	
Active C	g kg^{-1}	Sandy loam (Typic Ustochrept)	4.09	2.92	[95]
MWD	mm	Sandy loam (Typic Ustochrept)	1.21	0.92	
Total organic carbon	g kg^{-1}	Fluvisol (silty clay)	7.25	6.38	[64]
Saturated hydraulic conductivity ($\times 10^{-6}$)	m s^{-1}	Clay	7.32	2.13	[97]
MWD	mm	Clay	0.94	0.76	
MWD	mm	Silty loam (Typic Ustocrept)	1.86	0.95	[17]
WSA (>0.25 mm)	%	Silty loam (Typic Ustocrept)	96	84	
SOC	g kg^{-1}	Silty loam (Typic Ustocrept)	7.86	5.81	
Bulk density	Mg m^{-3}	Sandy loam	1.60	1.56	[98]
MWD	mm	Sandy loam	0.95	0.79	[99]
Porosity	%	Clay loam	42.40	42.62	[60]
Bulk density	Mg m^{-3}	Clay loam	1.43	1.40	
Soil moisture (%)		Non-calcareous brown sandy loam Haplaquept	18.6	7.4	[100]

EC, electrical conductivity; MWD, mean weight diameter; WSA, water stable aggregates; PAWC, plant available water capacity; BD, bulk density; SOC, soil organic carbon; ZT, zero tillage; PT, plow tillage.

Sowing wheat with ZT ensures early sowing and suppresses the obnoxious weed e.g., littleseed canarygrass (68–80% reduction in population) when compared with conventional farmers' practices [101]. Moreover, ZT facilitates the timeliness of wheat sowing [3], improves soil structure, fertility, soil biological activities [102], and water-stable aggregates [103], and reduces the costs of land preparation [73,104]. In ZT wheat, the activities of soil microbial biomass carbon [73,105], soil enzymes [106], soil respiration [66], and soil quality index [107] are higher than plow tillage. In no-till with permanent soil cover, water infiltration is usually higher than plow tillage [108].

The Happy Seeder is a zero-tillage seeder that sows wheat into large amounts of crop residue and saves \$136 ha^{-1}. Moreover, it facilitates timely wheat sowing, saves water, reduces air pollution, and enhances the sustainability of agriculture [24]. The use of Happy Seeder reduces the labor requirement for crop establishment by 80%, herbicide use by 50%, and irrigation by 20–25% [109]. Use of zero-tillage drill and Happy Seeder made it easy to plant wheat 2.7 days earlier (with earlier stand establishment) than that in CT wheat [24]. Many farmers in RWCS in South Asia are quickly shifting towards Happy Seeder wheat sowing due to the short turnover time between rice harvest and wheat sowing and imposition of huge penalties on the burning of rice residues. In conclusion, switching wheat sowing from conventional tillage to ZT ensures timely wheat sowing, saves production costs and improves soil health, yields, and yield sustainability.

4.3. Promotion of Precision Agriculture Practices for Nutrient Management

In the IGP, fertilizer recommendations are based on crop response data without considering the inherent nutrient supply capacity of the soil, causing over- or under-fertilization [68]. Improved nutrient management under CA improves yields and nutrient and water use efficiencies [110]. For RWCS in the IGP, a combination of macro- and micro-fertilizers with green manure, crop residues, and organic manures is a practical option for better nutrient management [111].

In maize–wheat–mungbean rotations, the adoption of ZT with site-specific nutrient management improved the soil physical, chemical, and biological properties, i.e., water-stable aggregates, saturated hydraulic conductivity, soil organic C, available N, P, and K, microbial biomass C, and enzyme activities (dehydrogenase, alkaline phosphatase, and β-glucosidase), relative to conventional and unfertilized treatments [112].

A recent study on N application rates in RWCS recommended N application rates of 120–200 kg ha^{-1} for rice and 50–185 kg ha^{-1} for wheat [26]. Zinc (Zn) application at 25 kg ha^{-1} as $ZnSO_4$ improved rice and wheat yields [113]. In another study, the application of Zn improved the grain yields in both DSR and PTR systems [114]. Likewise, the boron (B) application to soils deficient in B improved growth and grain yield of rice [115,116].

Leaf color charts and SPAD chlorophyll meters are good options for managing N application, with a strong correlation (0.84–0.91) reported between these and various rice and wheat genotypes. Moreover, net returns increased by 19–31% using a leaf color chart for N management rather than a fixed N application rate [68].

In conclusion, integrated nutrient management, crop rotations incorporating legumes, site-specific optimization of nutrients rates, and the use of SPAD chlorophyll meters and leaf color charts are the best options for nutrient management and enhanced nutrient use efficiencies in rice and wheat.

4.4. Planning Wise Crop Rotations

Continuous monocultures have caused nutrient imbalances and increased the risk of pest and disease occurrence [59]. Diversifying the area sown to rice to incorporate other remunerative crops sustains soil fertility and improves crop productivity and farmer income [66]. Rotating cereals and pulses help to maintain soil quality and soil microflora and fauna [66,107]. It has been reported that the inclusion of leguminous crops in the cereal system increased system productivity by 18% and net returns by 15% [117]. In another study, the CA-based rice–wheat–mungbean cropping system improved system productivity by 11% and profitability by 24%, and reduced energy inputs by 25%, relative to a conventional rice–wheat system [62].

Long-term crop rotations (2000–2004) in India revealed that the rice–potato–green gram rotation had the highest net returns, system productivity, production efficiency, benefit: cost ratio, and profitability. Moreover, the inclusion of summer grain/fodder legumes improved soil organic matter [118]. The addition of short-duration summer legumes (mungbean and cowpea) in RWCS enhanced system productivity and profitability and nutritional security [119]. A rice–fallow cropping system with the intensification of five winter crop rotations (chickpea, lentil, safflower, linseed, and mustard) resulted in higher productivity for grain legumes (chickpea and lentil) than oilseed crops (safflower, mustard, and linseed) [120].

The inclusion of legumes in the cereal system fixes atmospheric N and improves soil fertility through nutrient recycling from deeper soil layers and mycorrhizal colonization [86]. Legume residues contain 20–80 kg N ha^{-1} (70% derived from N fixation), depending on the crop type [121,122]. A long-term study (2001–2004) showed that rice–legume rotations improved rice yields more than a rice–fallow rotation. In conclusion, the inclusion of short-duration grain or forage legumes in rotation in RWCS improves soil fertility and the yield of succeeding crops.

4.5. Rice Harvesting with Advanced Rice Harvesters

Rice harvesting is the most expensive rice production field activity, as the timing, duration, and mode of conduct of the harvesting directly affect rice quality, efficiencies, and farmer incomes [123]. In developing countries, rice is manually harvested with hand tools (as sickles) and threshed by beating on a hard matter or durum. The harvesting of rice with modern rice harvesters saves time, costs, and labor and reduces grain losses when compared with conventional manual harvesting [124]. Modern crop-specific mechanical harvesters, such as combine and mini-combine harvesters and reapers, can save time and labor, reduce harvesting losses, and increase profit margins and rice quality [125]. A reaper saved 37% and mini-combine harvesters saved 52% of harvesting costs over manual harvesting [126]. On average, a mini-combine harvester saves 95.5% of the time, 61.5% of costs, and 4.9% of grain losses compared with manual harvesting [127].

Combine harvesters (mini, medium, and large) are a time-saving technology, saving 20–30% of operation time than ordinary machines [128]. The use of a mini-combine harvester or reaper saved 65% and 52% of the labor costs over manual harvesting [129]. A combine harvester increased the net benefit by 30.3%, relative to manual harvesting and threshing [130]. Likewise, a vertical conveyor reaper saved 44% of harvesting costs [131]. Mechanical harvesting can also save grain losses, which were 2.88–3.60% for a tractor-mounted combine harvester [125], compared with 6.36% for manual harvesting [132].

However, in Pakistan and many other countries of South Asia, rice crop is harvested through wheat combine harvesters through some modification in machines. The use of wheat combine harvester in rice cause substantial grain losses which affect farmer profitability. In a study, the use of rice specific harvester reduces harvest losses by 14% and an extra premium of 5% on the paddy harvested from rice harvesters which increased farmer profitability [133]. In conclusion, rice harvesting with specific rice harvesters improves grain quality, reduces grain losses, and increases profit. However, the price of advanced rice harvesters is not affordable for all farmers. But this problem can be solved through the subsidy by the governments or and through cooperative investment, where a group of farmers pool their resources to purchase such machinery. Provision of such machinery by the service providers, on rental basis, can be another option. However, the rental charges for rice harvesters are double than the old model combine wheat harvesters. Therefore, private investors are interested to invest in the purchase and provision of on-farm services to farmers in South Asia.

5. Conclusions

The RWCS is the major cereal-based cropping system in South Asia, providing food to millions of people. However, the sustainability and productivity of this system are at high risk due to climate change, deteriorating natural resources, yield stagnation, and the negative impacts of this system on the environment. Major issues with this system include GHG emissions, declining soil quality and health, and reduced profit margins. However, the adoption of alternative innovative and sustainable approaches, including smart seeding/DSR, ZT wheat, crop rotation, precision agriculture, and rice and wheat harvesting using advanced harvesters such as the reaper, mini, and combine harvesters are the best options for improving yield, grain quality, and soil health, reducing environmental pollution, and preserving the ecosystem and natural resources (i.e., water, air, and soil).

Author Contributions: Conceptualization: K.H.M.S., M.F.; methodology: all; formal analysis: all; investigation: all; writing—original draft preparation: K.H.M.S., M.F.; writing—review and editing: all. All authors have read and agreed to the published version of the manuscript.

Funding: This research received no external funding.

Institutional Review Board Statement: Not applicable.

Informed Consent Statement: Not applicable.

Data Availability Statement: Not applicable.

Conflicts of Interest: The authors declare no conflict of interest.

References

1. Ray, D.K.; Ramankutty, N.; Mueller, N.D.; West, P.C.; Foley, J.A. Recent patterns of crop yield growth and stagnation. *Nat. Commun.* **2012**, *3*, 1293. [CrossRef]
2. Jat, H.S.; Datta, A.; Sharma, P.C.; Kumar, V.; Yadav, A.K.; Choudhary, M.; Choudhary, V.; Gathala, M.K.; Sharma, D.K.; Jat, M.L.; et al. Assessing soil properties and nutrient availability under conservation agriculture practices in a reclaimed sodic soil in cereal-based systems of North-West India. *Arch. Agron. Soil Sci.* **2018**, *64*, 531–545. [CrossRef] [PubMed]
3. Nawaz, A.; Farooq, M.; Nadeem, F.; Siddique, K.H.; Lal, R. Rice–wheat cropping systems in South Asia: Issues, options and opportunities. *Crop Pasture Sci.* **2019**, *70*, 395–427. [CrossRef]
4. Humphreys, E.; Kukal, S.S.; Christen, E.W.; Hira, G.S.; Sharma, R.K. Halting the groundwater decline in north-west India—Which crop technologies will be winners? *Advan. Agron.* **2010**, *109*, 155–217.
5. Bhushan, L.; Ladha, J.K.; Gupta, R.K.; Singh, S.; Tirol-Padre, A.; Saharawat, Y.S.; Gathala, M.; Pathak, H. Saving of water and labor in a rice-wheat system with no-tillage and direct seeding technologies. *Agron. J.* **2007**, *99*, 1288–1296. [CrossRef]
6. Geethalakshmi, V.; Ramesh, T.; Palamuthirsolai, A.; Lakshmanan. Agronomic evaluation of rice cultivation systems for water and grain productivity. *Arch. Agron. Soil Sci.* **2011**, *57*, 159–166. [CrossRef]
7. Tuong, T.P.; Bouman, B.A.M. Rice production in water-scarce environments. In *Water Productivity in Agriculture: Limits and Opportunities for Improvement*; Kijne, J.W., Barker, R., Molden, D., Eds.; CABI: Wallingford, UK, 2003; pp. 53–67.
8. Brown, L.; Black, B.; Hassan, G.; Hussein, G. Aquifer depletion. In *Encyclopedia of Earth*; Cleveland, C.J., Ed.; Environmental Information Coalition and National Council for Science and the Environment: Washington, DC, USA, 2011.
9. Farooq, M.; Nawaz, A. Weed dynamics and productivity of wheat in conventional and conservation rice-based cropping systems. *Soil Tillage Res.* **2014**, *141*, 1–9. [CrossRef]
10. Ishaq, M.; Ibrahim, M.; Hassan, A.; Saeed, M.; Lal, R. Subsoil compaction effects on crops in Punjab, Pakistan: II. Root growth and nutrient uptake of wheat and sorghum. *Soil Tillage Res.* **2001**, *60*, 153–161. [CrossRef]
11. Yadav, A.; Malik, R.K. *Herbicide Resistant Phalaris Minorin Wheat—A Sustainability Issue*; Department of Agronomy and Directorate of Extension Education, CCS Haryana Agricultural University: Hisar, India, 2005.
12. Chhokar, R.S.; Singh, S.; Sharma, R.K. Herbicides for control of isoproturon-resistant Littleseed Canarygrass (Phalaris minor) in wheat. *Crop Prot.* **2008**, *27*, 719–726. [CrossRef]
13. Ranjit, J.D. Weeds and weed management in the rice–wheat system. In *Proceedings Rice-Wheat Research End-of-Project Workshop*; Hobbs, P.R., Rajbhandary, N.P., Eds.; NARC/CIMMYT/RWC: Kathmandu, Nepal, 1997; pp. 13–22.
14. Bhatt, R.; Kukal, S.S.; Busari, M.A.; Arora, S.; Yadav, M. Sustainability issues on rice–wheat cropping system. *Int. Soil Water Conserv. Res.* **2016**, *4*, 64–74. [CrossRef]
15. Nayyar, V.K. Soil micronutrient deficiencies in the rice–wheat cropping system. In *Addressing Resource Conservation Issues in Rice-Wheat Systems of South Asia: A Resource Book*; Rice–Wheat Consortium for the Indo-Gangetic Plains: New Delhi, India, 2003; pp. 157–162.
16. Alam, M.M.; Ladha, J.K.; Rahman, Z.; Khan, S.R.; Khan, A.; Buresh, R. Nutrient management for increased productivity of rice-wheat cropping system in Bangladesh. *Field Crop. Res.* **2006**, *96*, 374–386. [CrossRef]
17. Bhattacharyya, T.; Pal, D.K.; Ray, S.K.; Chandran, P.; Mandal, C.; Telpande, B.; Deshmukh, A.S.; Tiwary, P. Simulating change in soil organic carbon in two long term fertilizer experiments in India: With the RothC model. *Clim. Change Environ. Sustain.* **2013**, *1*, 104–117. [CrossRef]
18. Sharma, V.; Hussain, S.; Sharma, K.R.; Arya, V.M. Labile carbon pools and soil organic carbon stocks in the foothill Himalayas under different land use systems. *Geoderma* **2014**, *232*, 81–87. [CrossRef]
19. Bhatia, A.; Pathak, H.; Jain, N.; Singh, P.K.; Singh, A.K. Global warming potential of manure amended soils under rice-wheat system in the Indo-Gangetic plains. *Atmos. Environ.* **2005**, *39*, 6976–6984. [CrossRef]
20. Pathak, H.; Bhatia, A.; Prasad, S.; Singh, S.; Kumar, S.; Jain, M.C.; Kumar, U. Emission of nitrous oxide from rice-wheat systems of Indo-Gangetic plains of India. *Environ. Monit. Assess.* **2002**, *77*, 163–178. [CrossRef]
21. Van Groenigen, J.W.; Velthof, G.L.; Oenema, O.; Van Groenigen, K.J.; Van Kessel, C. Towards an agronomic assessment of N_2O emissions: A case study for arable crops. *Eur. J. Soil Sci.* **2010**, *61*, 903–913. [CrossRef]
22. Malla, G.; Bhatia, A.; Pathak, H.; Prasad, S.; Jain, N.; Singh, J. Mitigating nitrous oxide and methane emissions from soil in rice–wheat system of the Indo-Gangetic plain with nitrification and urease inhibitors. *Chemosphere* **2005**, *58*, 141–147. [CrossRef]
23. NAAS. Innovative Viable Solution to Rice Residue Burning in Rice-Wheat Cropping System through Concurrent Use of Super Straw Management System-Fitted Combines and Turbo Happy Seeder. Policy Brief No. 2. National Academy of Agricultural Sciences. Available online: https://naasindia.org/documents/CropBurning.pdf (accessed on 10 September 2019).
24. Keil, A.; Krishnapriya, P.P.; Mitra, A.; Jat, M.L.; Sidhu, H.S.; Krishna, V.V.; Shyamsundar, P. Changing agricultural stubble burning practices in the Indo-Gangetic plains: Is the Happy Seeder a profitable alternative? *Int. J. Agric. Sustain.* **2020**, 1–24. [CrossRef]
25. Lohan, S.K.; Jat, H.S.; Yadav, A.K.; Sidhu, H.S.; Jat, M.L.; Choudhary, M.; Peter, J.K.; Sharma, P.C. Burning issues of paddy residue management in north-west states of India. *Renew. Sustain. Energy Rev.* **2018**, *81*, 693–706. [CrossRef]

26. Sapkota, T.B.; Singh, L.K.; Yadav, A.K.; Khatri-Chhetri, A.; Jat, H.S.; Sharma, P.C.; Jat, M.L.; Stirling, C.M. Identifying optimum rates of fertilizer nitrogen application to maximize economic return and minimize nitrous oxide emission from rice–wheat systems in the Indo-Gangetic Plains of India. *Arch. Agron. Soil Sci.* **2020**, *7*, 1–6. [CrossRef]
27. Gupta, D.K.; Bhatia, A.; Kumar, A.; Das, T.K.; Jain, N.; Tomer, R.; Malyan, S.K.; Fagodiya, R.K.; Dubey, R.; Pathak, H. Mitigation of greenhouse gas emission from rice–wheat system of the Indo-Gangetic plains: Through tillage, irrigation and fertilizer management. *Agric. Ecosyst. Environ.* **2016**, *230*, 1–9. [CrossRef]
28. Liu, S.; Zhang, Y.; Lin, F.; Zhang, L.; Zou, J. Methane and nitrous oxide emissions from direct-seeded and seedling-transplanted rice paddies in southeast China. *Plant Soil* **2014**, *374*, 285–297. [CrossRef]
29. Saharawat, Y.S.; Ladha, J.K.; Pathak, H.; Gathala, M.K.; Chaudhary, N.; Jat, M.L. Simulation of resource-conserving technologies on productivity, income and greenhouse gas GHG emission in rice-wheat system. *J. Soil Sci. Environ. Manag.* **2012**, *3*, 9–22.
30. Pathak, H.; Chakrabarti, B.; Bhatia, A.; Jain, N.; Aggarwal, P.K. Potential and cost of low carbon technologies in rice and wheat systems: A case study for the Indo-Gangetic Plains. In *Low Carbon Technologies for Agriculture: A Study on Rice and Wheat Systems in the Indo-Gangetic Plains*; Pathak, H., Aggarwal, P.K., Eds.; Indian Agriculture Research Institute: New Delhi, India, 2012; pp. 12–40.
31. Tirol-Padre, A.; Rai, M.; Kumar, V.; Gathala, M.; Sharma, P.C.; Sharma, S.; Nagar, R.K.; Deshwal, S.; Singh, L.K.; Jat, H.S.; et al. Quantifying changes to the global warming potential of rice wheat systems with the adoption of conservation agriculture in northwestern India. *Agric. Ecosyst. Environ.* **2016**, *219*, 125–137. [CrossRef]
32. Pathak, H.; Saharawat, Y.S.; Gathala, M.K.; Mohanty, S.; Ladha, J.K. Simulating environmental impact of resource-conserving technologies in the rice–wheat system of the Indo-Gangetic Plains. In *Integrated Crop and Resource Management in the Rice-Wheat System of South Asia*; Ladha, J.K., Singh, Y., Erenstein, O., Hardy, B., Eds.; International Rice Research Institute: Los Baños, PH, USA, 2009; pp. 321–333.
33. Singh, S.K.; Bharadwaj, V.; Thakur, T.C.; Pachauri, S.P.; Singh, P.P.; Mishra, A.K. Influence of crop establishment methods on methane emission from rice fields. *Curr. Sci.* **2009**, *97*, 84–89.
34. Pathak, H.; Aggarwal, P.K.; Roetter, R.P.; Kalra, N.; Bandyopadhaya, S.K.; Prasad, S.; Van Keulen, H. Modelling the quantitative evaluation of soil nutrient supply, nutrient use efficiency, and fertilizer requirements of wheat in India. *Nutr. Cycl. Agroecosyst.* **2003**, *65*, 105–113. [CrossRef]
35. Ishibashi, E.; Yamamoto, S.; Akai, N.; Tsuruta, H. The influence of no-tilled direct seeding cultivation on greenhouse gas emissions from rice paddy fields in Okayama, Western Japan. 4. Major factors controlling nitrous oxide emission from rice paddy fields under no-till direct seeding cultivation. *Jpn. J. Soil Sci. Plant Nutr.* **2007**, *78*, 453–463.
36. Setyanto, P.; Makarim, A.K.; Fagi, A.M.; Wassmann, R.; Buendia, L.V. Crop management affecting methane emissions from irrigated and rainfed rice in Central Java (Indonesia). *Nutr. Cycl. Agroecosyst.* **2000**, *58*, 85–93. [CrossRef]
37. Corton, T.M.; Bajita, J.B.; Grospe, F.S.; Pamplona, R.R.; Assis, C.A.; Wassmann, R.; Lantin, R.S.; Buendia, L.V. Methane emission from irrigated and intensively managed rice fields in Central Luzon (Philippines). *Nutr. Cycl. Agroecosyst.* **2000**, *58*, 37–53. [CrossRef]
38. Jat, M.L.; Gathala, M.K.; Saharawat, Y.S.; Tetarwal, J.P.; Gupta, R. Double no-till and permanent raised beds in maize–wheat rotation of north-western IndoGangetic plains of India: Effects on crop yields, water productivity, profitability and soil physical properties. *Field Crop. Res.* **2013**, *149*, 291–299. [CrossRef]
39. Verburg, P.H.; van Bodegom, P.M.; van der Gon, H.A.; Bergsma, A.; van Breemen, N. Upscaling regional emissions of greenhouse gases from rice cultivation: Methods and sources of uncertainty. *Plant Ecol.* **2006**, *182*, 89–106. [CrossRef]
40. IPCC. The physical science basis. In *Contribution of Working Group I to the Fourth Assessment Report of the Intergovernmental Panel on Climate Change*; Cambridge University Press: Cambridge, UK, 2007.
41. Shindell, D.T.; Faluvegi, G.; Koch, D.M.; Schmidt, G.A.; Unger, N.; Bauer, S.E. Improved attribution of climate forcing to emissions. *Science* **2009**, *326*, 716–718. [CrossRef]
42. Pathak, H.; Bhatia, A.; Jain, N. *Greenhouse Gas Emission from Indian Agriculture: Trends, Mitigation and Policy Needs*; Approach Paper; Indian Agricultural Research Institute: New Delhi, India, 2014.
43. Tandon, H.L.S.; Sekhon, G.S. *Potassium Research and Agricultural Production in India*; Fertilizer Development and Consultation Organization: New Delhi, India, 1988.
44. Sharma, P.K.; Nayyar, V.K. *Diagnosing Micronutrient Related Constraints to Productivity in Muktsar, Patiala, Hoshiarpur and Ludhiana Districts*; Project Report for ITSAP (Information Technology for Sustainable Agriculture in Punjab)—UNDP-TIFAC Sponsored Project; Punjab Remote Sensing Centre: Ludhiana, India, 2004.
45. Chander, S. Annual review of fertilizer production and consumption. *Indian J. Fertil.* **2011**, *7*, 141–188.
46. Nayyar, V.K.; Arora, C.L.; Kataki, P.K. Management of soil micronutrient deficiencies in the rice-wheat cropping system. *J. Crop. Prod.* **2001**, *4*, 87–131. [CrossRef]
47. Singh, M.V. Micronutrient deficiencies in crops and soils in India. In *Micronutrient Deficiencies in Global Crop Production*; Alloway, B.J., Ed.; Springer: Dordrecht, The Netherlands, 2008; pp. 93–125.
48. Singh, V.K.; Dwivedi, B.S.; Shukla, A.K.; Chauhan, Y.S.; Yadav, R.L. Diversification of rice with pigeonpea in a rice-wheat cropping system on a Typic Ustochrept: Effect on soil fertility, yield and nutrient use efficiency. *Field Crops Res.* **2005**, *92*, 85–105. [CrossRef]

49. Sapkota, T.B.; Majumdar, K.; Jat, M.L.; Kumar, A.; Bishnoi, D.K.; McDonald, A.J.; Pampolino, M. Precision nutrient management in conservation agriculture based wheat production of Northwest India: Profitability, nutrient use efficiency and environmental footprint. *Field Crops Res.* **2014**, *155*, 233–244. [CrossRef]
50. Pandey, S.; Velasco, L.E. Economics of direct seeding in Asia: Patterns of adaptation and research priorities. *Int. Rice Res. Notes* **1999**, *24*, 6–11.
51. Farooq, M.; Siddique, H.M.K.; Rehman, H.; Aziz, T.; Lee, D.; Wahid, A. Rice direct seeding: Experiences, challenges and opportunities. *Soil Tillage Res.* **2011**, *111*, 87–98. [CrossRef]
52. Younas, M.; Rehman, M.A.; Hussain, A.; Ali, L.; Waqar, M.Q. Economic comparison of direct seeded and transplanted rice: Evidences from adaptive research area of Punjab Pakistan. *Asian J Agri. Biol.* **2015**, *4*, 1–7.
53. Tripathi, J.N.; Lorenzen, J.; Bahar, O.; Ronald, P.; Tripathi, L. Transgenic expression of the rice Xa21 pattern-recognition receptor in banana (Musa sp.) confers resistance to Xanthomonas campestris pv. musacearum. *Plant Biotech. J.* **2014**, *12*, 663–673. [CrossRef]
54. Kumar, R.; Batra, S.C. Economic and Constraint Analysis of Rice Cultivation by DSR Technology in Karnal District of Haryana. *Adv. Econ. Bus. Manag.* **2017**, *4*, 61–64.
55. Jagadish, S.V.; Craufurd, P.Q.; Wheeler, T.R. High temperature stress and spikelet fertility in rice (Oryza sativa L.). *J. Exp. Bot.* **2007**, *58*, 1627–1635. [CrossRef]
56. Morita, S.; Shiratsuchi, H.; Takanashi, J.; Fujita, K. Effect of high temperature on ripening in rice plants: Comparison of the effects of high night temperatures and high day temperatures. *Jpn. J. Crop Sci.* **2002**, *71*, 102–109. [CrossRef]
57. Kumar, N.; Kumar, N.; Shukla, A.; Shankhdhar, S.C.; Shankhdhar, D. Impact of terminal heat stress on pollen viability and yield attributes of rice (Oryza sativa L.). *Cereal Res. Commun.* **2015**, *43*, 616–626. [CrossRef]
58. Dikgwatlhe, S.B.; Chen, Z.D.; Lal, R.; Zhang, H.L.; Chen, F. Changes in soil organic carbon and nitrogen as affected by tillage and residue management under wheat–maize cropping system in the North China Plain. *Soil Tillage Res.* **2014**, *144*, 110–118. [CrossRef]
59. Jat, M.L.; Chakraborty, D.; Ladha, J.K.; Rana, D.S.; Gathala, M.K.; McDonald, A.; Gerard, B. Conservation agriculture for sustainable intensification in South Asia. *Nat. Sustain.* **2020**, *3*, 336–343. [CrossRef]
60. Jat, R.D.; Nanwal, R.K.; Jat, H.S.; Bishnoi, D.K.; Dadarwal, R.S.; Kakraliya, S.K.; Yadav, A.; Choudhary, K.M.; Jat, M.L. Effect of conservation agriculture and precision nutrient management on soil properties and carbon sustainability index under maize-wheat cropping sequence. *Int. J. Chem. Stud.* **2017**, *5*, 1746–1756.
61. Jat, R.D.; Jat, H.S.; Nanwal, R.K.; Yadav, A.K.; Bana, A.; Choudhary, K.M.; Kakraliya, S.K.; Sutaliya, J.M.; Sapkota, T.B.; Jat, M.L. Conservation agriculture and precision nutrient management practices in maize-wheat system: Effects on crop and water productivity and economic profitability. *Field Crop. Res.* **2018**, *222*, 111–120. [CrossRef]
62. Kumar, V.; Jat, H.S.; Sharma, P.C.; Gathala, M.K.; Malik, R.K.; Kamboj, B.R.; Yadav, A.K.; Ladha, J.K.; Raman, A.; Sharma, D.K.; et al. Can productivity and profitability be enhanced in intensively managed cereal systems while reducing the environmental footprint of production? Assessing sustainable intensification options in the breadbasket of India. *Agric. Ecosyst. Environ.* **2018**, *252*, 132–147. [CrossRef] [PubMed]
63. Yansheng, C.; Fengliang, Z.; Zhongyi, Z.; Tongbin, Z.; Huayun, X. Biotic and abiotic nitrogen immobilization in soil incorporated with crop residue. *Soil Tillage Res.* **2020**, *202*, 104664. [CrossRef]
64. Nandan, R.; Singh, V.; Singh, S.S.; Kumar, V.; Hazra, K.K.; Nath, C.P.; Poonia, S.; Malik, R.K.; Bhattacharyya, R.; McDonald, A. Impact of conservation tillage in rice-based cropping systems on soil aggregation, carbon pools and nutrients. *Geoderma* **2019**, *340*, 104–114. [CrossRef] [PubMed]
65. Kamboj, B.R.; Kumar, A.; Bishnoi, D.K.; Singla, K.; Kumar, V.; Jat, M.L.; Chaudhary, N.; Jat, H.S.; Gosain, D.K.; Khippal, A.; et al. *Direct Seeded Rice Technology in Western Indo-Gangetic Plains of India: CSISA Experiences*; CSISA, IRRI and CIMMYT: Patna, India, 2012; pp. 1–16.
66. Choudhary, K.M.; Jat, H.S.; Nandal, D.P.; Bishnoi, D.K.; Sutaliya, J.M.; Choudhary, M.; Sharma, P.C.; Jat, M.L. Evaluating alternatives to rice-wheat system in western Indo-Gangetic Plains: Crop yields, water productivity and economic profitability. *Field Crop. Res.* **2018**, *218*, 1–10. [CrossRef]
67. Gathala, M.K.; Kumar, V.; Sharma, P.C.; Saharawat, Y.S.; Jat, H.S.; Singh, M.; Kumar, A.; Jat, M.L.; Humphreys, E.; Sharma, D.K.; et al. Optimizing intensive cereal-based cropping systems addressing current and future drivers of agricultural change in the northwestern Indo-Gangetic Plains of India. *Agric. Ecosyst. Environ.* **2013**, *177*, 85–97. [CrossRef]
68. Jat, M.L.; Gathala, M.K.; Saharawat, Y.S.; Ladha, J.K.; Singh, Y. Conservation agriculture in intensive rice-wheat rotation of western Indo-Gangetic Plains: Effect on crop physiology, yield, water productivity and economic profitability. *Int. J. Environ. Sci. Nat. Resour.* **2019**, *18*, 555988.
69. Jat, M.L.; Gathala, M.K.; Ladha, J.K.; Saharawat, Y.S.; Jat, A.S.; Kumar, V.; Sharma, S.K.; Kumar, V.; Gupta, R. Evaluation of precision land leveling and double zero-till systems in the rice–wheat rotation: Water use, productivity, profitability and soil physical properties. *Soil Tillage Res.* **2009**, *105*, 112–121. [CrossRef]
70. Kumar, V.; Ladha, J.K. Direct seeding of rice: Recent developments and future research needs. *Adv. Agron.* **2011**, *111*, 297–413.
71. Hobbs, P.R.; Singh, Y.; Giri, G.S.; Lauren, J.G.; Duxbury, J.M. *Direct Seeding and Reduced Tillage Options in the Rice-Wheat Systems of the Indo-Gangetic Plains of South Asia*; Pandey, S., Mortimer, M., Wade, L., Tuong, T.P., Lopez, K., Hardy, B., Eds.; IRRI: Los Baños, PH, USA, 2002; pp. 201–215.

72. Sapkota, T.B.; Jat, M.L.; Aryal, J.P.; Jat, R.K.; Khatri-Chhetri, A. Climate change adaptation, greenhouse gas mitigation and economic profitability of conservation agriculture: Some examples from cereal systems of Indo-Gangetic Plains. *J. Integr. Agric.* **2015**, *22*, 1917. [CrossRef]
73. Nawaz, A.; Farooq, M.; Lal, R.; Rehman, A. Comparison of conventional and conservation rice-wheat systems in Punjab, Pakistan. *Soil Tillage Res.* **2017**, *169*, 35–43. [CrossRef]
74. Tuong, T.P.; Singh, A.K.; Siopongco, J.D.; Wade, L.J. Constraints to high yield of dry-seeded rice in the rainy season of a humid tropic environment. *Plant Prod. Sci.* **2000**, *3*, 164–172. [CrossRef]
75. Baghel, J.K.; Das, T.K.; Mukherjee, I.; Nath, C.P.; Bhattacharyya, R.; Ghosh, S.; Raj, R. Impacts of conservation agriculture and herbicides on weeds, nematodes, herbicide residue and productivity in direct-seeded rice. *Soil Tillage Res.* **2020**, *201*, 104634. [CrossRef]
76. Erenstein, O.; Laxmi, V. Zero tillage impacts in India's rice–wheat systems: A review. *Soil Tillage Res.* **2008**, *100*, 1–4. [CrossRef]
77. Singh, G.; Kumar, D.; Sharma, P. Effect of organics, biofertilizers and crop residue application on soil microbial activity in rice–wheat and rice-wheat mungbean cropping systems in the Indo-Gangetic plains. *Cogent Geosci.* **2015**, *1*, 1085296. [CrossRef]
78. Salahin, N.; Alam, K.; Mondol, A.T.; Islam, M.S.; Rashid, M.H.; Hoque, M.A. Effect of tillage and residue retention on soil properties and crop yields in Wheat-Mungbean-Rice Crop rotation under subtropical humid climate. *Open J. Soil Sci.* **2017**, *7*, 1. [CrossRef]
79. Bazaya, B.R.; Sen, A.; Srivastava, V.K. Planting methods and nitrogen effects on crop yield and soil quality under direct seeded rice in the Indo-Gangetic plains of eastern India. *Soil Tillage Res.* **2009**, *105*, 27–32. [CrossRef]
80. Bhattacharyya, R.; Das, T.K.; Das, S.; Dey, A.; Patra, A.K.; Agnihotri, R.; Ghosh, A.; Sharma, A.R. Four years of conservation agriculture affects topsoil aggregate associated 15nitrogen but not the 15nitrogen use efficiency by wheat in a semi-arid climate. *Geoderma* **2019**, *337*, 333–340. [CrossRef]
81. Dixit, J.; Gupta, R.S.R.; Behl, V.P.; Yadav, L.R. No-tillage and conventional tillage system evaluation for production of wheat-an analysis. *Indian J. Agric. Res.* **2003**, *37*, 199–203.
82. Singh, A.; Kaur, J. Impact of conservation tillage on soil properties in rice-wheat cropping system. *Agric. Sci. Res. J.* **2012**, *2*, 30–41.
83. Ghosh, P.K.; Das, A.; Saha, R.; Kharkrang, E.; Tripathy, A.K.; Munda, G.C.; Ngachan, S.V. Conservation agriculture towards achieving food security in north east India. *Curr. Sci.* **2010**, *99*, 915–921.
84. Das, T.K.; Bhattacharyya, R.; Sharma, A.R.; Das, S.; Saad, A.A.; Pathak, H. Impacts of conservation agriculture on total soil organic carbon retention potential under an irrigated agro-ecosystem of the western IndoGangetic Plains. *Eur. J. Agron.* **2013**, *51*, 34–42. [CrossRef]
85. Tomar, R.K.; Singh, D.; Gangwar, K.S.; Garg, R.N.; Chakraborty, G.V.K.; Sahoo, D.; Rajeev, R.N.R.; Chakravarty, N.V.K. Effect of tillage systems and irrigation schedules on soil cracking pattern, water requirement and performance of rice-wheat cropping system in inceptisols in semi-arid regions. *J. Soil. Water Conserv.* **2009**, *8*, 26–33.
86. Bandyopadhyay, K.K.; Mohanty, M.; Painuli, D.K.; Misra, A.K.; Hati, K.M.; Mandal, K.G.; Ghosh, P.K.; Chaudhary, R.S.; Acharya, C.L. Influence of tillage practices and nutrient management on crack parameters in a Vertisol of central India. *Soil Tillage Res.* **2003**, *71*, 133–142. [CrossRef]
87. Alam, M.K.; Salahin, N.; Islam, S.; Begum, R.A.; Hasanuzzaman, M.; Islam, M.S.; Rahman, M.M. Patterns of change in soil organic matter, physical properties and crop productivity under tillage practices and cropping systems in Bangladesh. *J. Agric. Sci.* **2017**, *155*, 216–238. [CrossRef]
88. Hati, K.M.; Chaudhary, R.S.; Mohanty, M.; Biswas, A.K.; Bandyopadhyay, K.K. Short-term tillage and fertilization impacts on soil organic carbon, aggregate stability and yield of soybean-wheat system in deep black soils of central India. *J. Indian Soc. Soil Sci.* **2015**, *63*, 1–12.
89. Samal, S.K.; Rao, K.K.; Poonia, S.P.; Kumar, R.; Mishra, J.S.; Prakash, V.; Mondal, S.; Dwivedi, S.K.; Bhatt, B.P.; Naik, S.K.; et al. Evaluation of long-term conservation agriculture and crop intensification in rice-wheat rotation of Indo-Gangetic Plains of South Asia: Carbon dynamics and productivity. *Eur. J. Agron.* **2017**, *90*, 198–208. [CrossRef] [PubMed]
90. Sapkota, T.B.; Jat, R.K.; Singh, R.G.; Jat, M.L.; Stirling, C.M.; Jat, M.K.; Bijarniya, D.; Kumar, M.; Yadvinder-Singh; Saharawat, Y.; et al. Soil organic carbon changes after seven years of conservation agriculture in a rice–wheat system of the eastern IndoGangetic Plains. *Soil Use Manag.* **2017**, *33*, 81–89. [CrossRef]
91. Singh, G.; Bhattacharyya, R.; Das, T.K.; Sharma, A.R.; Ghosh, A.; Das, S.; Jha, P. Crop rotation and residue management effects on soil enzyme activities, glomalin and aggregate stability under zero tillage in the Indo-Gangetic Plains. *Soil Tillage Res.* **2018**, *184*, 291–300. [CrossRef]
92. Dey, A.; Dwivedi, B.S.; Bhattacharyya, R.; Datta, S.P.; Meena, M.C.; Das, T.K.; Singh, V.K. Conservation agriculture in a rice-wheat cropping system on an alluvial soil of north-western Indo-Gangetic plains: Effect on soil carbon and nitrogen pools. *J. Indian Soc. Soil Sci.* **2016**, *64*, 246–254. [CrossRef]
93. Bhattacharyya, R.; Prakash, V.; Kundu, S.; Gupta, H.S. Effect of tillage and crop rotations on pore size distribution and soil hydraulic conductivity in sandy clay loam soil of the Indian Himalayas. *Soil Tillage Res.* **2006**, *86*, 129–140. [CrossRef]
94. Kumar, N.; Nath, C.P.; Hazra, K.K.; Das, K.; Venkatesh, M.S.; Singh, M.K.; Singh, S.S.; Praharaj, C.S.; Singh, N.P. Impact of zero-till residue management and crop diversification with legumes on soil aggregation and carbon sequestration. *Soil Tillage Res.* **2019**, *189*, 158–167. [CrossRef]

95. Mishra, G.; Kushwaha, H.S. Winter wheat yield and soil physical properties responses to different tillage and irrigation. *Eur. J. Biol. Res.* **2016**, *6*, 56–63.
96. Hati, K.M.; Chaudhary, R.S.; Mandal, K.G.; Bandyopadhyay, K.K.; Singh, R.K.; Sinha, N.K.; Mohanty, M.; Somasundaram, J.; Saha, R. Effects of tillage, residue and fertilizer nitrogen on crop yields, and soil physical properties under soybean–wheat rotation in vertisols of Central India. *Agric. Res.* **2015**, *4*, 48–56. [CrossRef]
97. Kumari, M.; Chakraborty, D.; Gathala, M.K.; Pathak, H.; Dwivedi, B.S.; Tomar, R.K.; Garg, R.N.; Singh, R.; Ladha, J.K. Soil aggregation and associated organic carbon fractions as affected by tillage in a rice–wheat rotation in North India. *Soil Sci. Soc. Am. J.* **2011**, *75*, 560–567. [CrossRef]
98. Bhattacharyya, P.; Roy, K.S.; Neogi, S.; Adhya, T.K.; Rao, K.S.; Manna, M.C. Effects of rice straw and nitrogen fertilization on greenhouse gas emissions and carbon storage in tropical flooded soil planted with rice. *Soil Tillage Res.* **2012**, *124*, 119–130. [CrossRef]
99. Talukder, H.; Meisner, C.A.; Baksh, M.E.; Waddington, S.R. Wheat-maize-rice cropping on permanent raised beds in Bangladesh. Permanent beds and rice-residue management for rice-wheat systems in the Indo-Gangetic Plain. In Proceedings of the Workshop, Ludhiana, India, 7–9 September 2006; pp. 111–123.
100. Mondal, S.; Kumar, S.; Haris, A.A.; Dwivedi, S.K.; Bhatt, B.P.; Mishra, J.S. Effect of different rice establishment methods on soil physical properties in drought-prone, rainfed lowlands of Bihar. *India. Soil Res.* **2016**, *54*, 997–1006. [CrossRef]
101. Sharma, P.C.; Jat, H.S.; Kumar, V.; Gathala, M.K.; Datta, A.; Yaduvanshi, N.P.S.; Choudhary, M.; Sharma, S.; Singh, L.K.; Saharawat, Y.; et al. *Sustainable Intensification Opportunities under Current and Future Cereal Systems of North-West India*; Central Soil Salinity Research Institute: Karnal, India, 2015; p. 46.
102. Alam, M.K.; Bell, R.W.; Haque, M.E.; Kader, M.A. Minimal soil disturbance and increased residue retention increase soil carbon in rice-based cropping systems on the Eastern Gangetic Plain. *Soil Tillage Res.* **2018**, *183*, 28–41. [CrossRef]
103. Gathala, M.K.; Ladha, J.K.; Saharawat, Y.S.; Kumar, V.; Kumar, V.; Sharma, P.K. Effect of tillage and crop establishment methods on physical properties of a medium-textured soil under a seven-year rice-wheat rotation. *Soil Sci. Soc. Am. J.* **2011**, *75*, 1851–1862. [CrossRef]
104. Shahzad, M.; Hussain, M.; Farooq, M.; Farooq, S.; Jabran, K.; Nawaz, A. Economic assessment of conventional and conservation tillage practices in different wheat-based cropping systems of Punjab, Pakistan. *Environ. Sci. Pollut. Res.* **2017**, *24*, 24634–24643. [CrossRef] [PubMed]
105. Lupwayi, N.Z.; Hanson, K.G.; Harker, K.N.; Clayton, G.W.; Blackshaw, R.E.; O'Donovan, J.T.; Johnson, E.N.; Gan, Y.; Irvine, R.B.; Monreal, M.A. Soil microbial biomass, functional diversity and enzyme activity in glyphosate-resistant wheat–canola rotations under low-disturbance direct seeding and conventional tillage. *Soil Biol. Biochem.* **2007**, *39*, 1418–1427. [CrossRef]
106. Sharma, P.; Singh, G.; Singh, R.P. Conservation tillage, optimal water and organic nutrient supply enhance soil microbial activities during wheat (Triticum aestivum L.) cultivation. *Braz. J. Microbiol.* **2011**, *42*, 531–542. [CrossRef]
107. Shaver, T.M.; Peterson, G.A.; Ahuja, L.R.; Westfall, D.G.; Sherrod, L.A.; Dunn, G. Surface soil physical properties after twelve years of dryland no-till management. *Soil Sci. Soc. Am. J.* **2002**, *66*, 1296–1303. [CrossRef]
108. Saunders, C.; Davis, L.; Pearce, D. *Rice-Wheat Cropping Systems in India and Australia, and Development of the Happy Seeder*; ACIAR Impact Assessment Series Report No.77; Australian Centre for International Agricultural Research: Canberra, Australia, 2012.
109. Jat, H.S.; Jat, R.K.; Singh, Y.; Parihar, C.M.; Jat, S.L.; Tetarwal, J.P.; Sidhu, H.S.; Jat, M.L. Nitrogen management under conservation agriculture in cereal-based systems. *Ind. J. Fert.* **2016**, *12*, 76–91.
110. Singh, A.P.; Sakal, R.; Sinha, R.B.; Bhogal, N.S. Use efficiency of applied zinc alone and mixed with biogas slurry in rice-wheat cropping system. *J. Ind. Soc. Soil Sci.* **1998**, *46*, 75–80.
111. Parihar, C.M.; Singh, A.K.; Jat, S.L.; Dey, A.; Nayak, H.S.; Mandal, B.N.; Saharawat, Y.S.; Jat, M.L.; Yadav, O.P. Soil quality and carbon sequestration under conservation agriculture with balanced nutrition in intensive cereal-based system. *Soil Tillage Res.* **2020**, *202*, 104653. [CrossRef]
112. Mahajan, A.; Gupta, R.D. Integrated nutrient management (INM) in a sustainable rice wheat cropping system. In *The Rice–Wheat Cropping System*; Mahajan, A., Gupta, R.D., Eds.; Springer: Dordrecht, The Netherlands, 2009; pp. 109–117.
113. Farooq, M.; Ullah, A.; Rehman, A.; Nawaz, A.; Nadeem, A.; Wakeel, A.; Nadeem, F.; Siddique, K.H. Application of zinc improves the productivity and biofortification of fine grain aromatic rice grown in dry seeded and puddled transplanted production systems. *Field Crops Res.* **2018**, *216*, 53–62. [CrossRef]
114. Atique-ur-Rehman, A.; Farooq, M.; Nawaz, A.; Ahmad, R. Influence of boron nutrition on the rice productivity, kernel quality and biofortification in different production systems. *Field Crop. Res.* **2014**, *169*, 123–131. [CrossRef]
115. Atique-ur-Rehman; Farooq, M.; Rashid, A.; Nadeem, F.; Stuerz, S.; Asch, F.; Bell, R.W. Boron nutrition of rice in different production systems. A review. *Agron. Sustain. Dev.* **2018**, *38*, 25. [CrossRef]
116. Kumar, S.; Karaliya, S.K.; Chaudhary, S. Precision farming technologies towards enhancing productivity and sustainability of rice-wheat cropping system. *Int. J. Curr. Microbio. App. Sci.* **2017**, *6*, 142–151. [CrossRef]
117. Singh, R.K.; Bohra, J.S.; Nath, T.; Singh, Y.; Singh, K. Integrated assessment of diversification of rice-wheat cropping system in Indo-Gangetic plain. *Arch. Agron. Soil Sci.* **2011**, *57*, 489–506. [CrossRef]
118. Yaqub, M.; Mahmood, T.; Akhtar, M.; Iqbal, M.M.; Ali, S. Induction of mungbean (Vigna radiata (L.) wilczek) as a grain legume in the annual rice-wheat double cropping system. *Pak. J. Bot.* **2010**, *42*, 3125–3135.

119. Kumar, R.; Mishra, J.S.; Rao, K.K.; Mondal, S.; Hazra, K.K.; Choudhary, J.S.; Hans, H.; Bhatt, B.P. Crop rotation and tillage management options for sustainable intensification of rice-fallow agro-ecosystem in eastern India. *Sci. Rep.* **2020**, *10*, 1–5. [CrossRef]
120. Wani, S.P.; Rupela, O.P.; Lee, K.K. Sustainable agriculture in the semi-arid tropics through biological nitrogen fixation in grain legumes. *Plant Soil* **1995**, *174*, 29–49. [CrossRef]
121. Giller, K.E. *Nitrogen Fixation in Tropical Cropping Systems*; CAB Int.: Wallingford, UK, 2001; p. 423.
122. Shah, Z.; Ahmad, S.R.; Rahman, H.U. Sustaining rice-wheat system through management of legumes I: Effect of green manure legumes on rice yield and soil quality. *Pak. J. Bot.* **2011**, *43*, 1569–1574.
123. Quick, G.R.; Quick, G.R. The Rice Harvesters Reference. Rural Industries Research and Development Corporation, 1999. RIRDC Rice Research and Development Program. RIRDC Publication No. 99/38. Available online: https://static1.squarespace.com/static/5a03c05bd0e62846bc9c79fc/t/5a4afdbfc83025f844b718de/1514864080791/99-046.pdf (accessed on 31 December 2020).
124. Hasan, K.; Tanaka, T.S.; Alam, M.; Ali, R.; Saha, C.K. Impact of Modern Rice Harvesting Practices over Traditional Ones. *Rev. Agric. Sci.* **2020**, *8*, 89–108. [CrossRef]
125. Hasan, M.K.; Ali, M.R.; Saha, C.K.; Alam, M.M.; Hossain, M.M. Assessment of paddy harvesting practices of Southern Delta Region in Bangladesh. *Progress. Agric.* **2019**, *30*, 57–64. [CrossRef]
126. Jones, M.; Alam, M.M.; Rahman, M.H.; Ali, M.R.; Hasan, M.K.; Pathan, M.S.I.A. Gender Technology Assessment of Harvesting Technologies in Bangladesh. Agrilinks, the Online Hub, Is a Part of the U.S. Government's Feed the Future. Available online: https://www.agrilinks.org/sites/default/files/resources/bangladesh_harvester_tech_profile_2019_08 (accessed on 31 December 2020).
127. Tiwari, P.S.; Gurung, T.R.; Sahni, R.K.; Kumar, V. *Agricultural Mechanization Trends in SAARC Region*; Gurung, T.R., Kabir, W., Bokhtiar, S.M., Eds.; Mechanization for Sustainable Agricultural Intensification in SAARC Agriculture Centre: Dhaka, Bangladesh, 2017; p. 302.
128. Ali, M.R.; Hasan, M.K.; Saha, C.K.; Alam, M.M.; Hossain, M.M.; Kalita, P.K.; Hansen, A.C. Role of mechanical rice harvesting in socio-economic development of Bangladesh. In Proceedings of the An ASABE Meeting Presentation, Detroit, MI, USA, 29 July–1 August 2018. Paper Number: 1800751.
129. Praweenwongwuthi, S.; Laohasiriwong, S.; Rambo, A.T. Impacts of rice combine harvesters on economic and social of farmers in a village of the Tung Kula Ronghai Region. *Res. J. Agri. Biol. Sci.* **2010**, *6*, 778–784.
130. Kurhekar, S.P.; Patil, S.R. Performance evaluation of self-propelled walking type vertical conveyor reaper. *Int. J. Process. Post Harvest Technol.* **2011**, *2*, 29–31.
131. Veerangouda, M.; Sushilendra, S.; Prakash, K.V.; Anantachar, M. Performance evaluation of tractor operated combine harvester. *Kamataka J. Agric. Sci.* **2010**, *23*, 282–285.
132. Ramesh, T.; Rathika, S. Evaluation of rice cultivation systems for greenhouse gases emission and productivity. *Int. J. Ecol. Environ. Sci.* **2020**, *2*, 49–54.
133. Islamic Republic of Pakistan; Asian Development Bank. *Final Report, TA 8578-PAK: Punjab Basmati Rice Value Chain*; Islamic Republic of Pakistan: Islamabad, Pakistan; Asian Development Bank: Metro Manila, Philippines, 2018.

sustainability

MDPI

Article

International Comparison of the Efficiency of Agricultural Science, Technology, and Innovation: A Case Study of G20 Countries

Xiangyu Guo [1], Canhui Deng [1], Dan Wang [1,*], Xu Du [2], Jiali Li [1] and Bowen Wan [1]

[1] College of Economics and Management, Northeast Agricultural University, Harbin 150030, China; guoxy@neau.edu.cn (X.G.); dengcanhuind@163.com (C.D.); LiLihehe135@163.com (J.L.); iwanbowen@163.com (B.W.)
[2] School of Economics, Harbin Normal University, Harbin 150025, China; neau711711@163.com
* Correspondence: wd@neau.edu.cn; Tel.: +86-0451-8713-2132

Abstract: An efficiency-oriented innovation analysis will enhance the understanding of the operational quality related to the transformation process of limited innovation investments for improving innovation outputs. The purpose of this study was to measure the static-dynamic efficiency of agricultural science, technology, and innovation (ASTI) and identify the efficiency determinants across the Group of Twenty (G20) countries. First, the static comprehensive efficiency of ASTI was measured employing the Data Envelopment Analysis (DEA)-BCC model, and some of the binding constraints to higher efficiency were investigated. Then, we applied the DEA-Malmquist index model to calculate the efficiency change of ASTI in certain periods and decomposed the sources of efficiency change. Finally, the G20 countries were classified into four-level clusters based on the rankings of efficiency measurement and capability evaluation of ASTI to locate the type of ASTI level and identify the type change in both the efficiency and capability. The empirical results indicate the following. (1) The efficiency range of the G20 developing countries was relatively larger than the G20 developed countries. The G20 developed countries showed a fluctuating downward trend, while the G20 developing countries showed an upward trend from the perspective of efficient proportion. The R&D expenditure redundancy and the agricultural journal papers deficiency were the main binding constraints to the higher efficiency of ASTI. (2) The total factor productivity change (TFPC) of ASTI showed an alternating trend of "decline–growth–continuous decline–growth recovery", where the G20 developed countries experienced "growth–decline–growth" and the G20 developing countries underwent a fluctuating upward trend. The TFPC of ASTI in most G20 countries was primarily due to technological change. (3) The G20 developed countries usually had advantages in capacity, while the G20 developing countries performed better in efficiency.

Keywords: agricultural science, technology and innovation; Innovation efficiency; DEA; G20

Citation: Guo, X.; Deng, C.; Wang, D.; Du, X.; Li, J.; Wan, B. International Comparison of the Efficiency of Agricultural Science, Technology, and Innovation: A Case Study of G20 Countries. *Sustainability* **2021**, *13*, 2769. https://doi.org/10.3390/su13052769

Academic Editor: Michael Blakeney

Received: 29 January 2021
Accepted: 28 February 2021
Published: 4 March 2021

Publisher's Note: MDPI stays neutral with regard to jurisdictional claims in published maps and institutional affiliations.

Copyright: © 2021 by the authors. Licensee MDPI, Basel, Switzerland. This article is an open access article distributed under the terms and conditions of the Creative Commons Attribution (CC BY) license (https://creativecommons.org/licenses/by/4.0/).

1. Introduction

The challenge of sustainable agriculture development in light of population growth, resource shortage, ecological deterioration, and climate change has led many governments to support agricultural science, technology, and innovation (ASTI). The investment of the United States government in agricultural research projects reached 3.03 billion dollars in 2018, 130 million dollars more than in 2017 [1]. The European Union has invested 10 billion euros in ASTI activities such as agriculture and forestry ecosystem restoration for the "Rural Development Project (2014–2020)" [2]. The UK adopted the "UK agricultural science and technology strategy" in 2013. In 2014, Germany's agricultural research funds reached 10% of the budget of the Federal Ministry of Food and Agriculture [3]. China issued the "Agricultural Science and Technology Development Plan (2006–2020)" [4] and "the National Agricultural Science, Technology and Innovation Capacity Building Plan (2012–2016)" [5].

However, the innovation performance is dependent not only on the available innovation resources but also and maybe most importantly on their efficient and productive use [6]. Innovation efficiency, which is "the ability to translate inputs into innovation outputs" by definition, has become very important and attractive to scholars and governments [7,8]. Because of the unique advantages in the efficiency evaluation of multi-input and multi-output [9], the Data Envelopment Analysis (DEA) has been widely used to measure the relative efficiency of Decision-Making Units (DMUs) by estimating the ratio of outputs to inputs [10–12]. Many studies investigated innovation efficiency at the national [13–15], regional [10,16,17], and institutional levels [18–20] by means of DEA. Several studies have been conducted to measure the efficiency of ASTI [21–27]. Most of these studies assess a particular nation [21,22,27] or a region [24–26], and very few studies attempt cross country comparisons for ASTI efficiency [23]. Moreover, the integration of static and dynamic ASTI efficiency analyses has been usually disregarded.

The limited attention to innovation efficiency at the national level could be a potentially significant omission from a policy-oriented perspective [28,29], since measuring the ASTI efficiency helps to both identify the best innovation practitioners for benchmarking and propose ways to improve efficiency by pinpointing areas of weakness [15]. The G20 countries account for 60% of global arable land and 80% of global agricultural trade [30]. Therefore, "G20 agriculture" has a significant effect on global agriculture development. In this context, this paper aims to address this gap by estimating the static-dynamic efficiency of ASTI for the G20 countries at the national level.

This paper proceeds as follows. Section 2 presents the DEA-BCC model and the DEA-Malmquist index model, as well as the input–output indicators and data sources. Section 3 shows the empirical results, including the static comprehensive efficiency and dynamic total factor productivity. In addition, we further classify the ASTI level of G20 countries through the results of efficiency measurement and capability evaluation. Section 4 is reserved for conclusions and implications.

2. Methodology

2.1. Definition of Efficiency of ASTI

According to Schumpeter's innovation theory, innovation is not only a technology and scientific research activity but also an economic activity [31]. In this paper, ASTI is defined as a complex innovation process in which a series of innovative actors transform input (personnel and expenditure) into output (new knowledge, new varieties, or new technologies) through cooperation and interaction to obtain economic benefits. Therefore, the efficiency of ASTI is the ability of transforming input into output in the above complex innovation process. The innovation efficiency reflects the effectiveness of innovation process from input to output. The maximum efficiency of ASTI is mainly reflected in the maximum innovation output at the given innovation input.

2.2. Data Envelopment Analysis

DEA is a non-parametric method proposed by Farrell [32] and developed by Charnes, Cooper, and Rhodes [33]. There are many unique advantages in the efficiency evaluation of multi-input and multi-output: First, the functional relationship between input and output indicators does not require a priori assumption [34]. Second, multi-input and multi-output are allowed to be processed simultaneously, without any input and output indicators dimensionless processing. Moreover, DEA does not need to verify in advance which input and output indicators are the most important in efficiency evaluation [35].

The CCR model and the BCC model are two basic DEA models. Both models are named after the author's initials. In 1978, Charnes, Cooper, and Rhodes created the first DEA model, which was named the CCR model [33]. Similarly, in 1984, Banker, Charnes, and Cooper proposed a new DEA model, which was named the BCC model [36]. The difference between BCC model and CCR model lies in the assumptions. The CCR model assumes that returns to scale are constant, while the BCC model assumes that returns to scale are variable. According to the efficiency measurement, the two models can be divided into input-oriented and output-oriented [37]. Input orientation emphasizes the degree to which the various input factors should be reduced to achieve technical efficiency without reducing output. In contrast, output orientation focuses on the extent to which all kinds of output should be increased for the purpose of achieving technical efficiency without increasing input. In practice, the ASTI in most countries is not in the optimal scale state, and ASTI will produce scale efficiency with the increasing input. This means that the measurement of the efficiency of ASTI meets the assumption of BCC model, that is, variable returns to scale. The fundamental purpose of increasing the input of ASTI is to expect more output, which is consistent with the output-oriented model. Therefore, we carried out the output-oriented BCC model to measure the comprehensive efficiency of ASTI in G20 countries. The linear form of the output-oriented BCC model is as follows:

$$\left(B^2C\right)^O \begin{cases} \max\left[\theta - \varepsilon\left(\hat{e}^t s^- + e^t s^+\right)\right], \\ \text{s.t.} \sum\limits_{j=1}^{k} x_{jl}\lambda_j + s^- = x_l^n, \\ \sum\limits_{j=1}^{k} y_{jm}\lambda_j - s^+ = \theta y_l^n, \\ \sum\limits_{j=1}^{k} \lambda_j = 1, \\ s^- \geq 0, s^+ \geq 0, \lambda_j \geq 0, j = 1, 2, \cdots k. \end{cases} \tag{1}$$

$\hat{e}^t = (1, 1, \ldots, 1) \in E^m, e^t = (1, 1, \ldots, 1) \in E^s$; x_{jl} represents the lth inputs of the jth DMU; y_{jm} represents the mth outputs of the jth DMU; ε is the non-Archimedes infinitesimal; λ_j is the weighting factor; s^- represents the relaxation variables; s^+ is the residual variable; and θ represents the relative efficiency of DMU.

If $\theta < 1$, DMU is inefficient.

If $\theta = 1, \hat{e}^t s^- + e^t s^+ > 0$, DMU is weakly efficient.

If $\theta = 1, \hat{e}^t s^- + e^t s^+ = 0$, DMU is efficient.

The BCC model can only use the cross-section data to reflect the efficiency value of DMU at a certain time statically. To show the dynamic changes of DMU in a specific time series, we need to use the DEA-Malmquist index model [38] to calculate the total factor productivity change (TFPC). The TFPC can be decomposed into the technical efficiency change (TEC) and technological change (TC) in two periods [39]. TEC can also be decomposed into the pure efficiency change (PEC) and scale efficiency change (SEC). The model is as follows:

$$\text{TFPC} = m_0\left(x_{t+1}, y_{t+1}; x_t, y_t\right) \tag{2}$$

$$= \left[\frac{d_0^t\left(x_{t+1}, y_{t+1}\right)}{d_0^t\left(x_t, y_t\right)} \times \frac{d_0^{t+1}\left(x_{t+1}, y_{t+1}\right)}{d_0^{t+1}\left(x_t, y_t\right)}\right]^{\frac{1}{2}} \tag{3}$$

$$= \text{TEC}\left(x_{t+1}, y_{t+1}; x_t, y_t\right) \times \text{TC}\left(x_{t+1}, y_{t+1}; x_t, y_t\right) \tag{4}$$

$$= \text{SEC}\left(x_{t+1}, y_{t+1}; x_t, y_t\right) \times \text{PEC}\left(x_{t+1}, y_{t+1}; x_t, y_t\right) \times \text{TC}\left(x_{t+1}, y_{t+1}; x_t, y_t\right) \tag{5}$$

where d_0 refers to the input and output matrix and x_t, x_{t+1} represent the input vectors of the t and $t+1$ periods, respectively. The relationship between variables satisfies the following conditions: TFPC = TEC × TC, EC = SEC × PEC. Thus, TFPC = SEC × PEC × TC.

2.3. Indicators Selection

The discriminatory power of DEA would be decreased when many input–output indicators are introduced [40]; the principle is as follows:

$$d \geq 3 * (m + n) \tag{6}$$

where d represents the number of DMUs, m represents the number of input indicators, and n represents the number of output indicators.

Following this restriction, only a few critical indicators can be selected. In this study, the number of DMUs is 19; therefore, the total number of indicators cannot be greater than 6.

As shown in Table 1, based on previous research experience [9,19,20,24,26,41,42], the definition of efficiency of ASTI and data availability, the input and output indicators are selected as follows.

Table 1. List of the innovation efficiency evaluation studies using DEA met.

Authors and Title	DEA Model	Input Indicators	Output Indicators
Chen, Z.; Zheng, R. et al. (2018) [24] Evaluation and analysis of agricultural science and technology innovation efficiency in Henan Province	CCR	Agricultural R&D expenditure; Agricultural R&D researchers; Total power of agricultural machinery	Number of agricultural journal papers; Total output value of Agriculture
Guo, X.Y.; Du, X. et al. (2020) [26] Evaluation and comparative analysis of the efficiency of provincial agricultural science, technology and innovation in China	BCC	Agricultural R&D expenditure; Agricultural R&D researchers	Number of agricultural patents; Number of new plant varieties; Added value of agriculture
Park, J.H. (2018) [20] Open innovation of small and medium-sized enterprises and innovation efficiency	BCC	The value of R&D expenditure divided by the total sales; The share of R&D staff in total employment	The percentage of sales from R&D activities
Shin, J.; Kim, C. (2018) [19] The Effect of Sustainability as Innovation Objectives on Innovation Efficiency	SBM	R&D Employee; R&D Expense	Patent Application; Innovation Sales
Zhang, C.; Wang, X.J. (2019) [41] The influence of ICT-driven innovation: a comparative study on national innovation efficiency between developed and emerging countries	BCC	Gross Domestic Expenditure on R&D; Total Researcher	Triadic Patent Families; Science & Engineering Articles; Value Added of Knowledge and Technology Intensive Industries
Fang, S.R.; Xue, X.S.; Yin, G. (2020) [42] Evaluation and Improvement of Technological Innovation Efficiency of New Energy Vehicle Enterprises in China Based on DEA-Tobit Model	two-stage DEA	Total assets R&D expenditure; Total number of employees; Technical asset rate	Number of patents; Operating income Net profit
Lin, Y.Y.; Deng, N.Q.; Gao, H.L. (2018) [43] Research on Technological Innovation Efficiency of Tourist Equipment Manufacturing Enterprises	DEA-Malmquist	Intensity of R & D personnel; Intensity of R & D expenditure	Number of patent applications; Profit ratio of sales; Total labor productivity

Input indicators: The innovation inputs mainly include the research and development (R&D) personnel and expenditure [9,19,20,44–46]. The R&D personnel served as the inputs in the brainwork for the upstream technological creation process in an innovation system, representing a basic element for the realization of the technological creation process. As a proxy for this indicator, we employ the number of agricultural researchers to measure R&D personnel [22,24,26,46]. As a supporting input, R&D expenditure is also needed, which is used to complete various R&D activities [47], including the payment of R&D employees' wages and the purchase of R&D equipment and facilities [48]. Percentage shares of R&D expenditure in agricultural value added is used as a proxy indicator to measure R&D expenditure.

Output indicators: The output indicators could be divided into two general categories: (1) scientific and technological output; and (2) economic performance. The scientific and technological output captured the extent to which a country produced some type of scientific and technological output. The commonly accepted measures of this are the number of agricultural journal papers [9,49] and the number of agricultural patents [19,25,47]. The agricultural value added (annual percent growth) is an appropriate proxy for economic performance generated by ASTI [49].

2.4. Data Sources

The study was limited to G20 countries and covered the period between 2008 and 2017. The G20 countries include Argentina, Australia, Brazil, Canada, China, France, Germany, India, Indonesia, Italy, Japan, the Republic of Korea, Mexico, the Russian Federation, Saudi Arabia, South Africa, Turkey, the United Kingdom, and the United States (note: the European Union (EU) is a political and economic union, and its major member states are already within the G20, so the EU was not included in this empirical analysis). The specific sources of each indicator are shown in Table 2 and its notes. The descriptive statistics of the input and output indicators are shown in Table 3.

Table 2. Index system for measuring ASTI efficiency.

Index	Sub-Index	Indicator	Data Sources
Input	R&D personnel	X_1: Number of agricultural researchers	UNESCO-UIS
	R&D expenditure	X_2: Percentage shares of R&D expenditure in agricultural value added	UNESCO-UIS, FAO
Output	Scientific and technological output	Y_1: Number of agricultural journal papers	WOS
		Y_2: Number of agricultural patents	WIPO
	Economic performance	Y_3: Agricultural value added (annual % growth)	WB

Notes: United Nations Educational, Scientific and Cultural Organization-Institute for Statistics (UNESCO-UIS): http://uis.unesco.org/ (accessed on 10 September 2020); The Food and Agriculture Organization (FAO): http://www.fao.org/home/en/ (accessed on 10 September 2020); World Intellectual Property Organization (WIPO): https://www3.wipo.int/ipstats/index.htm?lang=en (accessed on 10 September 2020); Web of Science (WOS): http://apps.webofknowledge.com/RAMore.do?product=WOS&search_mode=AdvancedSearch&SID=5BeAM2moXj26NR13wVH&qid=13&ra_mode=more&ra_name=CountryTerritory&colName=WOS&viewType=raMore (accessed on 10 September 2020); World Bank (WB): https://data.worldbank.org/indicator/NV.AGR.TOTL.ZS?view=chart (accessed on 10 September 2020).

Table 3. Descriptive statistics of the input and output indicators.

Indicator	Mean	Std. Dev.	Min	Max
X_1	14.81	19.48	0.12	95.92
X_2	42.17	27.39	5.92	89.94
Y_1	10.99	11.43	0.11	49.47
Y_2	11.89	19.07	0.01	100.00
Y_3	39.78	9.95	0.10	92.87

The empirical research framework of this paper is shown as Figure 1.

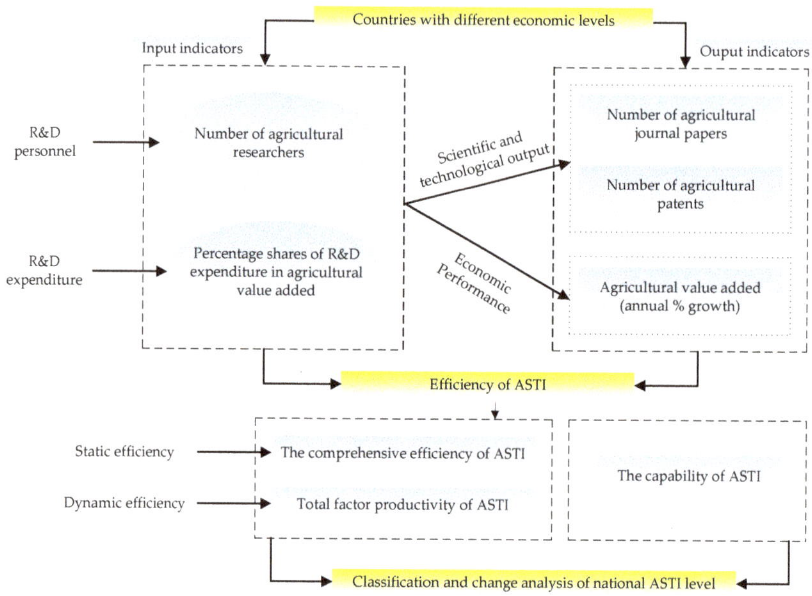

Figure 1. Empirical research framework.

3. Empirical Results and Discussion

This section analyzes and discusses the efficiency results of ASTI from the static and dynamic perspective. In addition, we further classify the ASTI level of the G20 countries through the results of efficiency measurement and capability evaluation. All of the computations were performed with the help of DEAP2.1.

3.1. Comprehensive Efficiency Analysis of ASTI

3.1.1. Overall Analysis of the Comprehensive Efficiency of ASTI

Figure 2 shows the measurement results for the static comprehensive efficiency of ASTI in the G20 countries from 2008 to 2017. A value equal to 1 represents that the ASTI of a country is efficient, while smaller values mean it is more inefficient. Brazil, China, India, Indonesia, Saudi Arabia, and the United States were all efficient from 2008 to 2017, while Italy, Japan, and the Republic of Korea were inefficient during those 10 years. The other 10 countries (Argentina, Australia, Canada, France, Germany, Mexico, the Russian Federation, South Africa, Turkey, and the United Kingdom) changed alternately between efficiency and inefficiency. It is worth noting that the countries that have been efficient during the 10 years include both developed countries (the United States) and developing countries (Brazil, China, India, Indonesia, and Saudi Arabia). This means that innovation efficiency is dependent not only on more innovation investments. Optimized translation from inputs into outputs will lead to high innovation efficiency.

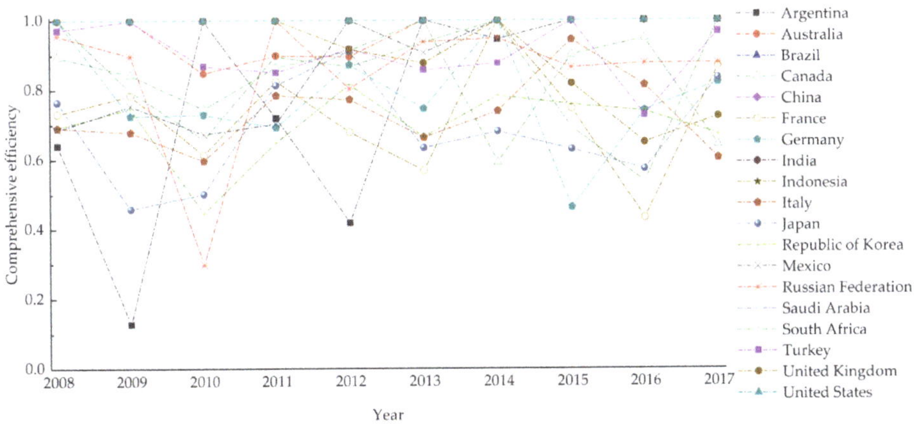

Figure 2. Comprehensive efficiency of ASTI in the period 2008–2017.

Due to the different levels of economic development and agricultural development, there are obvious differences in ASTI inputs between the G20 developed countries (Australia, Canada, France, Germany, Italy, Japan, the Republic of Korea, the United Kingdom, and the United States) and the G20 developing countries (Argentina, Brazil, China, India, Indonesia, Mexico, the Russian Federation, Saudi Arabia, South Africa, and Turkey). Figures 3 and 4 describe the efficiency range and efficient proportion of ASTI in the G20 developed countries and the G20 developing countries from 2008 to 2017.

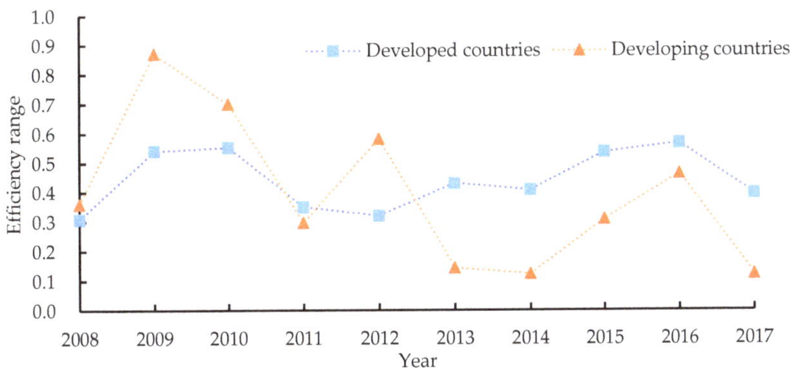

Figure 3. Efficiency range of ASTI.

The efficiency range refers to the difference value between the maximum and minimum values of the comprehensive efficiency, which is used to reflect the balance of the comprehensive efficiency development. The efficiency range of the G20 developed countries stabilized within 0.3–0.5, with no obvious change, reaching the highest value of 0.568 in 2016. The efficiency range of the G20 developing countries was relatively large (0.1–0.9), reaching the highest value of 0.872 in 2009. The above shows that, compared with developed countries, the development of comprehensive efficiency in developing countries is unbalanced. This is because developed countries generally attach importance to ASTI, while developing countries pay more or less attention to it.

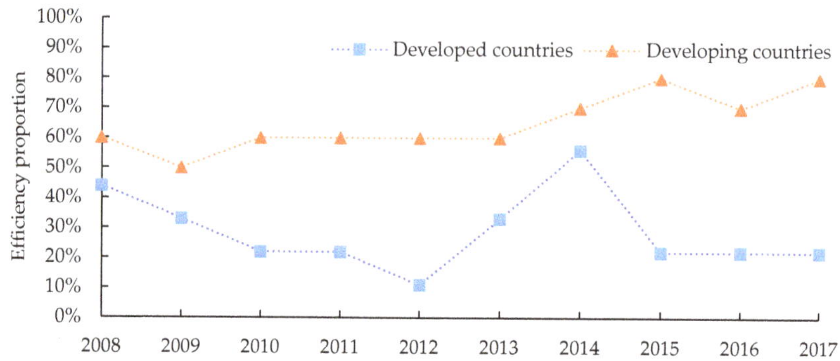

Figure 4. Efficient proportion of ASTI.

The efficient proportion represents the proportion of the number of countries that were efficient to the total number of countries, which is used to reflect the overall development level of the comprehensive efficiency. It could be seen that the G20 developed countries showed a fluctuating downward trend, with the highest proportion of 56% in 2014 and the lowest proportion of 11% in 2012. The G20 developing countries showed an upward trend, from the lowest proportion of 50% in 2009 to the highest proportion of 80% in 2015 and 2017. The reason for this result is that the investment of ASTI in developed countries has been at a high level for a long time. When the increase of inputs is lower than the increase of outputs, the efficiency will decline. However, with the emphasis on ASTI in developing countries, the output increases rapidly with the increasing input, which promotes the improvement of efficiency.

3.1.2. Input Redundancy and Output Deficiency of ASTI

If R&D resources are not used effectively, additional investment may be of little help in stimulating scientific and technological progress [29]. In this section, we analyze the input redundancy and output deficiency to investigate the binding constraints to higher efficiency. Table 4 shows the input redundancy frequency and output deficiency frequency of ASTI for 13 countries (Argentina, Australia, Canada, France, Germany, Italy, Japan, the Republic of Korea, Mexico, the Russian Federation, South Africa, Turkey, and the United Kingdom) during 2008–2017. Brazil, China, India, Indonesia, Saudi Arabia, and the United States all achieved efficiency in ASTI from 2008 to 2017, so there was no input redundancy and output deficiency. A frequency of 0 means that there is no input redundancy or output deficiency in the country during these 10 years; a frequency of 1–4 means that input redundancy or output deficiency happens occasionally; and a frequency of 5–10 means that input redundancy or output deficiency happens frequently.

There are input redundancy and output deficiency in both the G20 developed and developing countries. From the perspective of input indicators redundancy, R&D personnel redundancy occurred frequently (5–10) in Japan and the Russian Federation, while R&D expenditure redundancy occurred frequently (5–10) in Canada, France, Italy, the Republic of Korea, South Africa, and the United Kingdom. The output deficiency mainly focuses on the scientific and technological output, while the economic performance performs well. The countries with a higher frequency (5–10) of agricultural journal papers deficiency include Germany, Japan, the Republic of Korea, the Russian Federation, and Turkey. The countries with a higher frequency (5–10) of agricultural patents deficiency include Argentina, South Africa, and Turkey.

Table 4. Input redundancy frequency and output deficiency frequency of ASTI in the period 2008–2017.

Frequency	R&D Personnel	R&D Expenditure	Agricultural Journal Papers	Agricultural Patents	Economic Performance
0	Argentina, Australia, Canada, Mexico, South Africa, United Kingdom			Germany, United Kingdom	Argentina, Canada, Italy, Japan, Republic of Korea, Mexico, Russian Federation, South Africa, Turkey
1–4	France, Germany, Italy, Republic of Korea, Turkey	Argentina, Australia, Germany, Japan, Mexico, Russian Federation, Turkey	Argentina, Australia, Canada, France, Italy, Mexico, South Africa, United Kingdom	Australia, Canada, France, Italy, Japan, Republic of Korea, Mexico, the Russian Federation	Australia, France, Germany, United Kingdom
5–10	Japan, Russian Federation	Canada, France, Italy, Republic of Korea, South Africa, United Kingdom	Germany, Japan, Republic of Korea, Russian Federation, Turkey	Argentina, South Africa, Turkey	

We further investigated the factors reducing the static comprehensive efficiency of the inefficient countries over the years 2008–2017. Taking 2017 as an example (see Table 5), there are nine inefficient countries: Canada, France, Germany, Italy, Japan, the Republic of Korea, the Russian Federation, Turkey, and the United Kingdom. In terms of the input redundancy, France (0.30), Germany (0.13), the Republic of Korea (0.16), the Russian Federation (0.63), and Turkey (0.09) have R&D personnel redundancy; Canada (0.21), France (0.05), Italy (0.16), the Republic of Korea (0.26), and the United Kingdom (0.06) have R&D expenditure redundancy; therefore, these countries can appropriately reduce related R&D investments. In terms of the output deficiency, Germany (0.51), Japan (3.09), the Republic of Korea (0.11), the Russian Federation (7.00), and Turkey (2.46) are deficient in agricultural journal papers; the Russian Federation (2.90) and Turkey (0.88) are deficient in agricultural patents; therefore, these countries should pay more attention to the two scientific and technological outputs. In addition, 0.00 represents no input redundancy or output deficiency, which means these inputs and outputs of the country have already been optimized.

Table 5. Input redundancy and output deficiency of ASTI in G20 countries in 2017.

Country	R&D Personnel	R&D Expenditure	Agricultural Journal Papers	Agricultural Patents	Economic Performance
Canada	0.00	0.21	0.00	0.00	0.00
France	0.30	0.05	0.00	0.00	0.00
Germany	0.13	0.00	0.51	0.00	0.00
Italy	0.00	0.16	0.00	0.00	0.00
Japan	0.00	0.00	3.09	0.00	0.00
Republic of Korea	0.16	0.26	0.11	0.00	0.00
Russian Federation	0.63	0.00	7.00	2.90	0.00
Turkey	0.09	0.00	2.46	0.88	0.00
United Kingdom	0.00	0.06	0.00	0.00	0.00

3.2. *Total Factor Productivity Analysis of ASTI*

This section applies the DEA-Malmquist index model to calculate the dynamic total factor productivity change (TFPC) of ASTI for G20 countries from 2008 to 2017 and decomposes TFPC of ASTI under time dimension and spatial dimension to investigate the sources of efficiency change.

3.2.1. TFPC Decomposition of ASTI under Time Dimension

Table 6 shows the TFPC decomposition of ASTI for G20 countries at different stages. A value of less (more) than 1 represents decline (growth), and a value equal to 1 shows no change. The mean value of TFPC for G20 countries is 0.981 in the period 2008–2017, a 1.9% decline in the Total Factor Productivity (TFP). The mean value of technological change (TC) is 0.974, while the mean value of Technical Efficiency Change (TEC) is 1.008, indicating that the decline of Total Factor Productivity (TFP) of ASTI was caused by the decline of TC greater than the growth of Technical Efficiency (TE). The mean value of Pure Efficiency Change (PEC) is 1.000, that is, pure efficiency (PE) kept unchanged. Thus, a 0.8% increase in Scale Efficiency (SE) improves the TE of ASTI in G20 countries.

Table 6. TFPC decomposition of ASTI in the period 2008–2017 under time dimension.

Period \ Efficiency	Total Factor Productivity Change (TFPC)	Technical Efficiency Change (TEC)	Technological Change (TC)	Pure Efficiency Change (PEC)	Scale Efficiency Change (SEC)
2008–2009	0.826	0.871	0.949	0.875	0.995
2009–2010	1.217	1.063	1.144	0.970	1.096
2010–2011	1.017	0.889	1.144	1.158	0.768
2011–2012	0.945	0.973	0.971	0.998	0.975
2012–2013	0.937	1.351	0.694	1.003	1.347
2013–2014	0.890	0.964	0.923	1.051	0.917
2014–2015	0.958	0.931	1.030	0.950	0.980
2015–2016	0.976	1.023	0.954	0.944	1.083
2016–2017	1.117	1.079	1.035	1.077	1.002
Mean Value	0.981	1.008	0.974	1.000	1.008

The TFPC of ASTI in 2008–2017 could be divided into four stages. The first stage (2008–2009) was the decline stage: the TFP of ASTI fell by 17.4%, which was caused by the decline in TE (falling by 12.9%) and TC (falling by 5.1%) simultaneously. The second stage (2009–2011) is the growth stage with a 21.7% growth from 2009 to 2010 and a 1.7% growth from 2010 to 2011. The growth of TFP of ASTI during 2009–2010 was driven by the synchronous growth of TE (increasing 6.3%) and TC (increasing 14.4%), while it was driven by TC (increasing 14.4%) during 2010–2011. The third stage (2011–2016) was the continuous decline stage: the TFP of ASTI declined by 5.5%, 6.3%, 11%, 4.2%, and 2.4%, respectively. It was mainly caused by the decline in TC. The fourth stage (2016–2017) is to resume growth. Due to the simultaneous growth of TE (increasing by 7.9%) and TC (increasing by 3.5%), the TFP of ASTI for G20 countries rose by 11.7%. The above alternating trend of "decline–growth–continuous decline–growth recovery" of TFPC decomposition indicates that the efficiency of ASTI for G20 countries was in a period of constant adjustment.

The TFPC decomposition of ASTI in G20 developed and developing countries is presented in Table 7. The G20 developed countries experienced an increment during 2009–2012, a decline during 2013–2015, and an increment again during 2015–2017. The G20 developing countries showed a fluctuating upward trend of TFPC.

Table 7. TFPC decomposition of ASTI in G20 developed and developing countries under time dimension.

	G20 Developed Countries					G20 Developing Countries				
	TFPC	TEC	TC	PEC	SEC	TFPC	TEC	TC	PEC	SEC
2008–2009	0.917	0.957	0.958	0.919	1.042	0.752	0.800	0.940	0.837	0.955
2009–2010	1.080	0.951	1.136	0.887	1.073	1.354	1.176	1.152	1.052	1.117
2010–2011	1.001	0.770	1.301	1.192	0.645	1.032	1.013	1.018	1.128	0.898
2011–2012	1.014	1.043	0.972	1.037	1.005	0.888	0.915	0.971	0.964	0.949
2012–2013	0.851	1.563	0.544	0.906	1.726	1.022	1.184	0.863	1.098	1.078
2013–2014	0.867	0.928	0.934	1.093	0.849	0.911	0.997	0.914	1.014	0.983
2014–2015	0.859	0.868	0.990	0.924	0.939	1.058	0.991	1.067	0.973	1.019
2015–2016	1.025	1.073	0.955	0.943	1.138	0.933	0.980	0.952	0.946	1.036
2016–2017	1.092	1.047	1.043	1.057	0.990	1.140	1.109	1.028	1.095	1.013

3.2.2. TFPC Decomposition of ASTI under Spatial Dimension

As shown in Table 8, seven G20 countries (Saudi Arabia, Japan, China, Mexico, the Republic of Korea, the Russian Federation, and Argentina) (37%) have shown growth in TFP of ASTI (TFPC > 1) during 2008–2017. Among them, Saudi Arabia, Japan, and China saw the larger growth, with increases of 11.4%, 7.5%, and 6.6% respectively; Mexico and the Republic of Korea saw growth of 3.9% and 2.8%, respectively; and the Russian Federation and Argentina saw smaller growth, with increases of 0.6% and 0.4% respectively. The TFP of ASTI for 12 countries (Australia, Turkey, the United States, South Africa, France, Indonesia, India, Italy, Germany, Brazil, the United Kingdom, and Canada) (63%) declined from 2008 to 2017 (TFPC < 1). Australia, Turkey, the United States, South Africa, and Indonesia fell 0–5%; India, Italy, Germany, and Brazil fell 5–10%; and the United Kingdom and Canada fell by more than 10%.

Table 8. TFPC decomposition of ASTI in the period 2008–2017 under spatial dimension.

Country	TFPC	TEC	TC	PEC	SEC
Saudi Arabia	1.114	1.000	1.114	1.000	1.000
Japan	1.075	1.111	0.967	1.009	1.100
China	1.066	1.000	1.066	1.000	1.000
Mexico	1.039	1.054	0.986	1.043	1.011
Republic of Korea	1.028	1.077	0.955	0.997	1.080
Russian Federation	1.006	1.006	1.000	0.990	1.016
Argentina	1.004	1.078	0.931	1.051	1.026
Australia	0.994	1.000	0.994	1.000	1.000
Turkey	0.976	1.008	0.969	0.999	1.009
United States	0.976	1.003	0.973	1.000	1.003
South Africa	0.960	0.964	0.995	1.000	0.964
France	0.958	1.014	0.945	1.019	0.995
Indonesia	0.955	1.000	0.955	1.000	1.000
India	0.948	1.000	0.948	1.000	1.000
Italy	0.945	1.006	0.939	0.985	1.022
Germany	0.929	0.983	0.946	0.978	1.004
Brazil	0.923	1.009	0.915	1.000	1.009
United Kingdom	0.897	0.932	0.963	0.965	0.966
Canada	0.882	0.922	0.957	0.964	0.957

The influencing factors of TFPC of ASTI were identified, as shown in Tables 9 and 10. As shown in Table 8, the growth for China and Saudi Arabia in TFP of ASTI was due to the improvement in TC (TC > 1, TEC $\leq$ 1). The growth for Japan, Mexico, the Republic of Korea, the Russian Federation, and Argentina was attributed to the improvement in TE (TEC > 1, TC $\leq$ 1), where two countries (the Republic of Korea and the Russian Federation) were driven by SE (SEC > 1) only and three countries (Japan, Mexico, and Argentina) by the synchronous improvements of SE and PE (SEC > 1, PEC > 1).

Table 9. Sources for growth of TFPC of ASTI.

<table>
<tr><th></th><th>Source</th><th colspan="2">Country</th></tr>
<tr><td rowspan="3">TFPC > 1</td><td rowspan="2">TEC > 1, TC $\leq$ 1</td><td rowspan="2">Argentina, Japan, Republic of Korea, Mexico, the Russian Federation</td><td>SEC > 1 (Republic of Korea, the Russian Federation)</td></tr>
<tr><td>PEC > 1, SEC > 1 (Japan, Mexico, Argentina)</td></tr>
<tr><td>TC > 1, TEC $\leq$ 1</td><td colspan="2">China, Saudi Arabia</td></tr>
</table>

Table 10. Sources for decline of TFPC of ASTI.

<table>
<tr><th></th><th>Source</th><th colspan="2">Country</th></tr>
<tr><td rowspan="4">TFPC < 1</td><td rowspan="3">TC < 1, TEC < 1</td><td rowspan="3">South Africa, Germany, United Kingdom, Canada</td><td>SEC < 1 (South Africa)</td></tr>
<tr><td>PEC < 1 (Germany)</td></tr>
<tr><td>SEC < 1, PEC < 1 (United Kingdom, Canada)</td></tr>
<tr><td>TC < 1, TEC ≥ 1</td><td colspan="2">Australia, Turkey, United States, France, Indonesia, India, Italy, Brazil</td></tr>
</table>

The sources for decline of TFP of ASTI is shown in Table 10. Eight countries (Australia, Turkey, the United States, France, Indonesia, India, Italy, and Brazil) were because of the decrease in TC (TC < 1, TEC ≥ 1), while four countries (South Africa, Germany, the United Kingdom, and Canada) were attributed to the synchronous decrease in TE and TC (TC < 1, TEC < 1). Among them, the source for decline in South Africa is SE (SEC < 1), Germany is PE (PEC < 1), and the United Kingdom and Canada showed declines due to the decrease of SE and PE (PEC < 1, SEC < 1).

3.3. *Classification and Change Analysis of National ASTI level*

The innovation efficiency and innovation capability are two important aspects of national ASTI level [50,51]. The efficiency of ASTI focuses on the relationship of transformation from input to output, that is, whether more output can be obtained under the given input or less input can be invested under the given output. The capability of ASTI is a comprehensive performance from the joint influence of input and output [46]. The efficiency of ASTI concentrates on innovation quality, while the capability of ASTI focuses on innovation quantity. This study integrated the analysis of the efficiency and capability of ASTI to locate the type of ASTI level and identify the type change for each G20 country, from both the innovation "quality" and "quantity" aspects.

According to the rankings of efficiency measurement of ASTI, countries ranked 1–10 are called "efficiency superior" and countries ranked 11–19 are called "efficiency inferior". Similarly, based on to the rankings of capability evaluation of ASTI for G20 countries, countries ranked 1–10 are called "capability superior " and countries ranked 11–19 are called "capability inferior ". Combining the rankings of the efficiency and capability of ASTI, the ASTI level of G20 countries is divided into four categories: "double superior type" ("efficiency superior" and "capability superior"), "efficiency single-superior type" ("efficiency superior" and "capability inferior"), "capability single-superior type" ("efficiency inferior" and "capability superior"), and "double inferior type" ("efficiency inferior" and "capability inferior").

This study used the evaluation system and calculation model, which were referred from Wang's methodology [46], to evaluate the capabilities of ASTI for G20 countries. For the details of the efficiency measurement, please refer to Figure 2.

As shown in Figure 5, in 2017, Australia, China, and the United States with outstanding efficiency and capability of ASTI simultaneously belong to the double superior type; Brazil, India, Indonesia, Mexico, Saudi Arabia, and South Africa are efficiency single-superior type, indicating that the capabilities of ASTI for these six countries need to be improved; the capacity single-superior type included Canada, France, Italy, Japan, and the Republic of Korea, which means that these five countries are weak in the efficiency of ASTI; and the Russian Federation and Turkey are weak in both capability and efficiency of ASTI, so they fell in the double-inferior type.

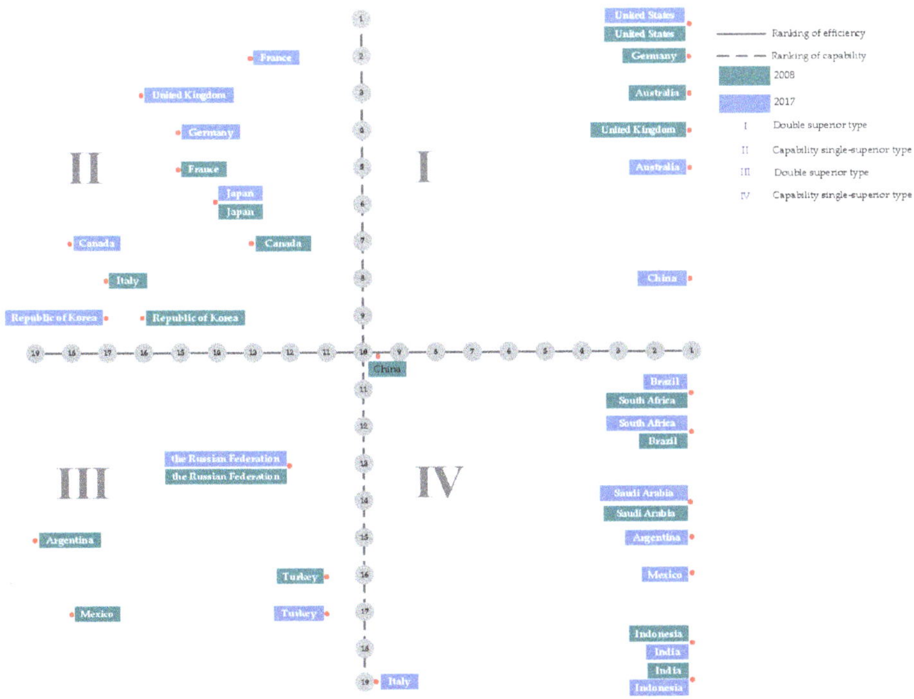

Figure 5. Classification for the national ASTI level for G20 countries.

Compared with 2008, in 2017, the types of ASTI levels in 15 countries, including Australia, Brazil, Canada, China, France, India, Indonesia, Italy, Japan, the Republic of Korea, the Russian Federation, Saudi Arabia, South Africa, Turkey, and the United States, remained unchanged. There were only four countries (Argentina, Germany, Mexico, and the United Kingdom) whose types of ASTI levels changed. Argentina and Mexico changed from the double inferior type to the efficiency single-superior type due to efficiency improvement. On the contrary, the ASTI types of Germany and the United Kingdom transformed from the double superior type to capacity single-superior type caused by a drop in efficiency. The results indicate that the ASTI levels of G20 countries were in a stable status on the whole during the period 2008–2017.

4. Conclusions and Implications

In this study, we estimated the static-dynamic efficiency of ASTI and identified the efficiency determinants across the G20 countries. First, we measured the static comprehensive efficiency of ASTI by means of the DEA-BCC model. The results show that one developed country (the United States) and five developing countries (Brazil, China, India, Indonesia, and Saudi Arabia) have been efficient in the period 2008–2017. The values of the G20 developing countries were relatively larger than the G20 developed countries from the perspective of efficiency range. The G20 developed countries showed a fluctuating downward trend, while the G20 developing countries showed an upward trend from the perspective of efficient proportion. The major binding constraints to the higher efficiency of ASTI included the R&D expenditure redundancy and the agricultural journal papers deficiency. Second, we applied the DEA-Malmquist index model to calculate the dynamic total factor productivity change (TFPC) of ASTI in the periods 2008–2017. The TFPC of ASTI for G20 countries showed "decline–growth–continuous decline–growth recovery" trend, where the G20 developed countries showed a "growth–decline–growth" trend, while the G20 developing countries experienced a fluctuating upward trend. The technological change (TC) was the main cause of the TFTC of ASTI in most G20 countries. Finally, we further classified the ASTI level of G20 countries based on efficiency and capability. The G20 developed countries usually had higher capacity, while the G20 developing countries had advantages in efficiency.

The theoretical contributions of this study are as follows. First, although measuring innovation efficiency is not a novel concept, the empirical evidence in ASTI is limited and most studies have taken a cross-sectional sample of one region or one country. In this study, we went beyond the range of single-country studies. We collected the G20 countries data during a 10-year period (2008–2017) and performed a comparative study of countries at different stages of development. Our second contribution is the integration of static and dynamic ASTI efficiency analyses, contrary to the previous studies only conducting static or dynamic efficiency analyses. Here, we investigated the efficiency determinants from static and dynamic perspectives. Third, this study is pioneering in classification and change analysis of the national ASTI level based on the rankings of efficiency measurement and capability evaluation of ASTI.

This study has important implications for the design and implementation of agricultural innovation strategies for policymakers. Firstly, developed countries should optimize the investment structure of ASTI, while developing countries should pay more attention to the scale of investment. The ASTI in developed countries started earlier, but most developed countries in G20 had R&D personnel and R&D expenditure investment redundancy during 2008–2017. Thus, policymakers in developed countries should pay more attention to the optimization of investment structure of ASTI, instead of focusing only on inputs. Although the efficiency of ASTI of developing countries is higher than that of developed countries on the whole, this is mainly based on the premise of low input–low output in ASTI of developing countries. Policymakers in developing countries should appropriately expand the scale of investment in ASTI while maintaining current efficiency. Secondly, the role of technological progress in promoting the efficiency of ASTI must be taken seriously. The empirical results show that the reason for the decline in the TFP of ASTI in G20 countries from 2008 to 2017 was that the decline in technological progress was greater than the increase in technological efficiency. All countries should strengthen the development and transformation of agricultural technology continuously. Finally, in 2008, there were four "double superior type" countries of G20 but only three in 2017. Therefore, the efficiency and capability of ASTI should be taken into consideration when implementing different policies simultaneously.

This research is not free of limitations and these could be addressed in future research. First, a national agriculture innovation system is a complex system, including various input and output indicators. Concerning the future work, it would be interesting and valuable to investigate more detailed indicators for a better simulation of the national agriculture innovation system, which can more effectively reflect the real process of ASTI activities at the national level. Second, due to the lack of data, our empirical research was not able to include more countries. Hence, a direction for our future research is to conduct comparative studies between more countries.

Author Contributions: Conceptualization, D.W., X.G., and C.D.; methodology, D.W., X.G., C.D., and X.D.; validation, D.W.; formal analysis, D.W.; investigation, D.W., C.D., X.D., J.L., and B.W.; writing—original draft preparation, D.W. and C.D.; writing—review and editing, D.W.; supervision, X.G.; and funding acquisition, D.W. All authors have read and agreed to the published version of the manuscript.

Funding: This research was funded by the Chinese Postdoctoral Science Foundation (2017M621239) and the Postdoctoral Science Foundation of Heilongjiang Province (LBH-Z17007).

Institutional Review Board Statement: Not applicable.

Informed Consent Statement: Not applicable.

Acknowledgments: We are grateful for the helpful insights and suggestions from the editor of this journal, as well as the anonymous referees.

Conflicts of Interest: The authors declare no conflict of interest.

References

1. Chen, Z.H.; Sun, T.T. American Agricultural Product: Analyzing Its International Competitiveness. *Acad. Exc.* **2019**, *10*, 99–113.
2. Ma, H.K.; Mao, S.P. Green-Ecological Transformation of EU Common Agricultural Policy: Policy Evolution, Reform Trend and Its Enlightenment. *Issues Agric. Econ.* **2019**, *9*, 134–144.
3. Gong, Y.T.; Sun, L.X.; Mao, S.P. Agricultural R&D policies in the UK and their implications to China. *Res. Agric. Modern.* **2018**, *39*, 559–566.
4. Ministry of Agriculture and Rural Affairs of the People's Republic of China: Notice of Ministry of Agriculture on Printing and Distributing "Agricultural Science and Technology Development Plan (2006–2020)". Available online: http://www.moa.gov.cn/nybgb/2007/dqq/201806/t20180614_6151989.htm (accessed on 16 November 2020).
5. *Ministry of Agriculture and Rural Affairs of the People's Republic of China: Notice of the Ministry of Agriculture on Printing and Distributing "The National Plan for the Construction of Agricultural Science and Technology Innovation Capacity (2012–2016)"*. Available online: http://www.moa.gov.cn/nybgb/2013/dliuq/201805/t20180509_6141642.htm (accessed on 16 November 2020).
6. Wennekers, S.; Thurik, R. Linking entrepreneurship and economic growth. *Small Bus. Econ. Group* **1999**, *13*, 27–56. [CrossRef]
7. Hollanders, H.; Celikel-Esser, F. *Measuring Innovation Efficiency*; European Commission: Brussels, Belgium, 2007.
8. Liu, Z.Y.; Chen, X.F.; Chu, J.F.; Zhu, Q.Y. Industrial development environment and innovation efficiency of high-tech industry: Analysis based on the framework of innovation systems. *Technol. Anal. Strateg. Manag.* **2018**, *30*, 434–446. [CrossRef]
9. Guan, J.C.; Chen, K.H. Modeling macro-R&D production frontier performance: An application to Chinese province-level R&D. *Scientometrics* **2010**, *82*, 165–173.
10. Zuo, K.R.; Guan, J.C. Measuring the R&D efficiency of regions by a parallel DEA game model. *Scientometrics* **2017**, *112*, 175–194.
11. Perović, S.; Radovanović, S.; Sikimić, V.; Berber, A. Optimal research team composition: Data envelopment analysis of Fermilab experiments. *Scientometrics* **2016**, *108*, 83–111. [CrossRef]
12. Campisi, D.; Mancuso, P.; Mastrodonato, S.L.; Morea, D. Efficiency assessment of knowledge intensive business services industry in Italy: Data envelopment analysis (DEA) and financial ratio analysis. *Meas. Bus. Excell.* **2019**, *23*, 484–495. [CrossRef]
13. Rousseau, S.; Rousseau, R. The scientific wealth of European nations: Taking effectiveness into account. *Scientometrics* **1998**, *42*, 75–87. [CrossRef]
14. Pan, T.W.; Hung, S.W.; Lu, W.M. DEA performance measurement of the national innovation system in Asia and Europe. *Asia Pac. J. Oper. Res.* **2010**, *27*, 369–392. [CrossRef]
15. Guan, J.C.; Chen, K.H. Modeling the relative efficiency of national innovation systems. *Res. Policy* **2012**, *41*, 102–115. [CrossRef]
16. Broekel, T.; Rogge, N.; Brenner, T. The innovation efficiency of German regions-A shared-input DEA approach. *Rev. Reg. Res.* **2018**, *38*, 77–109. [CrossRef]
17. Zemtsov, S.; Kotsemir, M. An assessment of regional innovation system efficiency in Russia: The application of the DEA approach. *Scientometrics* **2019**, *120*, 375–404. [CrossRef]
18. Wang, D.; Zhao, X.L.; Zhang, Z.Z. The time lags effects of innovation input on output in national innovation systems: The case of China. *Discrete Dyn. Nat. Soc.* **2016**, *8*, 1–12. [CrossRef]
19. Shin, J.; Kim, C.; Yang, H. The Effect of Sustainability as Innovation Objectives on Innovation Efficiency. *Sustainability* **2018**, *10*, 1966. [CrossRef]
20. Park, J.H. Open innovation of small and medium-sized enterprises and innovation efficiency. *Asian J. Technol. Innov.* **2018**, *26*, 115–145. [CrossRef]
21. Zhang, J.; Zhang, B.W. The empirical analysis of the innovation efficiency of Agricultural science and technology parks in China based on Malmquist index method. *Sci. Technol. Pro. Policy* **2011**, *28*, 129–133.
22. Du, J. Analysis of the Input-output Efficiency of Agriculture Science and Technology Innovation in China Based on Data Envelopment Analysis. *Sci. Technol. Pro. Policy* **2013**, *30*, 82–85.
23. Mekonnen, D.K.; Spielman, D.J.; Fonsah, E.G.; Dorfman, J.H. Innovation Systems and Technical Efficiency in Developing-Country Agriculture. *Agric. Econ.* **2015**, *46*, 689–702. [CrossRef]
24. Chen, Z.; Zheng, R.; Li, P.H.; Huang, S. Evaluation and analysis of agricultural science and technology innovation efficiency in Henan Province. *J. Henan Agric. Univ.* **2018**, *52*, 464–469.
25. Li, Y.H.; Bai, L.P. Study on Resource Allocation Efficiency and Influencing Factors of Agricultural Science and Technological Innovation in Yunnan Province. *Chin. J. Agri. Resour. Region. Plan* **2019**, *40*, 63–69.
26. Guo, X.Y.; Du, X.; Wang, D. Evaluation and comparative analysis of the efficiency of provincial agricultural science, technology and innovation in China. *Stud. Explor.* **2020**, *42*, 141–147.
27. Dong, M.T. A Study on Allocation Efficiency and Influencing Factors of Agricultural Science and Technology Innovation Resources in China. *East Chin. Econ. Manag.* **2014**, *28*, 53–58.
28. Wang, E.C. R&D efficiency and economic performance: A cross-country analysis using the stochastic frontier approach. *J. Policy Model* **2007**, *29*, 345–360.

29. Wang, E.; Huang, W. Relative efficiency of R&D activities: A cross-country study accounting for environmental factors in the DEA approach. *Res. Policy* **2007**, *36*, 260–273.
30. Wang, D.; Du, X.; Sun, J.; Guo, X.Y.; Chen, Y. Synergy of national agricultural innovation systems. *Sustainability* **2018**, *10*, 3385. [CrossRef]
31. Bloch, H.; Metcalfe, S. Innovation, creative destruction, and price theory. *Ind. Corp. Chang.* **2018**, *27*, 1–13. [CrossRef]
32. Farrell, M.J. The Measurement of Productive Efficiency. *J. R. Stat. Soc.* **1957**, *120*, 253–290. [CrossRef]
33. Charnes, A.; Cooper, W.W.; Rhodes, E. Measuring the efficiency of decision-making units. *Eur. J. Oper. Res.* **1978**, *2*, 429–444. [CrossRef]
34. Alirezaee, M.R.; Howland, M.; van de Panne, C. Sampling Size and Efficiency Bias in Data Envelopment Analysis. *J. Appl. Math. Decis. Sci.* **1998**, *2*, 51–64. [CrossRef]
35. Wober, K.W. Data Envelopment Analysis. *J. Travel. Tour. Mark.* **2007**, *21*, 91–108. [CrossRef]
36. Banker, R.D.; Charnes, A.; Cooper, W.W. Some Models for Estimating Technical and Scale Inefficiencies in Data Envelopment Analysis. *Manag. Sci.* **1984**, *30*, 1078–1092. [CrossRef]
37. Xu, K.; Bossink, B.; Chen, Q. Efficiency Evaluation of Regional Sustainable Innovation in China: A Slack-Based Measure (SBM) Model with Undesirable Outputs. *Sustainability* **2020**, *12*, 31. [CrossRef]
38. Fare, R.; Grosskopf, S.; Norris, M.; Zhang, Z.Y. Productivity growth, technical progress, and efficiency change in industrialized countries. *Am. Econ. Rev.* **1994**, *84*, 66–83.
39. Fare, R.; Grosskopf, S.; Kokkelenberg, E.C. Measuring Plant Capacity, Utilization and Technical Change: A Nonparametric Approach. *Int. Econ. Rev.* **1989**, *30*, 655–666. [CrossRef]
40. Ouenniche, J.; Carrales, S. Assessing efficiency profiles of UK commercial banks: A DEA analysis with regression-based feedback. *Ann. Oper. Res.* **2018**, *266*, 551–587. [CrossRef]
41. Zhang, C.; Wang, X.J. The influence of ICT-driven innovation: A comparative study on national innovation efficiency between developed and emerging countries. *Behav. Inf. Technol.* **2019**, *38*, 876–886. [CrossRef]
42. Fang, S.R.; Xue, X.S.; Yin, G. Evaluation and Improvement of Technological Innovation Efficiency of New Energy Vehicle Enterprises in China Based on DEA-Tobit Model. *Sustainability* **2020**, *12*, 7509. [CrossRef]
43. Lin, Y.Y.; Deng, N.Q.; Gao, H.L. Research on Technological Innovation Efficiency of Tourist Equipment Manufacturing Enterprises. *Sustainability* **2018**, *10*, 4826. [CrossRef]
44. Cruz-Cázares, C.; Bayona-Sáez, C.; García-Marco, T. You can't manage right what you can't measure well: Technological innovation efficiency. *Res. Policy* **2013**, *42*, 1239–1250. [CrossRef]
45. Suh, Y.; Kim, M.S. A taxonomy of service innovations based on the innovative activity efficiency of service firms: A DEA approach. *Int. J. Serv. Technol. Manag.* **2014**, *20*, 267–289. [CrossRef]
46. Wang, D.; Zhao, X.L.; Guo, X.Y.; Hu, Y.; Du, X. Theoretical framework and capability evaluation for the national agricultural science, technology and innovation—An empirical study based on the group of twenty. *Chin. Soft. Sci.* **2018**, *3*, 18–35.
47. Brown, M.G.; Svenson, R.A. Measuring R&D productivity. *Res. Technol. Manag.* **1998**, *41*, 30–35.
48. Kaihua, C.; Mingting, K. Staged efficiency and its determinants of regional innovation systems: A two-step analytical procedure. *Ann. Regional. Sci.* **2014**, *52*, 627–657. [CrossRef]
49. Wang, Q.W.; Hang, Y.; Sun, L.C.; Zhao, Z.Y. Two-stage innovation efficiency of new energy enterprises in China: A non-radial DEA approach. *Technol. Forecast. Soc. Chang.* **2016**, *112*, 254–261. [CrossRef]
50. Sun, H.; Yu, H.X. Research on Technology Innovation Efficiency and Innovation Capacity of Pharmaceutical Industry in Hebei Province. *Sci. Technol. Manage. Res.* **2012**, *32*, 85–90.
51. Zhang, Y.; Tang, L.B.; Jin, P.Z. Study on the Measurement and Influencing Factors of the BRICs' Innovation Capability. *Chin. Soft. Sci.* **2015**, *6*, 148–157.

sustainability MDPI

Review

Digitalization for Sustainable Agri-Food Systems: Potential, Status, and Risks for the MENA Region

Rachel A. Bahn [1], Abed Al Kareem Yehya [1] and Rami Zurayk [2,*]

[1] Food Security Program, American University of Beirut, Beirut 1107 2020, Lebanon; rb89@aub.edu.lb (R.A.B.); ay27@aub.edu.lb (A.A.K.Y.)

[2] Department of Landscape Design and Ecosystem Management and Food Security Program, American University of Beirut, Beirut 1107 2020, Lebanon

* Correspondence: rzurayk@aub.edu.lb

Abstract: Digital technologies offer a potential solution to improve sustainability—economic, social, and environmental—of agri-food systems around the globe. While developed countries have led the innovation and adoption of digital agriculture, the potential impact in developing countries—including in the Middle East and North Africa (MENA) region—is massive. This article synthesizes existing evidence to review the potential and current contribution of digital technologies to the agri-food sectors in MENA. Digital agriculture shows promise in addressing the key challenges facing the agri-food sector across MENA countries. Improvements in primary production, supply chain and logistics performance, and optimized use of scarce natural resources (notably agricultural water) could be notable, if digital technologies can be implemented as envisioned. Available evidence shows that adoption of digital agriculture is at early stages, generally led by high-value agricultural production targeting domestic markets in Gulf countries and export markets in Mashreq countries. Economic sustainability appears the strongest force for current adoption, with less focus on social or environmental sustainability. Public policies should not only foster the adoption of digital technologies in MENA but also ensure equity of access, transparency of use, data protections, and labor protections. Policymakers should move beyond traditional, production-centric views to deliver also on social and environmental sustainability.

Keywords: productivity; efficiency; sustainability; food security; agriculture; digital agriculture; smart farming; digitalization; digital technologies; Middle East and North Africa

Citation: Bahn, R.A.; Yehya, A.A.K.; Zurayk, R. Digitalization for Sustainable Agri-Food Systems: Potential, Status, and Risks for the MENA Region. *Sustainability* **2021**, *13*, 3223. https://doi.org/10.3390/su13063223

Academic Editor: Michael Blakeney

Received: 9 February 2021
Accepted: 11 March 2021
Published: 15 March 2021

Publisher's Note: MDPI stays neutral with regard to jurisdictional claims in published maps and institutional affiliations.

Copyright: © 2021 by the authors. Licensee MDPI, Basel, Switzerland. This article is an open access article distributed under the terms and conditions of the Creative Commons Attribution (CC BY) license (https://creativecommons.org/licenses/by/4.0/).

1. Introduction

The term "digitalization" refers to the socio-technical application of digital technologies or innovations [1]. Digital agriculture (sometimes also termed "smart farming") refers to the design, development, and use of digital technologies in agriculture and the wider agri-food sector. Digital agriculture encompasses a range of technologies that include sensors, robots, digital communication tools, blockchain, computational decision and analytical tools, and cloud-based technologies [2]. Controlled-environment agriculture (greenhouses, indoor farms, and vertical and hydroponic farms) increasingly applies digital technologies including sensors, robots, and digital communication. More advanced approaches leverage digital, mobile, internet of things (IOT), and cognitive technologies. For example, precision agriculture relies on tools including global positioning system (GPS)-enabled guidance, control systems, sensors, robotics, drones, autonomous vehicles, and variable rate technologies. Precision agriculture practices for livestock farming include sensors, radio frequency identification (RFID), and automated or robotic milking and feeding systems. Predictive analytics software and/or artificial intelligence (AI) use available data to provide farmers with guidance about crop rotation, optimal planting times, harvesting times, and soil management.

Digital technologies have been touted as a potentially revolutionary solution to improve agricultural production systems' performance and sustainability [3], defined to have economic, social, and environmental dimensions [4,5]. Digital technologies can make the agri-food sector more efficient, inclusive, and environmentally sustainable, thereby increasing benefits for farmers, consumers, and society at large [6]. Digital technologies can help to raise on-farm productivity, improve resource use efficiency, and support climate resilience [7]. Improvements in primary production, supply chain and logistics performance, and reductions in food losses and waste are especially notable, provided that digital technologies can be implemented. Moreover, the COVID-19 pandemic has increased attention to both the need for and utility of digital technologies including within the agri-food sector [8] and has catalyzed the introduction and adoption of digital technologies [9]. Despite the myriad benefits promised, digital agriculture—like other, major innovation breakthroughs—is not without its challenges or risks [6,10].

While developed countries have led much of the innovation and early adoption of digital technologies [11–14], the potential impact of these technologies on agriculture in developing countries is massive given the economic, social, and environmental roles played by their agricultural sectors [6]. The available literature has tended to focus on developing countries and regions that depend most highly on agricultural production, where agricultural production is viewed as a key economic sector or where the agri-food sector is a major employer.

The Middle East and North Africa (MENA) region has received little, albeit growing, attention within the field of digital agriculture [14] and vice versa [15]. This tendency to overlook the MENA region is despite the fact that agriculture continues to be practiced and that agri-food is an important, if relatively small, sector within most countries of the region and may reflect the fact that the sector is severely constrained by both natural and human factors [16], which may limit expectations for agriculture in MENA. Nevertheless, the sustainable development of the agri-food sector remains of critical importance for economic performance, employment, social stability, and food security.

The research questions motivating this article are as follows: (1) What are the challenges to the sustainability of agriculture in the MENA region? (2) To what extent can digital agriculture address the actual challenges facing the sector, across all three pillars of sustainability? (3) What is the current status of digital agriculture in the MENA region? (4) What policy action is needed to unlock the potential of digital agriculture and truly contribute to sustainability in the MENA region? Accordingly, this review article seeks to synthesize existing evidence to provide a tightly focused overview of the potential and current contribution of digital technologies to the agri-food sectors of MENA countries. This article does not offer new empirical evidence but rather synthesizes available knowledge to lay the groundwork for such future contributions to the literature. The article is structured as follows: Section 2 briefly describes the materials and methods used. Section 3 presents the results of the literature review, including a summary of the status, contributions, and challenges of agri-food in MENA countries; the potential contribution of digital technologies to the sector; and the current status of digital technologies within the MENA region. Section 4 presents the discussion and conclusions. Evidence shows that adoption of digital technologies within the agri-food sectors across the MENA region varies but is generally at early stages and led by high-value and export-oriented agricultural production. Policies are needed not only to foster the adoption of digital technologies in MENA but also to address concerns for equity of access, transparency of use, data protections, and protections against adverse labor impacts.

2. Materials and Methods

This research consists of a scoping review of the literature addressing the potential and current contribution of digital technologies to the agri-food sectors of MENA countries. The review draws on both academic and grey literature, including publications from institutions such as the World Bank and the Food and Agriculture Organization of the United Nations that play a significant role in policies and interventions targeting the application of digital technologies within the agricultural sector in developing countries [3,17]. News articles have also informed this review, to address the most recent developments not yet reflected in the academic press. Publications were identified through searches of academic databases (Web of Science) and open internet sources (Google search), and through the "snowball" referral to bibliographic references identified in relevant publications. Given the rapid evolution of digital technologies, this review has prioritized more recent studies (published since 2000), though has not strictly excluded earlier publications of relevance to the topic.

A systematic approach to the literature review was neither practical nor appropriate: A search for peer-reviewed academic articles addressing "digital agriculture" or "smart farming" within the MENA region or any of the corresponding countries (pairwise combinations, both within the TOPIC field in Web of Science) yielded precisely one result (Table S1). Moreover, the objectives of this research are to provide a synthesis of the evidence on the potential of technology application and its current status in MENA and to extrapolate lessons for policy relevance, rather than to provide an exhaustive examination of any single digital technology.

This research delimited the MENA region to include the 22 member countries of the Arab League. Given the nature of this review, the de facto focus is accordingly on those countries for which more evidence is available.

3. Results

3.1. *Agri-Food in MENA—Sustainability Considerations*

The agri-food sector makes a modest but important contribution to the economy of countries in the MENA region, particularly those countries that have already experienced the structural transformation of their economies. Nevertheless, the sustainable development of the agri-food sector can play a key role in informing economic performance and employment, social stability and inclusion, and environmental outcomes. Indeed, sustainability of agri-food systems needs to encompass the multiple pillars of sustainability: economic, social, and environmental [4,5,18].

3.1.1. Economic Contribution

The relative contribution of the agriculture sector to the total economy is a minor share in all MENA countries, ranging from less than 1% in Gulf countries like Bahrain, Qatar, and the United Arab Emirates to a high of more than 30% in Comoros (Table S2) [19]. Despite this small share, growth in agricultural production can stimulate growth in the wider economy and especially other areas of the agri-food sector, including production of inputs, food processing, logistics, and financial services. The economy-wide multiplier effect of the agricultural sector may be significant in developing countries, suggesting strong potential for investment in the agriculture sector to trigger wider economic growth [20,21]. In MENA, the downstream segments of the agri-food sector have demonstrated dynamism and faster growth in recent decades as compared to on-farm performance [22].

The agri-food sector can be an important contributor to the net trade performance of MENA countries, offsetting imports and increasing exports. Increasing agriculture's productive potential could help to reduce the dependence on food imports, which is high. For example, Iraq imports more than 80% of its food needs (including most cereals, meat, refined sugar, cooking oil, canned and processed foods, fruits and vegetables, and dairy products) [23]. Food import dependence in MENA countries is projected to rise in coming years due to population growth and shifting dietary patterns. Moreover, this dependence was illuminated during the COVID-19 pandemic, when shocks to international supply chains led to calls for shortened supply chains and sustained local agri-food production, including through digital technologies [8,24]. Improved agri-food performance can also help improve trade balances via the export channel. The expansion of agri-food exports can increase foreign currency inflows, which would be especially welcome in countries such as Lebanon where foreign currency is scarce [25].

3.1.2. Social Contribution

As a labor-intensive sector, agriculture plays an outsized role in employment and job creation in nearly every country of the MENA region. In fact, agricultural employment as a share of total employment is higher than agriculture as a share of GDP in every country in the region for which data are available, except Algeria and Jordan. Agriculture accounts for a minor share of employment in the Gulf countries and Jordan, but a relatively higher share in the other countries of the region, rising to half or more of all employment in Comoros, Mauritania, and Somalia (Table S2) [19]. The agricultural sector also has relatively strong potential for further job creation, as the long-term value-added elasticity of employment in the MENA region is higher for agriculture than the industrial or service sectors. For every percentage point of growth in the value added of the agriculture sector, employment increases by 0.36 percentage points (versus industry at 0.30 percentage points and service at 0.20 percentage points). Employment is created along the value chain, from production to processing, packaging, and distribution and with spillover effects on related sectors such as services, transport, and communications [26].

Poverty is concentrated in rural areas, where households disproportionately depend on agriculture as a primary source of income. The prevalence of poverty in rural areas is reported to be higher than in urban areas in every MENA country for which data is available (Table S2) [27]. As employer to many of the poorer and more marginalized segments of the population—such as migrants, refugees, and the internally displaced—agriculture can play a critical role in reducing poverty and vulnerability. For example, in Lebanon, the agriculture sector was the second-highest employer of displaced Syrians who were working in 2019 (17% of those who have a regular job); this figure was reported at 27% among children [28]. In Iraq, agriculture has proven critical to resettlement of internally displaced persons: 47% of people returned to rural areas, where agriculture, farming and animal husbandry ranks among the top three sources of income. Evidence showed that, as of 2019, returns were higher in locations where agriculture, livestock, and agricultural market activities had resumed [29]. Technologies that boost land and labor productivity may provide a direct path out of poverty. In terms of employment and labor demand, productivity growth in agriculture accumulates additional purchasing power among rural families, expanding job opportunities in off-farm sectors and thereby releasing labor to non-farm sectors [30]. The combination of appropriate investments and proper enforcement of social protections can allow agriculture to progress towards social inclusion and vulnerability reduction among its labor force.

The agri-food sector, as a component of the food system, has direct and indict impacts on food security [31]. Countries of the MENA region report significant food insecurity according to a range of indicators (Table S2). Undernourishment affected approximately 9.0% of the regional population as of 2017–2019, and 30.2% of the population suffered from moderate or severe food insecurity (figures are reported for the Western Asia and Northern Africa regional definition, which approximates but does not precisely correspond to the MENA region as defined within this article). Concurrently, some 27.7% of adults in the region were obese (2016 data) [32]. Support for the agri-food sector and improvements in its performance could improve the region's food security status by supporting agricultural development and incomes, as well as the production and consumption of more nutritious foods [33,34]. Insofar as food security contributes to social stability, the agri-food sector therefore also makes a contribution [35]. Moreover, food systems globally—including within the MENA region—must change fundamentally in the coming decades to align with the needs of healthy people, a healthy planet, and healthy economies, while sustainably increasing the food supply by 2050 [36].

3.1.3. Environmental Contribution

The principal environmental "contribution" of the agricultural sector in the MENA region is in fact as a major user—and polluter—of natural resources including water and land, as well as a contributor to biodiversity loss and climate change. The sector is thus both agent and victim of environmental harm [5].

The agricultural sector is the largest single user of water in multiple countries of the MENA region [37], a region which is one of the most water scarce on the planet. Most countries of the region are categorized as "extremely high" or "high" baseline water stress, and a number are ranked among the most water stressed in the world (Table S2) [38]. To manage this scarcity of water, irrigation is frequently employed for crop production; the use of modern technologies such as drip irrigation and hydroponics can multiply water productivity and produce more with less water. Though water is a scarce resource, agricultural water use efficiency remains low in many countries of the region [39]. Indeed, while water scarcity is a challenge at the national level, it may be less so for individual farmers who can access sufficient water through groundwater resources, even if these resources are threatened by over-extraction and declining quality. Widespread water pollution and substandard water infrastructure exacerbate water scarcity and complicate government efforts to manage the water demand of the agricultural sector. In agricultural areas, the runoff and infiltration of fertilizer and pesticide residues contributes to further environmental degradation. In light of water scarcity, expanding agricultural production and exports must be balanced with the sustainable use of scarce natural resources, specifically irrigation water. For example, experts have recommended against the expansion of irrigated agriculture within the Tigris–Euphrates watershed due to already excessive water use that has led to declining water tables and soil salinization [36]. The sustainable use of water resources is among the most pressing food production challenges in the Gulf [40].

Arable land is another key limitation to agriculture in MENA countries and is under threat from encroaching desertification. Desertification is estimated to affect some 75% of Iraq's land, for example, contributing to the displacement of farming activities [41]. Declining soil fertility is a pervasive problem and expected to limit future agricultural productivity in the region.

Unsustainable use of agricultural inputs, including fertilizers, pesticides, and herbicides, is damaging the current export potential and long-term sustainability of agriculture. Excessively high levels of fertilizer use can reduce yields over the long term. Excessive and inappropriate (ill-timed) application of pesticides may reduce the quality of production, limiting exports to markets that impose strict food safety and sanitary and phyto-sanitary (SPS) standards [26].

MENA's agri-food sector is a contributor to global climate change through its production of greenhouse gas emissions (GHGE) associated with every stage of the value chain. While the total GHGE attributed to agriculture (including agricultural land use) are not available at the regional level, these range widely from less than 1% in GCC countries where agricultural production is relatively limited and where energy production is significant (e.g., Bahrain, Kuwait, Qatar, Saudi Arabia, and United Arab Emirates), to more than 50% in Somalia and Mauritania (Table S2) [42].

3.1.4. Key Challenges

Agricultural production in the MENA region is constrained by challenges including scarcity and inefficient use of natural resources (notably water); small and fragmented farm structures; suboptimal agricultural practices and poor management due to weak extension services; poor integration of input and output markets; limited access to finance; weak infrastructure; and lagging food safety and traceability standards. These challenges disproportionately impact smallholder farmers (as compared to larger, commercial farmers), due to factors such as their lower production volumes, higher transaction costs, more limited capacity for investment, reduced market access, and weaker bargaining power [16,43].

In a number of countries of the MENA region, conflict and social unrest have challenged the performance of the agricultural sector in recent decades. Violent conflict such as that experienced in Iraq may destroy, disrupt, or deteriorate infrastructure, resources and productive assets, institutions, and value chains, e.g., [23,33,35].

Climate change exacerbates the existing challenges the agriculture sector is facing in the MENA region, undermining rural livelihoods and potentially leading to more fragility, displacement, and tensions. Climate change has and will continue to impact temperature and precipitation patterns, raising the risk of extreme weather events. The average temperature in the MENA region may increase by up to 4.8 °C by 2100, with a higher rate of warming in summer than winter. Crop yields in MENA could fall by 10–20% by 2050 under worst-case scenario models. Livestock productivity will fall due to less available animal feed, increased animal heat stress, and a higher risk of infection and disease. Climate change is expected to disproportionately affect small-scale farmers, who are more vulnerable to its effects and may be more susceptible to pressures to migrate out of agriculture and rural areas as a result [44]. Fortunately, adaptation to climate change in the agriculture sector and promotion of climate-smart agriculture practices may yield economic, social, and environmental benefits [45].

Despite the economic, social, and environmental importance of the agri-food sector in the MENA region, and its potential to deliver more benefits, the sector generally does not receive adequate attention or funding from policymakers. The agri-food sector is typically underfunded and rarely seen as a priority in major economic policy reform and investment programs. The agricultural orientation index—the ratio of agriculture share of government expenditures to agriculture share of GDP—is less than 1.0 in every country for which data is available, excepting Qatar (Table S2) [46]. The neglect of the agricultural sector in terms of public expenditure and foreign direct investment (FDI) has contributed to weak investment in agricultural research and poor performance of public extension systems in many MENA countries.

MENA countries range in the extent to which their respective agricultural sectors enjoy the support of favorable government policies or face legal and regulatory constraints to growth. According to the Enabling the Business of Agriculture index, the best performer in the region is Morocco, trailed by Jordan and Egypt; Tunisia, Sudan, and Iraq perform less favorably, though all countries have room for improvement to enable the growth and performance of their respective agricultural sectors. Data is unavailable for the remaining countries of the region (Table S2) [47].

3.2. *Contribution of Digital Technologies to Agri-Food in MENA*

Applied within the agri-food sector, digital technologies can promote long-term development by enhancing the sector's economic, social, and/or environmental sustainability. Below, the most significant potential contributions of digital technologies to agri-food in MENA are reviewed. Current, illustrative examples from the MENA region are noted according to their primary contribution; although digital technologies may indeed be multi-functional and contribute to more than one sustainability pillar simultaneously, this is necessary for organizational purposes. Selected illustrative examples are presented in Table 1; Table 2 within the appropriate sub-sections below; a fuller tabulation of examples of digital agriculture within MENA is presented in Table S3.

3.2.1. Economic Contribution

Digital technologies may enhance the economic contribution of MENA's agri-food sector via greater productivity and/or efficiency of resource use, whether physical, natural, or intangible such as data. The potential applications of digital technology are found on-farm, downstream in value chains, and within supporting e-government and public services.

Access to Land: Digital land information systems or e-registries can improve cadastral data and facilitate its access and use. Cadastral maps often differ from reality, impeding agricultural land market development and farm consolidation. Governments can improve existing cadastral maps via assessment, re-surveying land parcels, and updating digital cadastral maps. Digital technologies can also improve the functioning of land markets: Digitized land-use permits/registration and automated application processes can reduce times for filing and approval and increase transparency. Publicly accessible cadastral databases can increase efficiency of land markets by identifying available land and reducing registration costs [48]. In MENA, Morocco's National Land Registry now incorporates digital technologies to execute its functions [49].

Access to Machinery and Services: Digital platforms can improve on-farm access to agricultural machinery and services, particularly those imposing a high purchase price. Alternatives to direct purchase (e.g., leasing arrangements, contract-based provision, and/or cooperative approaches) may facilitate adoption, especially by smaller farmers [50]. Digital platforms can create efficient new markets for machinery rental by providing more affordable access to physical capital for smallholder farmers and employing underused assets by matching suppliers of machinery rental services with potential customers. The marginal cost of matching buyers and sellers through digital platforms is extremely low; therefore, platforms have the potential to reduce unit costs of machinery rental services through saved transaction and search costs [6,51]. Examples include mobile platforms to deliver tractor services and to connect animal health and veterinary service providers to small livestock operators. Within MENA, the Moroccan Agricultural Development Agency has since 2020 expanded the functionality of Attaisir—a farmer-focused system that aims to ensure sugar production for the domestic market—to enable remote monitoring of a fleet of 2000 GPS-linked agricultural machines and respect social distancing rules in light of COVID-19 [52].

Input Efficiency: Digital technologies can improve the quantity and quality of agricultural output while using less inputs (e.g., water, energy, fertilizers, and pesticides) and increase efficiency by performing farming practices remotely [6]. Crop yields could rise significantly by 2050 with the introduction of precision agriculture technologies, with an 18% yield increase due to precision fertilizer application, 13% due to precision planting, 4% due to precision spraying, and 10% due to precision irrigation [53].

Opportunities for improved water efficiency are critical for MENA. More water-efficient agricultural production is an important goal for water-scarce countries, but the use of partial productivity measures ("crop per drop") can be misleading. Moreover, farmers rarely seek to maximize the crop per drop measure in their production decisions. However, efforts to improve the allocation of water to more efficient uses could, even under conditions of climate change, yield better economic performance [45]. Smart irrigation systems that incorporate IoT and remote sensing technologies are one option to achieve more water-efficient agricultural production that accounts for farmer incentives. High-resolution satellite images combined with specific algorithms can determine spatial and temporal variability of agricultural water and land productivity. In field, remote sensors can measure water use and monitor net groundwater withdrawals; the resulting data can inform targets for sustainable irrigation water management and allocations to system users. Farmers' or landholders' water pumps can then be integrated into smart irrigation systems, allowing for real-time metering and remote shut-off if a water user exceeds a water allocation. Digital applications can allow farmers to use their mobile phones to control their irrigation pumps remotely and save time, energy, and water [6,54]. More efficient use of irrigation systems can also reduce the energy needed to operate those systems, with corresponding GHGE savings [11]. Smart irrigation systems with optimized fertigation could be particularly beneficial for MENA countries in view of current water scarcity, inefficient water use, and soil degradation. In fact, such systems are already in use in some areas. For example, in Jordan, digital technologies have been adopted to estimate evapotranspiration and monitor agricultural water use, including within irrigation systems [15]. This and similar examples are presented in Table 1.

Farm Management and Decision-Making: Higher quality agronomic data, weather data, and price information can improve farmer decision-making and management, ultimately increasing farmer profits and incomes, and this data can be recorded, analyzed, and disseminated more rapidly and more widely through the use of digital technologies. Access to information can also promote the inclusion of marginalized rural producers in markets, as when female farmers benefit from tools that encourage knowledge and information sharing [6].

Table 1. Precision agriculture for input use efficiency: illustrative examples of current application in Middle East and North Africa (MENA).

Country	Illustrative Examples of Precision Agriculture for Input Use Efficiency
Egypt	• Tomatiki is a start-up in the area of precision agriculture applying smart irrigation solutions [15,49]. • Innovation is a start-up producing large-scale, automated fish feeders for the fishery sector [15].
GCC	• ICARDA has developed digital technologies for precision management of date palm plantations [15].
Iraq	• Digital technologies have been adopted to support the application of drip irrigation [15].
Jordan	• Digital technologies have been adopted to estimate evapotranspiration and monitor agricultural water use, including within irrigation systems [15].
Lebanon	• The wine industry has adopted digital technologies including drone- and sensor-based IoT to assess growing conditions, water stress, vine performance, and disease [15].
Saudi Arabia	• The public Sustainable Agriculture Research Centre conducts applied research on innovative techniques for protected agriculture, including water use efficiency and pest management [15].
Tunisia	• Start-up companies including Ezzayra offer precision agriculture technologies [15,49]. Ezzayra is a private company offering management solutions in the form of sensors to monitor and regulate soil conditions (including irrigation and fertilization), as well as control irrigation leaks [55].
United Arab Emirates	• Responsive Drip Irrigation is a start-up company developing self-regulating irrigation technologies [56].

Extension Services: Across MENA, digital platforms for knowledge dissemination could play a useful role in improving the quality of extension services. Existing extension services are currently insufficient to meet farmers' needs in many countries, offering limited farm visits and outdated information that excludes new production technologies [26]. E-extension services (public or private) may address constraints of traditional extension services and provide cost-effective ways to reach more farmers. E-extension can supply farmers with real-time access to relevant data and information to support sustainable farming practices, climate-smart solutions, and market access [6,51], with the ability to customize services based on farmer characteristics such as location, educational background, or financial situation. Digital technologies that incorporate use of photographs or video—for example, to deliver remote pest and disease diagnostics—may complement to field advisory visits but may not yet offer a complete substitute. In the context of COVID-19, e-extension services could expand to include new information on personal hygiene and safe handling of food products [57]. In Lebanon, the Facebook group *Izraa* was established by agricultural engineers to offer free advice, tutorials, and questions-and-answers for a variety of agricultural contexts (rural or urban, high- or low-technology) during the pandemic and has since reached more than 100,000 members [58].

Meta-analyses report that the transmission of agricultural information through mobile technologies can increase yields by 4% and adoption of recommended agrochemical inputs by 22% [59]. Benefits of e-extension services significantly exceed the costs, and this gap should widen further as the cost of information transmission continues to fall. Nevertheless, evaluation is still ongoing as to which of the various forms of agricultural e-extension services work best and under what conditions [6].

Resilient Production and Risk Mitigation: Digital technologies may support improved agri-food sector performance throughout or in response to shocks or crises, whether natural or manmade. The combination of remote sensing and big data applications is increasingly used to enhance agri-food management and to mitigate risk. These applications include automated early warning systems for crop or livestock health linked to threats from weather, pests, and diseases, which can facilitate proactive and timely management responses as well as generate crop yield projections [51]. Applications can be used by farmers and along the entire value chain including input suppliers, logistics providers, market actors, and policymakers [26,60].

Digital technologies can also mitigate financial risks. Access to tools targeting weather-related risks is limited or out of reach for many farmers and small-to-medium size businesses in MENA. Remote sensing technologies can reduce the monitoring costs of traditional insurance contracts, improve the contracts available to smallholder farmers, and reduce their costs [26]. Elsewhere, smart contracts—blockchain-enabled arrangements that contain all information about the contract terms and execute all envisaged actions automatically—can provide a risk mitigation tool that is flexible, low-cost, secure (traceable and irreversible), and highly customizable to a variety of risks and payouts, with minimal transaction costs and without the need for third parties. Smart contracts are already in use for weather-based crop insurance, with a number of digital tools reaching the market in recent years [61]. The continued development and spread of these technology-backed product offerings will require legal-regulatory reforms (both agricultural and financial) in many MENA countries [26].

Access to Finance: The digitization of financial transactions and payments can foster inclusion of unbanked farmers. Digital payment systems, mobile phone-based financial and insurance products, and big data analysis can reduce the cost of credit. A minor share of smallholder farmers has access to formal credit in many MENA countries, but big data and advanced analytics can reduce the cost of establishing their creditworthiness and assessing insurance risk [51]. These reduced costs may translate into lower interest rates for farmers, expanding access to financial services. Digital solutions can also help smallholder farmers to access family and friends' savings in times of need [26]. In addition, digital financial solutions have been shown to have a positive and significant effect on household input use, agricultural commercialization, and household income (e.g., [51,62,63]). Over the long term, farmer use of e-wallet services can pave the way for adoption of other digital financial services such as input credit, agricultural insurance, and savings accounts [64]. As interoperability (the ability to both pay and receive funds across banks, internet, and mobile platforms) becomes standard, the use of digital payments is facilitated. Within MENA, digital applications are already targeting access to finance, with digital payments systems reported in Jordan, Lebanon, and West Bank and Gaza (see Table S3). In Tunisia, AHMINI targets rural women's inclusion in social protections and expands their access to insurance against work accidents, health insurance, and retirement income [15].

Market Access and Information: Digital marketplaces and e-platforms for agricultural products can link producers directly to consumers, shorten agri-food value chains, expand producers' access to new markets (usually within national borders), reduce food loss, and create new business opportunities for small agricultural producers and small- and medium-sized enterprises (SMEs) [6,65]. Lowering information-related transaction costs through digital technologies helps to expand access to input and output markets and reduce information asymmetries and inefficiencies caused by reliance on market intermediaries [66,67]. Digital marketplaces may improve price transparency, which is particularly needed in markets marked by obscure pricing that is subject to manipulation, for example wholesale markets where traders can exploit information asymmetries to the detriment of farmers [26]; however, digital marketplaces may not prove a panacea if AI and data analytics in fact facilitate price targeting or manipulation [68]. Because e-platforms can match producers and consumers at nearly no cost, they have tremendous potential to overcome past market failures, expand market access, and restructure value chains. To date, the application of e-platforms or e-commerce for international transactions for agri-food products has been hampered by a lack of explicit international guidance regarding e-commerce, food trade, and consumer protection to assist national regulators [69].

An illustrative example comes from Oman. In partnership with the Oman Technology Fund, the government of Oman introduced the Behar digital auction platform to sustain aquaculture production despite the disruption of COVID-19 to traditional, in-person market operations. The platform also facilitates electronic payments for registered accounts. The platform has sustained market activity during COVID-19, while ensuring proper health and safety measures are respected [70]. This and other examples of digital technologies supporting expanded market access and information in MENA countries, including digital marketplaces and e-commerce, are presented in Table 2.

Table 2. Market access and information: illustrative examples of current application in MENA.

Country	Illustrative Examples of Digital Technologies for Market Access and Information
Egypt	• Freshsource is an e-commerce, business-to-business platform linking horticulture producers to vendors in Cairo [15,49]. • Freshfarm is a start-up working as an intermediary between farmers and consumers, facilitating direct purchases to raise farmer incomes and reduce consumer prices [15].
Jordan	• Websites operated by large wholesale and producer associations (Wholesale and Vegetable Market of Greater Amman Municipality, Jordanian Exporters and Producers Association for Vegetables and Fruits) share market price information, as well as export standards, to support both farmer incomes and export performance [15].
Lebanon	• The AgVisor application allows farmers to compare crop market prices and to access a director of actors across value chains [15].
Oman	• The Behar digital auction platform provides an alternative to traditional, in-person market operations during the COVID-19 pandemic and facilitates electronic payments for registered accounts [70].
Tunisia	• Herundo delivers e-commerce services to the National Olive Oil Exporter Association. Its business-to-consumer platform seeks to expand access to export markets, reduce transaction costs, capture economies of scale, and improve branding [15,49].

Transaction Costs: Digital technologies can improve agricultural supply chain organization and management by optimizing aggregators' performance. Smallholder farms increasingly turn to aggregators such as farmer cooperatives and organizations that use digital tools to improve collection, transportation, and quality control [65]. Gathering producers, aggregators, and buyers on one platform increases transaction volumes, raises prices for farmers, creates enterprise opportunities for aggregators, and delivers traceable and better-quality produce to buyers. Digital platforms have been developed to collect and aggregate data on farm production, which networks of village aggregators and buyers then use to market and sell produce. Other applications use competitive bidding to link transport providers and producers at optimal prices [26].

Transparency and Traceability: Food sensing technologies, blockchain (distributed ledger), and e-platforms can be used to improve value chain transparency and traceability [6]. Barcodes and digital scanners to improve traceability have been in increasing use since the 1970s, enhanced by the introduction of higher-capability coding systems (e.g., DataBar, RFID chips) [71]. More recently, distributed ledger technologies such as blockchain have allowed for improved product traceability and integrity, contract certainty, and compliance with SPS requirements [61]. Blockchain technology prevents data from being improperly altered, which helps to ensure transparency and thereby build trust among retailers and consumers located at the end of decentralized supply chains [72]. Improved traceability of food products can reduce the lost productivity and medical expenses associated with inadequate food safety, which imposes costs estimated at $110 billion annually across low- and middle-income economies [51,73].

Food Loss and Waste: Digital technologies have the potential to reduce food loss and waste in food systems. For example, blockchain-enabled traceability could reduce food loss by 1–2% within the global food system by 2030 [74]. Online, blockchain-supported marketplaces and smart food loss management systems have been designed to facilitate the sale of surplus food directly to end-consumers either as food or for non-edible uses (e.g., compost, alternative processing) and avoid its disposal [75]. In Egypt, e-commerce technologies connecting farmers to modern markets and supporting sustainable agriculture practices have enhanced their efforts to tackle food loss and empower farmers during the COVID-19 pandemic [76].

Food Safety Compliance and Control: Digitally enhanced traceability could help producers to expand exports to more demanding and more lucrative markets that impose more demanding food safety standards, such as European Union countries. Producers can adopt digital technologies to ensure compliance with maximum residue levels for pesticides and with standards preventing microbial contamination. If maintained over the longer-term, digital traceability systems could assist and guide producers to undertake needed improvements in the quality, food safety, and sustainability standards of their production. Moreover, moving directly from rudimentary solutions to digital technologies for improved traceability may be an opportunity to leapfrog intermediary technologies [26].

Public actors can use digital technologies including DLT to improve food control systems. Food safety data collected via remote technologies can inform the development of risk-based strategies for inspection along food chains, which is a more cost-effective approach to food safety control than universal inspection. The use of e-certification for electronic transmission of SPS data can improve the accuracy and reduce the cost of international food trade, as compared to traditional paper-based systems [69].

Data Collection and Dissemination: The use of digital technologies to collect and disseminate available data on a timely basis has been identified as an opportunity in multiple MENA countries, especially where data gaps are large. Remote sensing may offer a powerful and cost-effective means to assess agricultural production in inaccessible areas, such as conflict-affected areas, across a wide scale and over time. Efforts to use digital technologies to improve both the collection and the publication of accurate and timely data specific to the agriculture sector could support the development of evidence-based policies. The timely provision of accurate data could also help identify specific challenges facing the agriculture sector. In the context of COVID-19, digital delivery of market information and pricing platforms could allow for transparency while reducing physical contact. Frequent digital monitoring and/or real-life crowdsourcing of food prices can allow taking legal action against any vendor manipulating prices during the crisis [26]. For example, Qatar's Ministry of Commerce and Industry maintains a website sharing daily price information for major food commodities (fish, fruit, and vegetables) [77] and has operated a multi-channel consumer complaints line as a means to monitor and control price hikes linked to both trade embargoes as well as the COVID-19 pandemic [78].

Monitoring and Evaluation: The adoption of digital technologies for agricultural data generation can support more cost-effective monitoring and evaluation of government programs. For example, satellite imagery provides timely, consistent, and unbiased information on whether agricultural investments and development projects are sustainable and effective [51]. Enhanced aerial imagery allows plot-level monitoring; drone-based inspections; and sensor-based plant, water, and soil analyses. Remote sensing technologies significantly reduce the cost and time required to monitor land cover and land use, elevation, soils, and watersheds, especially in remote areas [2]. In conflict-affected areas, satellite data can support intervention targeting, analysis, and advance planning [26].

3.2.2. Social Contribution

Employment: Digital technologies—notably automation and robotics—have a wide range of applications in the agri-food sector, ranging from surveying to planting to livestock monitoring to food delivery. These applications could deliver a dizzying array of social impacts including safer working conditions but could also reduce labor requirements with corresponding impacts on employment [10]. Indeed, the application of digital technologies is often assumed to displace labor within the agriculture sector. This concern may be overstated as displacement may apply only to selected technologies, specifically digital technologies that can replace manual labor within capital-intensive agricultural operations. For example, robotic technology could replace manual labor for tasks such as weeding, harvesting, and milking. Other digital technologies may be labor enhancing, increasing the efficiency of agricultural labor and the productivity of agricultural operations in terms of lower production costs, higher yields, reduced loss, and higher revenues. Digital technologies can also enable (smallholder) farmers to upgrade their skills, encouraging inclusion of low-skilled farmers and helping enhance their productivity. Additionally, the knowledge needed to develop and operate digital technologies for agriculture may generate new employment opportunities. Thus, the modernization of the agriculture sector could create more productive, skill-intensive, and remunerative jobs in the agricultural sector and along value chains [6].

Digital Identification: The digital identification of farmers offers several opportunities for improved social outcomes. Identification provision can support the transition from informality to formality, linking farmers to their assets (land, livestock) and increasing access to financial services. Moreover, farmers' identification can support agricultural entrepreneurs and digital solution providers who currently invest roughly half of their initial business development efforts in profiling and identifying target farmers [6,26].

Subsidy Distribution and Social Safety Nets: Governments can use digital identification systems to more effectively target input and cash subsidies to farmers, create digital farmer profiles to improve service delivery, and open up new economic opportunities for the poor. For example, the Nigerian government applied an e-wallet digital payment program for subsidized fertilizers that reduced leakage compared to its previous subsidy system [79], increasing efficiency significantly. International organizations are similarly applying digital payments for efficient distribution of humanitarian and development assistance. In Jordan, a partnership between Making Cents International and BanQu has applied blockchain to deliver digital identification to vulnerable populations for facilitate delivery of social, financial, and health services [15]. Digital technologies can also facilitate the distribution of social safety net payments with improved efficiency. In a context such as COVID-19, the use of digital payments or e-vouchers can further reduce human contact and thereby preserve the health of beneficiaries and government employees.

3.2.3. Environmental Contribution

Reduced Use of Scarce Natural Resources: Through the application of digital technologies, the more efficient use of natural resources including water and land can, in theory, lead to use reductions and thereby more sustainable outcomes. However, this is not guaranteed: The optimization of resources such as groundwater may perversely lead to their expanded use if limits are not imposed, a phenomenon known as rebound use [80]. While precision agriculture and digitally-enhanced controlled agricultural production in MENA are reportedly motivated by water scarcity concerns [15], evidence of their net water impact is not currently available.

Reduced Use of External Inputs: On-farm, digital technologies including precision agriculture applications can achieve positive environmental effects through reducing excessive or inappropriate input (fertilizer, pesticides) use [6] that contributes to unnecessary soil and water pollution, energy use, or GHGE. However, a critical perspective of digital agriculture argues that it cannot deliver sustainable agriculture rooted in agroecological approaches, insofar as it still relies on the use of agrochemicals [12]. Despite multiple examples of precision agriculture across MENA (see Table S3), evidence on their ability to reduce the use of external inputs is not currently available.

Reduced GHGE: On-farm, precision agriculture can reduce in-field fuel use, use of nitrogen fertilizers, and land tillage, the latter two of which can have positive effects on carbon sequestration in soils. All three shifts may contribute to lower GHGE [11,12]. Downstream, ICT and sensors may optimize transport logistics within agri-food chains, reducing fuel usage and delivering environmental benefits via reduced carbon footprint. Whether e-ecommerce can similarly deliver GHGE savings depends on the efficiency of transportation, and the available empirical evidence of net impacts is mixed [11].

Biodiversity and Ecosystem Services Impacts: Proponents of precision agriculture argue that the improved efficiency offered by these technologies can reduce pressure on natural resources and limit the clearing of land [12], which can have positive effects on the maintenance of biodiversity and ecosystem services [18]. Additionally, technology-assisted reductions in the use of chemical pesticides and herbicides may slow the development of resistance among animal and plant pests [13]. Moreover, empirical evidence shows that precision agriculture can deliver higher crop yields with positive or at least neutral impacts on ecosystem services (notably water flow regulation and soil structure and fertility enhancement), versus conventional intensive farming techniques [81].

Aggregate View of Agriculture and Evidence-Based Policies: Insofar as sustainable agri-food systems are knowledge-intensive [11], they may rely on extensive data and information derived from a range of sources, including digital technologies such as ICT and remote sensing. Over time, the accumulation of significant data from digital technologies could provide a clearer understanding of the impacts of intensive agriculture and the current industrial food system on the environment, as well as the effects of climate change on agricultural production. This, in turn, could inform a shared understanding of sustainable agriculture and policies to promote it, or could motivate action to combat climate change [82]. For example, monitoring through satellite technology could enable governments to assess how agricultural practices affect the ecosystem, develop better regulations, enforce sustainable land management practices, and address vulnerability to climate change [51].

3.2.4. Summary of Contributions within MENA Sub-Regions

Gulf Cooperation Council (GCC): The GCC includes Bahrain, Kuwait, Oman, Qatar, Saudi Arabia, and the United Arab Emirates. According to a recent analysis by the FAO, the GCC region is most advanced in the adoption of (largely but not only public) e-extension services with respect to digital agriculture technologies, though farmer uptake may be limited. Public authorities in Qatar similarly offer several e-government solutions to facilitate agricultural service delivery, such as for veterinary health certificates. Remote sensing (satellite imagery) and crop modelling have been adopted for high-value export crops including date palms, with a focus on improved farm management, pest and disease prevention, and irrigation management. Another key area for the application of digital technologies in GCC countries is within controlled-environment agriculture (e.g., greenhouse production) [15], reflecting the challenging natural conditions for agricultural production.

The COVID-19 pandemic has pushed GCC countries to boost digital agriculture, with notional and concrete advances reported in 2020 [83]. Significant investments in agricultural technologies—including digital agriculture—have been announced across the GCC [84,85]. Notable are sizeable investments by the sovereign wealth funds of Kuwait and the United Arab Emirates [56]. ICARDA has established more than 2000 plots for soilless agriculture using digital applications for remote monitoring and control across the GCC sub-region [86]. Despite the limited advance of digital agriculture thus far, public actions have signaled support for further adoption. Selected GCC countries have hosted events to exchange knowledge, to launch collaborative efforts (e.g., a joint FAO-KSA-UAE regional kickoff workshop for the Research Technology and Innovation project), and to jump-start investment in emerging digital technologies (e.g., UAE hosted an international conference for agricultural technology innovation) [87].

Hurdles limiting the adoption of digital agriculture by GCC farmers include a lack of awareness, ill-suited digital solutions to farmer needs, and farmers' disinterest in digital agriculture [15].

Maghreb: A recent diagnostic of digital agriculture in the Maghreb determined that it has a limited presence, though there have been interesting developments [49]. The FAO has reported that Egypt, Morocco, and Tunisia have begun to adopt digital agriculture technologies including precision agriculture (particularly for irrigation), e-extension services (both public and private), e-finance services for farmers, and e-commerce applications to expand market access, although their development is not consistent across countries. For example, while Morocco has well-developed e-finance services, e-commerce is nearly non-existent [15]. As for e-government, Egypt's public efforts have met with limited citizen adoption (only 2% by 2019) [7]. Little or no information has been reported regarding digital agriculture in other Maghreb countries. In Algeria, precision agriculture has been touted as an important contribution to large-scale farming projects in the south of the country [88], but details about current implementation are unavailable. Similarly, Sudan's Gedaref Digital City Organization has announced ambitious plans to apply digital technologies to improve agricultural production [89], but the status of these plans is unknown. In the case of Mauritania, the gap in information is unsurprising given barriers including exceptionally high costs of mobile and internet services and low levels of adult literacy [7].

Maghreb countries have accelerated this adoption during the COVID-19 pandemic. For example, in mid-2020 Egypt announced a cooperation protocol between its Ministry of Agriculture and Land Reclamation and the Ministry of Communications and Information Technology to develop irrigation practices incorporating AI, ICT, and remote control for enhanced water resource use and management [90].

Hurdles to the adoption of digital agriculture in Maghreb countries include farmers' resistance to innovative technologies [55], a lack of digital services and broadband in rural areas, legal and regulatory obstacles, and logistical and administrative barriers to entry [49,76].

Mashreq: Evidence shows that adoption of digital technologies within the agri-food sector in the Mashreq sub-region (specifically Iraq, Jordan, and Lebanon) is at early stages, generally led by high-value and export-oriented agricultural production. Nevertheless, digital technologies show promise to advancing the transformation of the agri-food sector in Mashreq. Indeed, governments have at least acknowledged this potential, and the digital transformation agenda already features in the respective countries' economic and sector development strategies [26].

As in other sub-regions, the advance of digital agriculture has accelerated since COVID-19. For example, Jordanian companies and start-ups are increasingly offering technological services to meet the needs of the agricultural sector [91].

The adoption and development of digital agriculture in Iraq, Jordan, and Lebanon are limited by gaps in awareness and understanding of digital technologies and their potential use within the agriculture sector, low levels of digital connectivity among farmers, mismatches in the language used by digital technologies (English) and by some farmers, and inadequate access to finance and data [26].

Conflict-Affected States: Little information is available on the status of digital agriculture in Somalia, Syria, West Bank and Gaza, and Yemen. Notable examples demonstrate the ability of digital technologies to respond to conflict-related challenges, such as the application of remote sensing to estimate agricultural production and conflict-related impacts on agricultural performance in Syria [92,93] and e-learning solutions for agricultural education [94].

Small Countries: Little information is available to suggest a robust adoption of digital agriculture in Comoros or Djibouti. For example, a recent, comprehensive review of the agricultural sector in Comoros made a single mention of digital technologies, highlighting the lack of mobile money or digital credit, which limits access to finance in rural areas [95].

3.3. Challenges to and Risks of Digital Technology Adoption in Agriculture in MENA

3.3.1. Challenges

The adoption of digital technologies within the agri-food sector is not automatic, and a number of hurdles have been identified. An exhaustive review of the conditions and drivers of successful technology adoption and application is beyond the scope of this article, but readers are referred to existing works in the areas of precision agriculture [50,96,97]; and big data [98,99]. Several of the key hurdles within the MENA region are reviewed here.

Cost and Profitability: High purchase, operating, and/or maintenance costs can impede the adoption of digital technologies, particularly by smaller farmers or SMEs who may lack access to credit or other financial resources [100]. Moreover, the financial benefits of adoption must be clear: Evidence shows that profitability is a fundamental condition for the adoption of new technologies by farmers, particularly smallholder farmers [96,101].

Technical Knowledge or Expertise: A lack of technical expertise to understand or utilize digital technologies (or the data that they generate) can restrict their adoption, particularly by smaller farmers or those on the weaker side of the digital divide [82].

Suitability to Local Conditions: Digital agriculture resources have often been developed outside of the MENA region, adapted to the needs of farmers in other areas or production systems, and are often available (only) in English, which can limit their use and uptake in MENA countries. For successful adoption and application, digital technologies and services need to be well suited to application in local conditions, including languages [102]. For example, e-extension services must provide timely, localized, and customized information addressing specific farming concerns in a comprehensible format and in Arabic or another relevant local language [51]. Low-performance, frugal innovations may be developed to meet the capacities and needs of less sophisticated or smaller actors [96].

Complementary Technologies: The absence of complementary technologies—notably minimum data system or internet capacities—can challenge the adoption of digital technologies in agriculture [98]. For this reason, digital agriculture applications that can operate at a low and medium internet connectivity are especially well suited to connect smallholders along agri-food value chains. Even in low-connectivity, rural environments, many (off-line) digital agriculture technologies can be deployed to support poor or illiterate farmers or marginalized groups with limited access to information and markets [51].

Underlying Constraints: The application of digital technologies alone may have a limited effect on expanding market access to farmers, unless underlying constraints are simultaneously are addressed. For example, in Lebanon, the creation of digital platforms for pricing information did not address the fact that farmers were engaged in complex relationships with both input dealers and wholesale traders. Crop farmers benefitted from informal credit arrangements with both input suppliers, who provided seeds, fertilizers, etc., on credit and received payment at the end of the agricultural season and with wholesale purchasers, who paid up-front upon receipt of the produce. These informal financial arrangements limited the ability of a farmer to shift among input suppliers or wholesale purchasers, regardless of the ability of farmers to gain price advantages elsewhere in the market [26].

3.3.2. Risks

As any major technology change, the digital transformation of the agri-food sector incurs risks—or even negative future outcomes—that must be understood, acknowledged, and managed [3,17]. The various risks may relate to the use of the internet, including inequality in access and affordability; concentration of market power of e-commerce platforms, social networks, and search engines; data privacy and consumer protection; and cybersecurity [103]. Several of the potentially most critical risks for the MENA region are highlighted below.

Labor Displacement: Digital technologies may displace agricultural labor with labor saving technologies [1,104]. Such technologies may include harvesting robots, driverless tractors, sprayer drones, AI to manage chemical and fertilizer application, and precision dairy farming. Conversely, the adoption of digital technologies to support automation within the agriculture sector could create high-skilled, high-paying jobs such as managing and maintaining robots or analyzing and interpreting data collected from digital sources and AI [104].

Adoption Gaps and Widening Inequality: Given the differences in the challenges they face, or the extent to which they face them, smallholder and larger, commercial farmers have different potential for adoption of digital technologies. It is important that small and large farmers and businesses alike should benefit from digital innovation [82]. Furthermore, digital innovations should not deepen the inequalities that typically disadvantage women, youth, refugees, and other vulnerable groups who may not have equal access to technologies and skills.

Non-Adoption: Conversely, there are implications for MENA countries associated with the non-adoption or low adoption of digital technologies [3] as their agriculture sectors may lose competitiveness and fall further behind other countries. Already the lack of competitiveness (not solely due to non-adoption of digital technologies) has limited the ability of agriculture producers in Iraq, Jordan, and Lebanon to sell their products to lucrative export markets [26]. In MENA countries that will pursue digital agriculture, public and private actors alike will need to partner and keep pace with fast-moving technology actors [49].

Data Ownership, Privacy, and Security: Farmers and operators may have valid concerns over the ownership, privacy, security, and equitable use (and re-use) of the data produced from digital technologies, and these concerns can limit their willingness to adopt them for use on their farms [82,99,105].

Resource Use and Waste: While digital technologies may offer important environmental benefits, including through the optimization of scarce inputs including water and energy, they may also generate new streams of resource use and waste, including GHGE related to increased energy use [10], energy-intensive data storage or the waste of electronic or digital materials. The environmental impacts of digital technologies adopted by the agriculture sector in MENA are as-yet unknown. The importance of this impact on the development of digital agriculture and opportunities to mitigate these impacts remain areas for future inquiry [26].

4. Discussion

The evidence reviewed above amply demonstrates that the digital transformation is applicable to the agri-food system at all stages of the value chain. Digital technologies such as precision agriculture, e-extension services, and digital markets matching inputs, producers, and/or consumers have tremendous potential to increase productivity on farm and within agri-food value chains, improve resource use efficiency and support adaptation to climate change, and thereby promote sustainable development within MENA. Moreover, digital technologies can help to reduce cost and increase efficiency and transparency of all public services, including those related to the agri-food system, for example by digitalizing land tenure mapping and registration, subsidy distribution, weather forecasting, and water resource management [6].

The potential economic contribution of digital technologies is presented in the literature more than either the social or the environmental contribution. While there is significant concern that digital technologies will exacerbate inequalities or contribute to labor displacement, the evidence appears to be mixed, perhaps due to the relatively early stage of digital technologies and the longer-term scale on which social phenomena play out. Similarly, the examination of potential applications of digital technologies for improved environmental sustainability—to reduce if not eliminate damaging environmental impacts as well as to generate benefits such as ecosystem services [106]—within the agriculture sector is only emerging. Moreover, evidence on whether this potential environmental contribution will materialize is in its early stages [17]. This observed lag of the social and environmental dimensions is consistent with previous research [3,17].

Neither the adoption nor the benefits of digital technologies are guaranteed, and digital technologies may not fully deliver the promised transformation for the agricultural sector [17]. Digital technologies could accelerate the depletion of natural resources and increase the absolute value of greenhouse gas emissions from agricultural production—the so-called "rebound effect" because efficiency gains could lead to increased use [80]. In terms of equity, the use of advanced digital technologies in the agriculture sector may widen the digital divide, if smallholders are unable to make use of these technologies due to a high cost and specialized skills needs [82].

Across MENA, there are some notable examples of where digital technologies have been introduced into the agri-food sector, marked by clustering around public services, high-value, export-oriented value chains (in Maghreb and Mashreq), and intensive production to satisfy local market demand (in GCC). The lack of evidence and reports of adoption suggests, unsurprisingly, that the digital agriculture transformation is less advanced in conflict-affected and less developed countries.

4.1. Policy Implications and Priorities

With the appropriate policy support, the adoption and application of digital technologies within MENA's agri-food sector could help to transform it into a source of improved economic growth, social inclusion, and environmental sustainability. Without the appropriate policy support, digitalization may disrupt the sector in adverse ways such as reducing employment, widening inequalities, and further exploiting already scarce resources. Policymakers should take prompt, comprehensive, and thoughtful action to ensure that the digital transformation of agriculture is to the collective benefit of both stakeholders and societies. Policies must not only foster the adoption of digital technologies in MENA, but also address concerns around equity of access, transparency of use, data protections, and protections against adverse labor impacts. Public action should respond to the actual and urgent needs and should build on the most promising opportunities for digital transformation of agriculture in the region as reviewed above.

Vision and Strategy for Digital Agriculture: Few if any countries in the MENA region have a comprehensive vision or strategy for digital agriculture [7]. Accordingly, they those that have not yet done so, should develop a vision and strategy for the role of digital technologies within their respective agricultural sectors and wider agri-food systems. Such visions and strategies should highlight the potential contribution of digital agriculture to outcomes like food security but also acknowledge that technology-based, production-focused solutions are only partial and will not resolve parallel concerns such as problematic food access and distribution [3,5,18]. Moreover, responsible policymaking should acknowledge and articulate the trade-offs associated with innovative technologies such as digital agriculture [10]. These visions and strategies should reflect stakeholders' and the wider society's ethical perspectives on the application of digital agriculture [17,82].

Strategies aiming to promote the adoption of digital technologies at the production level should consistently consider the role of both upstream and downstream contexts and actors. Such an approach would reflect the shifting agri-food landscape: "Increasingly, the urban market, the food industry firms that mediate access to the urban market, input supply chains, and agribusiness firms that determine the development of input supply chains, set the market incentives and conditions for the affordability and profitability of new farm technologies, and thus their adoption" [107] (p. 48). This consideration is particularly relevant in MENA, given the greater dynamism of downstream segments of the agri-food sector in recent decades [22].

Strategies should be complemented with appropriate action plans and targets for implementation. Strategies, action plans, and targets should be rooted in the concrete needs of potential users and in a clear understanding of both the barriers and incentives to the adoption of digital technologies by farmers and other actors along the agri-food value chain. Within these strategies, interventions must help ensure that small farmers and small businesses benefit from digital innovation in the same way as larger operators. Similarly, gender-sensitive approaches are needed to ensure that disparities and women's lack of empowerment within the agricultural sector are addressed or, at a minimum, not widened [26].

Knowledge and Skills Development: Public actions to expand and enhance appropriate knowledge and skills can facilitate the adoption and/or development of digital agriculture technologies, including via improved education [101]. Farmers and agricultural stakeholders may require support around basic digital literacy, business and farm operations management, and the use of customer-facing technologies like digital marketplaces. Within agricultural universities and vocational training programs, exposure to digital technologies within the curriculum can improve adoption, adaptation, and development of locally appropriate technologies. Elsewhere, entrepreneurs and start-ups innovating in the digital technology space may require knowledge and skills around business development and finance, as well as a better understanding of the challenges facing the agri-food sector [26]. Education initiatives to facilitate the adoption of digital technologies must balance the optimization of resources (e.g., targeting actors who show higher likelihood to adopt) [97] with attention to widening disparities.

Data Ownership, Use, and Privacy Protections: To address serious and valid concerns over data ownership, use, and privacy linked to digital technologies, policies can help to establish and clarify the rights of different stakeholders. While such policies are typically not specific only to agricultural technologies or data [99,108], there is a need to balance the privacy and legal protections for the entities (farmers and firms) that generate data, with the potentially tremendous benefits to be gained from the aggregation and analysis of data and application of the resulting information. Indeed, open access to large datasets could facilitate critical evaluation of smart farming, or even enhance entrepreneurship and economic activity [82]. Policies can help to set the frame for positive feedback loops, in which good governance builds user trust in digital technologies. Protections for data integrity can result in greater trust in that data and its application in evidence-based policies and programs [26]. Due to the rapid evolution of digital technologies, public actors should track sector developments to ensure that policies and regulation around digital agriculture reflect current challenges and needs.

Support and Enabling Environment for Private Investment: Both the public and the private sectors will play a role in the digital transformation of agriculture, with private actors playing a particularly important role in developing and adapting digital technologies. Public policies can facilitate private investment in digital agriculture—as well as the mobile and internet infrastructure and services necessary for digital technologies—by offering a supportive enabling environment, investment incentives, and partnership opportunities. Public policies can also foster an innovation ecosystem for digital agriculture technologies and innovation, providing information, networking opportunities, and incubation and acceleration services where the private sector has yet to do so [7,26]. Policies should seek to balance an environment that supports profitable investment opportunities with protections against market power concentration that can stifle innovation [7].

Non-Digital Reforms and Investments: Policy action that is specific to digital technologies may not be sufficient to unlock the transformation of the agricultural sector, if complementary reforms and investments addressing non-digital barriers to the sector are not also addressed [7]. For example, digital technologies that facilitate logistics and target reduced food loss will offer only partial improvements without investments to improve transportation, storage, and power infrastructure, which are badly needed in many lower-income MENA countries. To deliver on the promise of digital technologies to expand access to finance to small farmers or rural areas, improvements to financial sector laws and regulations may be required [26]. Extending market access via e-commerce platforms may require the establishment or improvement of digital payment systems, which build on banking and telecommunications policies as well as consumer protection frameworks [109].

Public Supply of Digital Agriculture Services—E-Government and E-Extension: The public sector can not only foster the demand for digital technologies but can also adopt or supply them directly. Digital technologies may improve efficiency, reduce cost, and increase transparency and accountability of many public services—not only agriculture. Within the agri-food sector, e-government could deliver statistical data collection and dissemination; open data platforms for land, soil, weather, and market price data; e-extension services; digital identification for farmers and other stakeholders; land e-registries; and public subsidy payments [26]. Successful e-government services require administrative capacity for design and management, as well as commitment to the often slow pace of user adoption [7]. Of note, through e-extension, the public sector can support the use of digital technologies that enhance the efficient use of scarce resources (e.g., irrigation water) and optimize the use of inputs (e.g., fertilizer, pesticides) to support productivity, efficiency, and environmental sustainability. E-extension can particularly target the needs of smallholders, addressing their risk aversion, information gaps, and mistrust of technologies or their promised benefits. Techniques including digital piloting and knowledge exchange may be well-suited to the needs of small farmers [26].

4.2. Research Contributions and Conclusion

This article has sought to contribute to the topic of digital agriculture in the MENA region in several ways. First, this article briefly reviewed the key features of the agri-food sector and key challenges to its sustainability including scarcity and inefficient use of natural resources, suboptimal agricultural management practices, poor market functioning, inefficient markets, and limited access to finance. These challenges are compounded by conflict and social unrest, climate change, and insufficient public support for the agri-food sector.

Second, we reviewed the potential contribution of digital agriculture to address these challenges and found significant evidence of positive impacts across all three dimensions of sustainability. While the literature has more clearly mapped and established the economic contribution of digital agriculture, the social and environmental contributions promise to be significant and highly needed within the region.

Third, a review of the current status of digital agriculture in the MENA region generally finds an early stage of adoption, though with some notable sub-regional patterns. Digital agriculture in the GCC is oriented to ensuring a minimum level of domestic food production to ensure social needs, while in Mashreq and Maghreb digital technologies are being applied for both local and export market destinations including EU and GCC. Conversely, digital agriculture in conflict-affected countries appears to be more practically oriented to conduct or deliver basic services at a distance, while maintaining human safety. This review further highlights a gap in knowledge around the state of digital agriculture in several countries of the region, including lower-income countries (e.g., Mauritania, Sudan), small states (e.g., Comoros, Djibouti), and conflict-affected states (e.g., Syria, Yemen). While the absence of information would suggest that digital agriculture is not well developed in these parts of the MENA region, the status of digital agriculture in these countries nevertheless remains an area for future research.

From a thematic perspective, the application and introduction of digital technologies to the agri-food sector—globally and in MENA countries—have tended to focus on economic objectives, with less contribution to addressing the social and environmental challenges facing the agri-food sector. Indeed, evidence of the environmental contribution of digital agriculture in the MENA region is as-yet missing. Though expected, this gap suggests that the application of digital technologies to address the significant social and environmental challenges facing the agri-food sector in the MENA region requires prompt and careful consideration. Insofar as the countries of the MENA region have a responsibility to determine for themselves the future of digital agriculture that they wish for themselves, this consideration is timely.

COVID-19 has catalyzed the introduction of digitalization along the agri-food system, including in the MENA region. Whether this transition will prove temporary or permanent remains to be seen [9]. This introduction was not structured; rather, it was an organic response from market-based actors, rather than a public-led policy decision. This allowed for faster penetration and market-led response to needs, but the disadvantage is that this has been driven by profitability considerations. As a result, we need now to take stock of these experiences—to which this paper contributes—in order to mainstream these experiences into polices that foster inclusive economic performance as well as social and environmental goals, for a true contribution to sustainability.

Finally, this article proposes a number of policy implications and priorities for relevant stakeholders across MENA countries, to unlock the potential of digital agriculture towards sustainability. Policymakers are urged to expand their traditional, production-centric views of the agri-food sector to account for social and environmental considerations and to work towards a digital future in which the agri-food sector not only generates productive value but also delivers on social and environmental sustainability.

Supplementary Materials: The following are available online at https://www.mdpi.com/2071-1050/13/6/3223/s1. Table S1: Search results from Web of Science for "digital agriculture" and "smart farming" in MENA countries. Table S2: Selected economic, social, environmental, and policy indicators. Table S3: Evidence of digital agriculture across MENA countries.

Author Contributions: Conceptualization, R.A.B. and R.Z.; methodology, R.A.B. and R.Z.; analysis, A.A.K.Y., R.A.B., and R.Z.; writing—original draft preparation, R.A.B.; writing—review and editing, A.A.K.Y., R.A.B., and R.Z. All authors have read and agreed to the published version of the manuscript.

Funding: This research received no external funding.

Institutional Review Board Statement: Not applicable.

Informed Consent Statement: Not applicable.

Data Availability Statement: Not applicable.

Acknowledgments: The authors acknowledge the work of the Mashreq Agricultural Initiative—an institutional collaboration between the American University of Beirut, the World Bank Group, the Food and Agriculture Organization/Near East North Africa Regional Office, the United Nations Economic and Social Commission for Western Asia, and the Center for Mediterranean Integration—in inspiring this research and in laying the groundwork for this review.

Conflicts of Interest: The authors declare no conflict of interest.

References

1. Klerkx, L.; Jakku, E.; Labarthe, P. A review of social science on digital agriculture, smart farming and agriculture 4.0: New contributions and a future research agenda. *NJAS Wagening. J. Life Sci.* **2019**, *90–91*, 100315. [CrossRef]
2. OECD. *Digital Opportunities for Better Agricultural Policies*; OECD Publishing: Paris, France, 2019.
3. Klerkx, L.; Rose, D. Dealing with the game-changing technologies of Agriculture 4.0: How do we manage diversity and responsibility in food system transition pathways? *Glob. Food Secur.* **2020**, *24*, 100347. [CrossRef]
4. Velten, S.; Leventon, J.; Jager, N.; Newig, J. What Is Sustainable Agriculture? A Systematic Review. *Sustainability* **2015**, *7*, 7833–7865. [CrossRef]
5. Béné, C.; Oosterveer, P.; Lamotte, L.; Brouwer, I.D.; de Haan, S.; Prager, S.D.; Talsma, E.F.; Khoury, C.K. When food systems meet sustainability—Current narratives and implications for actions. *World Dev.* **2019**, *113*, 116–130. [CrossRef]
6. World Bank Group. *Future of Food: Harnessing Digital Technologies to Improve Food System Outcomes*; World Bank: Washington, DC, USA, 2019.
7. Trendov, N.M.; Varas, S.; Zeng, M. *Digital Technologies in Agriculture and Rural Areas—Status Report*; FAO: Rome, Italy, 2019.
8. FAO. *COVID-19 and Its Impact on Food Security in the Near East and North Africa: How to Respond?* FAO: Cairo, Egypt, 2020.
9. Mittra, B. COVID-19 Pandemic Presents Opportunities for Innovation. In *TCI Blog*; Tata-Cornell Institute for Agriculture and Nutrition: Ithaca, NY, USA, 2020; Volume 2021.

10. Herrero, M.; Thornton, P.K.; Mason-D'Croz, D.; Palmer, J.; Bodirsky, B.L.; Pradhan, P.; Barrett, C.B.; Benton, T.G.; Hall, A.; Pikaar, I.; et al. Articulating the effect of food systems innovation on the Sustainable Development Goals. *Lancet Planet. Health* **2021**, *5*, e50–e62. [CrossRef]
11. El Bilali, H.; Allahyari, M.S. Transition towards sustainability in agriculture and food systems: Role of information and communication technologies. *Inf. Process. Agric.* **2018**, *5*, 456–464. [CrossRef]
12. Clapp, J.; Ruder, S.-L. Precision Technologies for Agriculture: Digital Farming, Gene-Edited Crops, and the Politics of Sustainability. *Glob. Environ. Politics* **2020**, *20*, 49–69. [CrossRef]
13. Srinavasan, A. Precision Agriculture: An Overview. In *Handbook of Precision Agriculture: Principles and Applications*, 1st ed.; Srinavasan, A., Ed.; CRC Press: Boca Raton, FL, USA, 2006; pp. 3–18. [CrossRef]
14. Ronaghi, M.H.; Forouharfar, A. A contextualized study of the usage of the Internet of things (IoTs) in smart farming in a typical Middle Eastern country within the context of Unified Theory of Acceptance and Use of Technology model (UTAUT). *Technol. Soc.* **2020**, *63*, 101415. [CrossRef]
15. FAO. Executive Summary. In Proceedings of the FAO Regional Conference for the Near East: Digital Innovation for Promoting Agriculture 4.0 in the Near East and North Africa, Muscat, Oman, 2–4 March 2020.
16. UNESCWA. *Arab Horizon 2030: Prospects for Enhancing Food Security in the Arab Region*; UNESCWA: Beirut, Lebanon, 2017.
17. Lajoie-O'Malley, A.; Bronson, K.; van der Burg, S.; Klerkx, L. The future(s) of digital agriculture and sustainable food systems: An analysis of high-level policy documents. *Ecosyst. Serv.* **2020**, *45*, 101183. [CrossRef]
18. Garnett, T.; Appleby, M.C.; Balmford, A.; Bateman, I.J.; Benton, T.G.; Bloomer, P.; Burlingame, B.; Dawkins, M.; Dolan, L.; Fraser, D.; et al. Sustainable Intensification in Agriculture: Premises and Policies. *Science* **2013**, *341*, 33–34. [CrossRef]
19. World Bank. Databank: World Development Indicators. Available online: https://databank.worldbank.org/source/world-development-indicators# (accessed on 26 January 2021).
20. Haggblade, S.; Hammer, J.; Hazell, P. Modeling Agricultural Growth Multipliers. *Am. J. Agric. Econ.* **1991**, *73*, 361–374. [CrossRef]
21. Haggblade, S.; Hazell, P.; Dorosh, P. Sectoral Growth Linkages between Agriculture and the Rural Nonfarm Economy. In *Transforming the Rural Nonfarm Economy—Opportunities and Threats in the Developing World*; Haggblade, S., Hazell, P., Reardon, T., Eds.; Johns Hopkins University Press: Baltimore, MD, USA, 2007; pp. 141–173.
22. IFPRI. *2020 Global Food Policy Report: Building Inclusive Food Systems*; IFPRI: Washington, DC, USA, 2020.
23. FAO. *Iraq: Restoration of Agriculture and Water Systems Sub-Programme 2018–2020*; FAO: Rome, Italy, 2018.
24. HLPE. *Impacts of COVID-19 on Food Security and Nutrition: Developing Effective Policy Responses to Address the Hunger and Malnutrition Pandemic*; FAO: Rome, Italy, 2020.
25. FAO. *Special Report—FAO Mission to Assess the Impact of the Financial Crisis on Agriculture in the Republic of Lebanon*; FAO: Rome, Italy, 2020.
26. Bahn, R.; Juergenliemk, A.; Zurayk, R.; Debroux, L.; Broka, S.; Mohtar, R. *Harnessing the Power of Digital Agriculture Transformation in Mashreq*; American University of Beirut; World Bank; FAO; UNESCWA; CMI: Washington, DC, USA, 2021.
27. World Bank. Databank: Poverty and Equity. Available online: https://databank.worldbank.org/source/poverty-and-equity (accessed on 26 January 2021).
28. UNHCR; UNICEF; WFP. *VASyR 2019: Vulnerability Assessment of Syrian Refugees in Lebanon*; UNHCR: Beirut, Lebanon, 2019.
29. International Organization for Migration. *Integrated Location Assessment III*; IOM: Baghdad, Iraq, 2018.
30. Yeboah, F.K.; Jayne, T.S. Africa's Evolving Employment Trends. *J. Dev. Stud.* **2018**, *54*, 803–832. [CrossRef]
31. HLPE. *Nutrition and Food Systems. A Report by the High Level Panel of Experts on Food Security and Nutrition of the Committee on World Food Security*; FAO: Rome, Italy, 2017.
32. FAO; IFAD; UNICEF; WFP; WHO. *The State of Food Security and Nutrition in the World: Transforming Food Systems for Affordable Healthy Diets*; FAO: Rome, Italy, 2020.
33. World Bank. *Iraq Economic Monitor, Fall 2019: Turning the Corner—Sustaining Growth and Creating Opportunities for Iraq's Youth*; World Bank: Washington, DC, USA, 2019.
34. FAO; IFAD; UNICEF; WFP; WHO. *Regional Overview of Food Security and Nutrition in the Near East and North Africa 2019—Rethinking Food Systems for Healthy Diets and Improved Nutrition*; FAO: Cairo, Egypt, 2020.
35. Bahn, R.A.; Zurayk, R. Agriculture, conflict, and the agrarian question in the 21st century. In *Crisis and Conflict in Agriculture*; Zurayk, R., Woertz, E., Bahn, R.A., Eds.; CABI: Boston, MA, USA, 2018; pp. 3–27.
36. Willett, W.; Rockström, J.; Loken, B.; Springmann, M.; Lang, T.; Vermeulen, S.; Garnett, T.; Tilman, D.; DeClerck, F.; Wood, A.; et al. Food in the Anthropocene: The EAT–Lancet Commission on healthy diets from sustainable food systems. *Lancet* **2019**, *393*, 447–492. [CrossRef]
37. FAO. *AQUASTAT Main Database*; FAO: Rome, Italy, 2016.
38. World Resources Institute. 17 Countries, Home to One-Quarter of the World's Population, Face Extremely High Water Stress. Available online: https://www.wri.org/blog/2019/08/17-countries-home-one-quarter-world-population-face-extremely-high-water-stress (accessed on 26 January 2021).
39. FAO. *Progress on Water Use Efficiency—Global Baseline for SDG 6 Indicator 6.4.1*; FAO/UN-Water: Rome, Italy, 2018.
40. Ben Hassen, T.; El Bilali, H. Food Security in the Gulf Cooperation Council Countries: Challenges and Prospects. *J. Food Secur.* **2019**, *7*, 159–169. [CrossRef]
41. Gibson, G. Effects of the Islamic State of Iraq and Syria on Cropland Area. *J. Food Secur.* **2019**, *7*, 20–27. [CrossRef]

42. FAO. FAOStat: Emissions Shares—Agriculture Total—Share of total emissions (CO2 eq) (AR5). Available online: Http://www.fao.org/faostat/en/#data/EM (accessed on 27 January 2021).
43. AFED. *Arab Environment: Food Security. Annual Report of the Arab Forum for Environment and Development, 2014*; AFED: Beirut, Lebanon, 2014.
44. Lewis, P.; Monem, M.A.; Impiglia, A. *Impacts of Climate Change on Farming Systems and Livelihoods in the Near East and North Africa—With a Special Focus on Small-Scale Family Farming*; FAO: Cairo, Egypt, 2018.
45. Adaptation, G.C. *Adapt Now: A Global Call for Leadership on Climate Resilience*; Global Center on Adaptation & World Resources Institute: Rotterdam, The Netherlands, 2019.
46. FAO. Sustainable Development Goals: The Agriculture Orientation Index for Government Expenditures. Available online: http://www.fao.org/sustainable-development-goals/indicators/2a1/en/ (accessed on 26 January 2021).
47. World Bank. Enabling the Business of Agriculture (Dataset). Available online: https://datacatalog.worldbank.org/dataset/enabling-business-agriculture (accessed on 26 January 2021).
48. Woetzel, L.; Remes, J.; Boland, B.; Lv, K.; Sinha, S.; Strube, G.; Means, J.; Law, J.; Cadena, A.; von der Tann, V. *Smart Cities: Digital Solutions for a More Livable Future*; McKinsey Global Institute: Shanghai, China, 2018.
49. Bravi, C.; Sylvester, G. Digital Agriculture in North Africa in Time of COVID-19 Crisis. In Proceedings of the Innovation Zoominar I: How Can Digital Innovation Help Smallholder Farmers Cope with COVID-19 Impact in NENA? Cairo, Egypt, 7 June 2020.
50. Kutter, T.; Tiemann, S.; Siebert, R.; Fountas, S. The role of communication and co-operation in the adoption of precision farming. *Precis. Agric.* **2011**, *12*, 2–17. [CrossRef]
51. Kim, J.; Shah, P.; Gaskell, J.C.; Prasann, A.; Luthra, A. *Scaling Up Disruptive Agricultural Technologies in Africa*; World Bank: Washington, DC, USA, 2020.
52. North Africa Post. *FAO Designates Moroccan Innovation as MENA's Best COVID-19 Protection for Farmers*; The North Africa Post: Rabat, Morocco, 2020.
53. Revich, J.; Koort, R.; Archambault, P.; Samuelson, A.; Nannizzi, M.; Moawalla, M.; Bonin, A. *Precision Farming: Cheating Malthus with Digital Agriculture*; Goldman Sachs Investment Research: New York, NY, USA, 2016.
54. World Bank Group. *Mashreq 2.0: Digital Transformation for Inclusive Growth and Jobs (Vol. 2) (English)*; World Bank Group: Washington, DC, USA, 2018.
55. Blaise, L. *In Tunisia, a Digital Revolution for Agriculture Takes Root*; Le Monde: Paris, France, 2020.
56. Pothering, J. Amid Covid-19, the Gulf region commits $200M to safeguard its food security with agtech. *AgFunderNews*, 15 April 2020.
57. FAO. *Extension and Advisory Services: At the Frontline of the Response to COVID-19 to Ensure Food Security*; FAO: Rome, Italy, 2020.
58. Izraa. Group of agricultural engineers on Facebook. In Facebook: 2020. 2020. Available online: https://www.facebook.com/groups/izraa.group/ (accessed on 5 February 2021).
59. Fabregas, R.; Kremer, M.; Schilbach, F. Realizing the potential of digital development: The case of agricultural advice. *Science* **2019**, *366*, eaay3038. [CrossRef]
60. Sharma, R.; Shishodia, A.; Kamble, S.; Gunasekaran, A.; Belhadi, A. Agriculture supply chain risks and COVID-19: Mitigation strategies and implications for the practitioners. *Int. J. Logist. Res. Appl.* **2020**, 1–27. [CrossRef]
61. Sylvester, G. *E-Agriculture in Action: Blockchain for Agriculture—Opportunities and Challenges*; FAO & ITU: Bangkok, Thailand, 2019.
62. Kirui, O.K.; Okello, J.J.; Nyikal, R.A.; Njiraini, G.W. Impact of Mobile Phone-Based Money Transfer Services in Agriculture: Evidence from Kenya. *Q. J. Int. Agric.* **2013**, *52*, 141–162. [CrossRef]
63. Jack, W.; Suri, T. Mobile Money: The Economics of M-PESA. *Natl. Bur. Econ. Res. Work. Pap. Ser.* **2011**, 16721. [CrossRef]
64. Tsan, M.; Totapally, S.; Hailu, M.; Addom, B.K. *The Digitalisation of African Agriculture Report 2018–2019*; CTA: Wageningen, The Netherlands, 2019.
65. Deichmann, U.; Goyal, A.; Mishra, D. Will Digital Technologies Transform Agriculture in Developing Countries? World Bank Group: Washington, DC, USA, 2016.
66. Aker, J.C.; Fafchamps, M. Mobile Phone Coverage and Producer Markets: Evidence from West Africa. *World Bank Econ. Rev.* **2015**, *29*, 262–292. [CrossRef]
67. Nakasone, E.; Torero, M.; Minten, B. The Power of Information: The ICT Revolution in Agricultural Development. *Annu. Rev. Resour. Econ.* **2014**, *6*, 533–550. [CrossRef]
68. Babin, B.J.; Feng, C.; Borges, A. As the wheel turns toward the future of retailing. *J. Mark. Theory Pract.* **2021**, 1–14. [CrossRef]
69. World Health Organization. *Digitalization, Food Safety and Trade*; WHO: Geneva, Switzerland, 2019.
70. FAO Regional Office for Near East and North Africa. From bustling Omani Fish Markets to Online Auctions: Digital Innovation is Boosting the Fisheries Supply Chain in Oman during COVID-19. Available online: http://www.fao.org/neareast/news/view/en/c/1294228/ (accessed on 28 January 2021).
71. GS1. How We Got Here. Available online: https://www.gs1.org/about/how-we-got-here (accessed on 26 January 2021).
72. The Seam. Blockchain and Food Safety. In *Agritech Talk*; The Seam: Memphis, TN, USA, 2019.
73. Jaffee, S.; Henson, S.; Unnevehr, L.; Grace, D.; Cassou, E. *The Safe Food Imperative: Accelerating Progress in Low- and Middle-Income Countries*; World Bank: Washington, DC, USA, 2019.
74. World Economic Forum. *Innovation with a Purpose: Improving Traceability in Food Value Chains through Technology Innovations*; World Economic Forum: Geneva, Switzerland, 2019.

75. Annovazzi-Jakab, L. Cutting Food Loss Where It Matters: Leveraging Digital Solutions for Greener Trade and Less Waste—UNECE's Impact Initiatives. In Proceedings of the UN-ESCWA Regional Meeting on Promoting Food and Water Security in the Arab Region, Amman, Jordan, 27–28 November 2019.
76. IFPRI. IFPRI Egypt Webinar "Fostering Digitalization for a Future-Proof Food System in Egypt". Available online: https://egyptssp.ifpri.info/2021/01/19/ifpri-egypt-webinar-fostering-digitalization-for-a-future-proof-food-system-in-egypt-january-26th-2021-330-pm-500-pm-eet/ (accessed on 19 January 2021).
77. Ministry of Commerce and Industry. Commodities Daily Prices. Available online: https://www.moci.gov.qa/en/our-services/consumer/commodities-daily-prices/ (accessed on 5 February 2021).
78. How to Complain about Price Hikes Amid the Coronavirus (COVID) Crisis in Qatar. Available online: https://www.iloveqatar.net/news/general/how-to-complain-about-price-hikes-in-qatar (accessed on 5 February 2021).
79. Grossman, J.; Tarazi, M. Serving Smallholder Farmers: Recent Developments in Digital Finance. *Cgap Focus Note* **2014**, *94*.
80. Perry, C. Accounting for water use: Terminology and implications for saving water and increasing production. *Agric. Water Manag.* **2011**, *98*, 1840–1846. [CrossRef]
81. Garbach, K.; Milder, J.C.; DeClerck, F.A.J.; Montenegro de Wit, M.; Driscoll, L.; Gemmill-Herren, B. Examining multi-functionality for crop yield and ecosystem services in five systems of agroecological intensification. *Int. J. Agric. Sustain.* **2017**, *15*, 11–28. [CrossRef]
82. Van der Burg, S.; Bogaardt, M.-J.; Wolfert, S. Ethics of smart farming: Current questions and directions for responsible innovation towards the future. *NJAS Wagening. J. Life Sci.* **2019**, *90–91*, 100289. [CrossRef]
83. FAO Regional Office for Near East and North Africa. Strengthening Research, Technology and Innovation to Enable Food Security in Saudi Arabia. Available online: http://www.fao.org/neareast/news/view/en/c/1366767/ (accessed on 28 January 2021).
84. Nair, M. *Kuwait's The Sultan Center Ties up with Abu Dhabi's Pure Harvest for 'Hi-Tech' Farm*; Gulf News: Dubai, United Arab Emirates, 2020.
85. Oman Daily Observer. *RO 6.5 m Agreement to Set up Smart Agriculture Project*; Oman Daily Observer: Muscat, Oman, 2021.
86. Ould Belgacem, A. Zoominar II: Innovative Success Stories in the NENA Region: Tools to Support Smallholders under COVID-19 Disruptive Impact. In Proceedings of the International Forum on Innovation in Agri-Food Systems to Achieve the SDGs, Cairo, Egypt, 30 April 2020.
87. FoodTech Challenge. About FoodTech Challenge. Available online: https://foodtechchallenge.com/about-foodtech-challenge/ (accessed on 28 January 2021).
88. Oxford Business Group. Algeria Focuses on Mega-Farms to Speed up Agricultural Development. Available online: https://oxfordbusinessgroup.com/analysis/go-large-focus-mega-farms-speed-development (accessed on 7 February 2021).
89. Eisa, A. Gedaref Digital City Organization Sudan e-Agriculture Project. In My Blog; UniteIT e-Inclusion Network. 2015, Volume 2021. Available online: http://www.unite-it.eu/profiles/blog/list?user=21w449uixv4i (accessed on 28 January 2021).
90. Daily News Egypt. *Egypt Uses AI to Improve Water-Efficient Irrigation*; Daily News Egypt: Giza, Egypt, 2020.
91. Magnitt. *Jordanian Agritech Startup TWIG Secures Seed Funding*; Magnitt: Dubai, United Arab Emirates, 2021.
92. Jaafar, H.H.; Ahmad, F.A. Crop yield prediction from remotely sensed vegetation indices and primary productivity in arid and semi-arid lands. *Int. J. Remote Sens.* **2015**, *36*, 4570–4589. [CrossRef]
93. Jaafar, H.H.; Zurayk, R.; King, C.; Ahmad, F.; Al-Outa, R. Impact of the Syrian conflict on irrigated agriculture in the Orontes Basin. *Int. J. Water Resour. Dev.* **2015**, *31*, 436–449. [CrossRef]
94. Abdullateef, S.; Parkinson, T.; Sarmini, I. Cross border connected learning in northern Syria: An agricultural pilot study. *Int. J. Educ. Res. Open* **2020**, *1*, 100005. [CrossRef]
95. World Bank Group. *The Union of the Comoros: Jumpstarting Agricultural Transformation*; World Bank: Washington, DC, USA, 2019.
96. Feder, G.; Umali, D.L. The Adoption of Agricultural Innovations: A Review. *Technol. Forecast. Soc. Chang.* **1993**, *43*, 215–239. [CrossRef]
97. Fuglie, K.; Gautam, M.; Goyal, A.; Maloney, W.F. *Harvesting Prosperity: Technology and Productivity Growth in Agriculture*; World Bank: Washington, DC, USA, 2019.
98. Pierpaoli, E.; Carli, G.; Pignatti, E.; Canavari, M. Drivers of Precision Agriculture Technologies Adoption: A Literature Review. *Procedia Technol.* **2013**, *8*, 61–69. [CrossRef]
99. World Bank. *ICT in Agriuclture (Updated Edition): Connecting Smallholders to Knowledge, Networks, and Institutions*; World Bank: Washington, DC, USA, 2017.
100. Turland, M.; Slade, P. Farmers' willingness to participate in a big data platform. *Agribusiness* **2020**, *36*, 20–36. [CrossRef]
101. World Bank. *World Development Report 2016: Digital Dividends*; World Bank: Washington, DC, USA, 2016.
102. Gallardo, R.K.; Sauer, J. Adoption of Labor-Saving Technologies in Agriculture. *Annu. Rev. Resour. Econ.* **2018**, *10*, 185–206. [CrossRef]
103. Wolfert, S.; Ge, L.; Verdouw, C.; Bogaardt, M.-J. Big Data in Smart Farming—A review. *Agric. Syst.* **2017**, *153*, 69–80. [CrossRef]
104. Wiseman, L.; Sanderson, J.; Zhang, A.; Jakku, E. Farmers and their data: An examination of farmers' reluctance to share their data through the lens of the laws impacting smart farming. *NJAS Wagening. J. Life Sci.* **2019**, *90–91*, 100301. [CrossRef]
105. Pretty, J.; Benton, T.G.; Bharucha, Z.P.; Dicks, L.V.; Flora, C.B.; Godfray, H.C.J.; Goulson, D.; Hartley, S.; Lampkin, N.; Morris, C.; et al. Global assessment of agricultural system redesign for sustainable intensification. *Nat. Sustain.* **2018**, *1*, 441–446. [CrossRef]

106. Reardon, T.; Echeverria, R.; Berdegué, J.; Minten, B.; Liverpool-Tasie, S.; Tschirley, D.; Zilberman, D. Rapid transformation of food systems in developing regions: Highlighting the role of agricultural research & innovations. *Agric. Syst.* **2019**, *172*, 47–59. [CrossRef]
107. Tey, Y.S.; Brindal, M. Factors influencing the adoption of precision agricultural technologies: A review for policy implications. *Precis. Agric.* **2012**, *13*, 713–730. [CrossRef]
108. Martinez, J. Setting the scene—Opening address. In Proceedings of the Digital Transformation of the Agricultural Value Chain—Opportunities, Challenges and the Role of Science, Berlin, Germany, 2–3 December 2020.
109. Klapper, L. How Digital Payments Can Benefit Entrepreneurs. *Iza World Labor* **2017**, *396*. [CrossRef]

sustainability

MDPI

Article

On Spatially Dependent Risk Preferences: The Case of Nigerian Farmers

Omotuyole Isiaka Ambali [1,2,*], Francisco Jose Areal [2,3] and Nikolaos Georgantzis [2,4,5]

1 Department of Agricultural Economics and Farm Management, Ayetoro Campus, College of Agricultural Sciences, Olabisi Onabanjo University, Ago Iwoye PMB 0012, Nigeria
2 Department of Agricultural and Applied Economics, School of Agriculture, Policy and Development, University of Reading, Reading RG6 6AR, UK; Francisco.Areal-Borrego@newcastle.ac.uk (F.J.A.); nikolaos.georgantzis@bsb-education.com (N.G.)
3 Centre for Rural Economy, School of Natural and Environmental Sciences, Newcastle University, Newcastle Upon Tyne NE1 7RU, UK
4 School of Wine and Spirits Business, Burgundy School of Business, 21000 Dijon, France
5 Laboratorio de Economía Experimental, University Jaume I—LEE-UJI, 12072 Castellón de la Plana, Spain
* Correspondence: tuyoleambali@gmail.com; Tel.: +44-074-2437-4990

Abstract: Rice farmers' attitudes toward risk-taking have been identified as one of the factors affecting investment decisions and wealth accumulation. While existing studies have identified the socio-economic factors driving farmers' risk attitudes, spatial variables that may correlate with decisions are often ignored in the risk models due to difficulties in measurement. We studied unobserved spatial heterogeneity in farmer's risk preferences by incorporating spatial dependency into a farmer's risk preference model. We used data from a survey conducted with Nigerian farmers between March and May 2016. The survey collected information on 2016 farmers' socio-demographic characteristics and farm attributes including its geographical location as well as information on the quality of roads. In addition, a set of experiments design to elicit famers' attitudes toward risk were conducted. We estimated a spatial autoregressive model using the instrumental variable method. We found that unobserved spatial heterogeneity (e.g., soil, topographic farmers emulating each other) was present in farmer's risk preferences along with socio-demographic variables such as age, gender, marital status, and religion and farm characteristics such as farm size and road quality. These results are relevant for policy decision-making processes.

Keywords: decision making; instrumental variable; neighbourhood effects; rice farmers; risk attitudes; spatial dependence

Citation: Ambali, O.I.; Areal, F.J.; Georgantzis, N. On Spatially Dependent Risk Preferences: The Case of Nigerian Farmers. *Sustainability* **2021**, *13*, 5943. https://doi.org/10.3390/su13115943

Academic Editor: Michael Blakeney

Received: 25 March 2021
Accepted: 21 May 2021
Published: 25 May 2021

Publisher's Note: MDPI stays neutral with regard to jurisdictional claims in published maps and institutional affiliations.

Copyright: © 2021 by the authors. Licensee MDPI, Basel, Switzerland. This article is an open access article distributed under the terms and conditions of the Creative Commons Attribution (CC BY) license (https://creativecommons.org/licenses/by/4.0/).

1. Introduction

Farmers' adoption of agricultural practices and technologies that contribute to achieving sustainable intensification, sustainable development, and food security require some degree of risk taking and risk management by farmers. Farmers face climatic shock risks (e.g., flood, drought) and pest and disease risks, as well as market input and output price fluctuations [1]. In addition, the inadequate access to insurance and other risk mitigation strategies by smallholder farmers in developing countries means that risks associated with agricultural production are relatively important in their decision-making process. Despite smallholder farmers usually being thought to have homogeneous risk averse attitudes [2], there is evidence that this is not the case. Hence, identifying and understanding the heterogeneity in farmers' risk preferences is crucial to guide policy formulation and implementation on risk management and investment decisions (e.g., technology, the adoption of new crop varieties, or the adoption of sustainable agricultural practices). However, few studies have analysed heterogeneity in risk preferences [3–11]. Institutional and non-institutional factors have been associated with farmers' risk attitudes [12–16]. Importantly, farmer's risk preferences may be associated with the climatic, soil, topographic, and economic conditions of the farm's geographic location and the farmer's economic situation. For instance, farmers in the coastal areas of Vietnam are reportedly less risk averse than farmers in non-coastal areas [14,15,17]. Additionally, a negative correlation is reported between low rainfall areas and farmers' risk aversion in Uganda [16]. Farmers' risk preferences may also be influenced by the risk preferences of other farmers who are in geographical proximity, or by the availability or otherwise of infrastructural or institutional facilities such as roads, schools, and markets [18]. For example, neighbourhood effects are observed in farmers' agricultural technology adoption patterns [19–23], partly because culturally, farmers living closely often rely on their friends and neighbours to acquire and share information on improved farm practices. The social composition of farmers may reveal neighbourhood effects [24]. Such an influence may lie within or extend beyond the current agricultural zones/land divisions. The degree of heterogeneity in farmer's risk preferences may therefore reflect the existing economic reality of farmers within and across agricultural zones in Nigeria.

Hence, there may be the presence of spatial unobserved heterogeneity when analysing farmer's risk preferences. Ignoring this spatial unobserved heterogeneity in farmers' risk preference models may lead to biased coefficient estimates [25–29]. Despite advances in the spatial econometrics [30,31], there is no attempt to examine the role of spatial dependence in risk preference.

The aim of this paper is to investigate how unobserved spatial heterogeneity in farmers' risk preferences may affect farmers' risk preferences. We investigate the heterogeneity of farmer's risk preferences (i.e., the extent to which a decision maker (DM thereafter) is willing to take risky decisions [32,33]) and the determinants of these preferences by incorporating spatial unobserved heterogeneity into a farmer's risk preference model along with farm- and farmer-specific factors (age, education, religious beliefs, household size, farm size, gender, marital status) and infrastructure quality (bad roads). Our approach differs from past studies in terms of the elicitation method used in this setting. We hypothesize that rice farmers' risk preferences are spatially dependent, a novel hypothesis in the field of agricultural and applied economics. That is, farmers living closely have similar risk attitudes relative to distant ones due to spatially determined conditions.

2. Materials and Methods

2.1. The Theoretical Model

A structural autoregressive (SAR) model was employed to account for spatial heterogeneity in risk preferences in line with past studies [25–27]. The application of a spatial model is driven by the nature of the data and theory [30,31]. Spatial dependence is a tendency for random variables to correlate with one another due to geographical proximity. It is hypothesized that the observed variation in DM risk preferences may be associated with spatially unobserved conditions such as infrastructure, cultural values, climatic conditions, etc. These are accounted for through an SAR model where the association between the distance-weighted average of neighbouring DM's willingness to take risks and the DM's own willingness to take risks can be investigated. Equation (1) is based on the assumption that the DM maximizes the payoff or expected payoff in the panel risk lotteries. Panel lotteries with four treatments are applied in this study. Detailed formulations are presented in the data sub-section. The treatments are defined as small gain one (SG_1), small gain two (SG_2), large gain one (LG_1), and large gain two (LG_2) to capture heterogeneity in risk preferences.

$$Max\ U(y_i, y_j; X) \tag{1}$$

where U is a utility function, y_i represents the utility derived by a DM from the lottery in location i, y_j implies the utility derived by the DM from the lottery in location j, and X is the vector of farmers' exogenous (and endogenous) socio-economic variables. The maximization objective produces a spatial reaction function, $y_i = F(y_{ij}, X)$ which forms the SAR (Equation (3)). This captures the dependency between the observational units [34]. The data generating process (DGP) reveals a global spillover, since $(I - \rho W)^{-1}$ links y_i

to all X through a multiplier, the spatial weights matrix (W). The power weights matrix adopted from Roe, Irwin and Sharp [18] is shown in Equation (2). This is adapted from Areal, Balcombe and Tiffin [28]. The distance based power weights function has many advantages. First, unlike the binary contiguity method, neighbours may be assigned with different weights. Second, more weights are attached to shorter distances, implying that the closer proximity the neighbours are, the more the influence. In other words, the weights are closer to one when the distance (d) is less than the cut-off distance (s), but tend towards zero when the distance is greater than the cut-off distance. In addition, assuming an equal number of neighbours may be inappropriate, since the number of sampled farmers is not equal across all locations or agricultural zones.

$$W_{ij} = exp\left(-{d_{ij}}^2/s^2\right) \tag{2}$$

where d_{ij} is the distance between DM in locations i and j, estimated from the recorded farmers' GPS coordinates (latitude and longitude), s is the cut-off distance that sets the spatial dependency limit distance after which the spatial effect is decreasing at a slower pace. Different cut-off distances were tested to determine the limit of spatial dependence in line with past studies [18,28,29]. W is often row-standardized, in which the sum of each row of the matrix equals one to facilitate the interpretation of the spatial coefficient results [19,22,24]. Only the diagonal elements of the weights matrix are set to zero to prevent rice farmers from being a neighbour to themselves.

$$y_r = \rho W y_r + X\beta + \varepsilon \tag{3}$$

In Equation (3), y_r is a column vector of willingness towards risk taking (risk avoidance is used interchangeably with willingness towards risk taking to refer to risk aversion because the parameter of the curvature of the utility function is not estimated. This is because risk preference has been previously defined as the extent to which an individual is willing to take risky decisions [2]). This is a probability index corresponding to farmers' choices in the panel lotteries and it ranges between 0 and 1 with an index of 1 indicating being highly unwilling to take risks. The ρ measures the strength of spatial dependence or spatial correlation between the risk preference of a DM and the adjusted-by-distance mean risk preference of their neighbours. W is the $N \times N$ weights matrix (Equation (2)). X is the $N \times K$ vector of exogenous explanatory variables. β is a $K \times 1$ vector of estimated parameters. Wy_r is a spatial lag, which is the weighted average of risk willingness in the neighbourhood locations. The disturbance term is assumed to be independently and identically distributed, $\varepsilon \sim N(0,\ I\sigma^2)$. The rho (ρ) is not restricted between -1 and 1 [35]. This suggests that it cannot be linearly interpreted as a conventional correlation between decision makers' willingness to take risks (y_r) and the adjusted-by-distance willingness to take risks of the neighbours (Wy_r). Equation (3) also suggests that the expected value of DM willingness to take risks, y_r, depends on $X\beta$ plus the neighbouring values of DM scaled by the dependence parameter, ρ.

The potential endogenous problem of spatial lag variables (the correlation between the spatial lag (W) and the disturbance error, ε) is addressed using the instrumental variable (IV) method. The application of IV requires the choice of an instrument, Z, which must satisfy two conditions. First, an instrument must be exogenous, which may be mathematically represented as $Cov = 0$. Second, an instrument must correlate with the endogenous explanatory variable (that is relevant), $Cov \neq 0$. Thus, X are assumed to be exogenous variables, and we use as an instrument the spatial lag of education, *Weducation*.

We used the R package ivreg [36] to estimate the model in Equation (3) using an instrument variable regression where the instrumental variable used is *Weducation*. The R package provides three different tests to ascertain the relevance of the instruments, the endogenous nature of an explanatory variable, and the validity of the instrument. The test of instrument relevance involves examining the significant of the Wald statistic. The Wu–Hausman test, a test of restriction, was adopted to test the endogenous nature of the spatial lag variables. This test is important since IV may produce estimates with larger standard errors relative to OLS if the spatial lag variable is not endogenous. Thus, it is referred to as the test of the consistency of OLS. Lastly, the test of validity of the instrument, often called the Sargan test. This tests over-identification restriction, but is not usually reported in an exactly identified model.

As part of the explanatory variables, we considered farm and farmer characteristics, as well as the quality of the infrastructure around the farm (road quality). We accounted for farm size, which can be a proxy for income in developing countries where livelihoods largely depend on farming. Studies from Ethiopia showed that farm size and risk aversion were negatively correlated [8] and positively related [5], although some studies found no significant relationship [10,37]. We included the farmer's level of education as an explanatory variable for farmer's risk preferences. The direction of the relationship between education and risk preferences has been mixed. Educated farmers are reported as showing aversion to risk taking in developing countries [10,15,38]. However, a positive relationship was also reported between risk aversion and education in Southern Peru [39] and West Africa [40]. The farmer's age was also considered as an explanatory variable for farmer's risk preferences. Research showed that younger farmers are less risk averse [9,15], while others indicated that older farmers are more risk averse [10,40]. The debate on whether women are more risk averse relative to men is inconclusive. For example, Schubert [41] found contrary results when compared to studies that provided strong statistical evidence that males are less averse to risk. In finance and investment, for instance, women are less financially tolerant and more financially risk averse compared to men [42–44]. On the other hand, Harris, Jenkins and Glaser [45] attributed the gender differences in perceptions about outcomes and risky decision making to less desire for enjoyment among women. Research also shows that the social status of individuals may drive risk aversion [46]. Although results on gender have been mixed, in agricultural settings, women are reported to be more risk averse than men [16,37,47]. Consequently, we have included gender as an explanatory variable in the analysis. Marital status was also included as an explanatory variable since it is important in a farmer's decision making process. On one hand, married individuals may be risk takers to cope with the financial burden. On the other hand, they may be more risk averse than the singles because of the fear of income loss when under intense financial pressure. Another variable that has received less attention in the literature is a farmer's religious beliefs. Religious farmers were found to be more risk averse than non-religious people, although it is difficult to know the degree of how risk averse religious people are [37]. Since religion relates to belief, it may affect farmers' perceptions and risk preferences. Notwithstanding, there is no expectation on the direction of this variable. Like other variables, mixed results have been reported between risk aversion and family size. For example, Liebenehm and Waibel [40] reported a positive correlation in West Africa. Large family sizes may prompt action towards taking risky decisions. Thus, farmers with a large family size are expected to be more willing to take risky decisions.

2.2. *Source of Data*

This study used experimental and survey data collected between March and May 2016 from Ogun State Nigeria. Following Binswanga [3,4], a number of studies have experimentally examined farmers' risk attitudes using different methods. As earlier stressed, the term risk avoidance is introduced in place of risk aversion to refer to an individual farmer who is strongly less willing to take risky decisions since the parameter of the curvature of the utility function is not estimated. The DM's risk preferences were elicited using panel lotteries originally proposed by [48], given the name S-GG. The S-GG has been applied in different contexts and countries, but we follow the specifications in [49], with modifications to the nomenclatures. Other applications of this risk attitude elicitation method can be found in some European studies [50,51]. The panel lotteries have four treatments each, with the nomenclature being small gain one (SG_1), small gain two (SG_2), large gain one (LG_1), and large gain two (LG_2). Each treatment has four panels each. A recently published working paper highlighted the advantages and limitations of this risk elicitation method [52]. One unique feature of the panel lottery is that each panel has ten separate lotteries from which the DM chooses one option. We adapted the original S-GG lottery that was presented in Euro to Naira with an exchange rate of 1 Euro to 225 Naira in 2016. Most risk preference elicitation methods in the literature are categorized into laboratory or field, but our risk experiment belongs to lab experiments in the field [2,53].

For SG_1 (and other stakes), DM is faced with a probability (P) to win a payoff (X), or nothing otherwise. Both the payoffs and the probabilities vary across the rows in each panel. Note that the probabilities are the same for each panel of each treatment. The payoffs increase while the probability associated with winning a reward decreases as we move from row (option) one to row (option) ten. The panel lotteries have four treatments with four panels each. The summary of the payoffs is presented in Table 1.

Table 1. Panel lotteries' payoffs.

Panel Lotteries for Four Treatments (Currency in Naira)										
P	1	0.9	0.8	0.7	0.6	0.5	0.4	0.3	0.2	0.1
					$X\ (SG_1)$					
Panel 1	225	251	282	322	376	451	563	751	1126	2251
Panel 2	225	251	282	322	376	451	564	753	1129	2259
Panel 3	225	251	283	324	379	455	570	762	1145	2295
Panel 4	225	252	284	326	382	460	578	774	1165	2340
					$X\ (SG_2)$					
Panel 1	0	26	57	97	151	226	338	526	901	2026
Panel 2	0	26	57	97	151	226	339	528	904	2034
Panel 3	0	26	58	99	154	230	345	537	920	2070
Panel 4	0	27	59	101	157	235	353	549	940	2115
					$X\ (LG_1)$					
Panel 1	22,500	25,002	28,128	32,148	37,507	45,010	56,265	75,024	112,540	225,090
Panel 2	22,500	25,012	28,150	32,186	37,567	45,100	56,400	75,234	112,900	225,900
Panel 3	22,500	25,056	28,250	32,358	37,834	45,500	57,000	76,167	114,500	229,500
Panel 4	22,500	25,112	28,375	32,572	38,167	46,000	57,750	77,334	116,500	234,000
					$X\ (LG_2)$					
Panel 1	0	2502	5628	9648	15,007	22,510	33,765	52,524	90,040	202,590
Panel 2	0	2512	5650	9686	15,067	22,600	33,900	52,734	90,400	203,400
Panel 3	0	2556	5750	9858	15,334	23,000	34,500	53,667	92,000	207,000
Panel 4	0	2612	5875	10,072	15,667	23,500	35,250	54,834	94,000	211,500

Note: P is the probability (10 represents 100% while 1 represents 10%), X is the payoff. The payoffs are rounded to the nearest ten and thousand. Source: authors' compilation.

Rice farmers who avoided risky decisions are more likely to choose from the first few rows (top five options), while risk neutral and risk loving subjects may prefer payoffs that are closer to the bottom (last five rows). Thus, the avoidance of zero earning by not picking higher rewards implies risk aversion. In other words, a DM with a uniformly concave utility function may choose extreme options, sure choices (with 100 per cent probability), while those with uniformly convex utility functions may choose the last or risky options (when the probability is 10 per cent). In addition, the lotteries expose subjects to the entire range of the probabilities and monetary rewards. In fact, a DM who avoids taking more risky options in the first and second panels of each treatment are attracted to risky decisions in the third and fourth panels which have relatively higher rewards. The choice of one (1) out of the ten (10) options in each panel results in sixteen (16) observations per subject. In other words, unlike most risk elicitation methods, the panel lotteries give four observations per treatment. Thus, different models were estimated for SG_1, SG_2, LG_1, and LG_2 to observe/compare the heterogeneity in rice farmers' risk attitudes. The payoff associated with each probability in the SG_1 treatment is constructed using Equation (4).

$$EV_{ij}(SG_1) = P_{ij}X_{ij} = C + (1 - P_{ij})t_j, \\ X_{ij}(SG_1) = \frac{C+(1-P_{ij})t_j}{P_{ij}} \tag{4}$$

where $EV_{ij}(SG_1)$ is the expected value of SG_1. $X_{ij}(SG_1)$ is the payoff associated with (SG_1). i varies from 1 to 10 corresponding to the lottery rows; j varies from 1 to 4, representing panels 1, 2, 3, and 4, respectively, P is the winning probability, which varies from 1 to 0.1, with 1 representing 100 per cent while 0.1 stands for 10 per cent. C is a constant fixed at N 225 for each of the panels in the SG_1. This is the Naira equivalent of the 1 Euro used in the original S-GG lottery. Therefore, all the four panels under SG_1 began with a sure amount (225), which is responsible for a linear large increment in the expected values down the vertical rows. t_j = 0.1, 1, 5, 10 is a panel-specific risk premium corresponding to panels 1, 2, 3, and 4, respectively. The risk premium is responsible for the increment in the expected values as we move from panel one to panel four. Other treatments are calculated from the SG_1. SG_2 is SG_1 less 225 ($SG_2 = SG_1 - 225$), as defined in Equation (5). On the other hand, LG_1 is a product of SG_1 and a constant, $LG_1 = SG_1 \times 100$, as defined in Equation (6). This is done to bring about a large increment in small gain one to examine the variation in subjects' risk attitudes. Lastly, LG_2 is expressed as LG_1 less than 22,500, ($LG_2 = LG_1 - 22,500$), as illustrated in Equation (7).

$$X_{ij}(SG_2) = X_{ij}(SG_1) - 225 \tag{5}$$

$$X_{ij}(LG_1) = (X_{ij}(SG_1))100 \tag{6}$$

$$X_{ij}(LG_2) = (X_{ij}(LG_1)) - 22{,}500 \tag{7}$$

Order of Presentation

A total number of 329 rice farmers were interviewed during the survey period with 328 fully completed questionnaires. All data were electronically collected using open data kit (ODK collect) with the aid of two smart android phones. This technology was used to record the GPS coordinates (latitude and longitude) of the locations of individual rice farmers. Notwithstanding the poor quality or absence of mobile networks in most villages, the locations (towns or villages) of each DM were manually recorded and later used to obtain the coordinates. Prior to the commencement of the survey, three postgraduate students were trained as research assistants on the use of the technology for data collection in late February 2016. The enumerators were also illustrated on how to fill in the record sheets. The record sheet used for visualising the lottery to farmers is presented in Figure 1.

Rice farmers were individually interviewed by contacting them at their homes and/or on their farms. In all cases, subjects' consents were sought before participating in the experiments/survey in line with the University of Reading regulations on research. In addition, respondents were informed about the voluntary participation and that they can withdraw from the experiment and survey at any stage. In all, no participant indicated interest in withdrawing from the experiments and survey. The risk experiment was conducted first, and lastly, questions were asked on the socio-economic factors. Respondents' minds were equally prepared for the need to use smart phones, because most farmers were not familiar with such technology for data collection. Subjects were presented first with the panel lotteries, starting from panel 1 to panel 4 of SG_1, then SG_2, LG_1, and LG_2, respectively. In addition, each DM was shown a bag containing 10 mixed blue and red balls, which represented the winning and losing probability. For the payment, only one of the panels in each treatment determined the earnings. However, this task was not incentivized for two reasons: first, due to the relatively high rewards involved, and second, it prevented non-rice farmers from participating in the experiment.

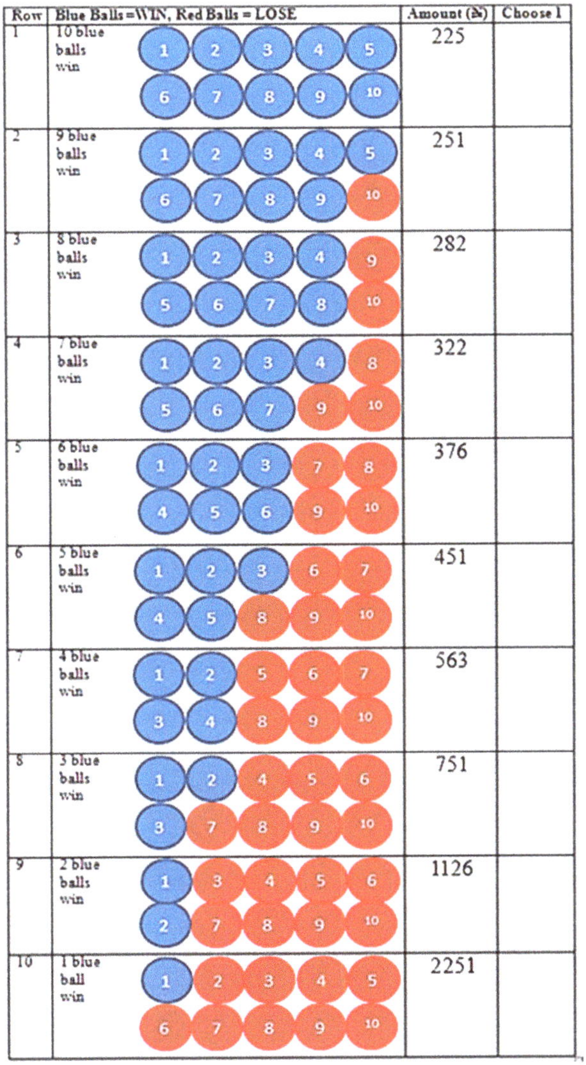

Figure 1. Record sheet for SG_1 panel 1.

- Instruction for small gain one and large gain one treatments

After the welcoming rice farmers with brief explanation on the importance of the survey, the experiments, and the likely impact of the study, the instructions for SG_1 were read to the farmers as follows: "The following 4 panels have 10 options each, the winning prize in each panel is the amount of Naira shown under the heading amount". The blue balls represent the chances of winning; 10 blue balls imply one hundred per cent chances (sure), while 1 blue ball means a ten percent chance of winning a payoff (Figure 1). Conversely, the red balls imply a loss. The subject earned nothing if they did not win the lottery. The earning was determined by tossing a four-sided die. That is, any of the numbers 1, 2, 3, or 4 occurring from a toss of a four-sided die determines the payment panel. For instance, if a subject chose option 7, and one appeared during the die toss, they would win N 563 if any of the blue balls 1, 2, 3, or 4 was drawn from the bag, but nothing if otherwise. Lastly, the record sheet was shown to the DM to make their choice. Similar instructions were given for LG_1.

- Instruction for small gain two and large gain two treatments

The instructions for SG_2 were read as follows. "The following 4 panels have ten options each. The winning prize in each panel is the amount of Naira shown under the heading "amount". The blue balls indicate the chances of winning; 10 blue balls imply hundred per cent chance (sure), while 1 blue ball means ten per cent chance. Conversely, the red balls imply loss. If you do not win the lottery, you will earn nothing or lose the sure amount. Your earning would be determined by tossing a die; any of the number 1, 2, 3, or 4 occurring from a toss of four-sided die determines the payment panel. For instance, chosen option seven and one appears during die toss earn you N 338 if any of the balls 1, 2, 3, or 4 is drawn from the bag. Kindly choose one option from each panel". Then, the record sheet was given to the DM to make a choice. Similar instructions applied for LG_2.

3. Results

The summary statistics of the variables included in the model are presented in Table 2. The average values suggest that rice farmers are risk avoidant with respect to SG_1, SG_2, LG_1, and LG_2, respectively. This is because the higher the probability values (the closer to 1) associated with the choices, the more averse an individual farmer is.

Table 2. Definition and Summary Statistics of the Variables used in the SAR Model.

Variables	Definition	Mean (SD)	Min	Max
SG_1	Small gain one probability index	0.80 (0.15)	0.10	1.00
SG_2	Small gain two probability index	0.60 (0.13)	0.10	1.00
LG_1	Large gain one probability index	0.70 (0.15)	0.10	1.00
LG_2	Large gain two probability index	0.60 (0.16)	0.10	1.00
Age	Age of the farmer in years	47.00 (12.50)	20.00	80.00
Education	Years of formal schooling	4.60 (4.50)	0.00	16.00
Male	1 if male, 0 if female	0.68	0.00	1.00
Christian	1 if Christian, 0 otherwise	0,56	0.00	1.00
Married	1 if married, 0 otherwise	0.94	0.00	1.00
Household size	Number of household members	6.00 (3.00)	1.00	21.00
Farm size	Rice farm area in hectares	1.90 (1.50)	0.20	16.00
Bad road	1 if farmers live in untarred, poorly accessible road areas, 0 otherwise	0.37	0.00	1.00

Source: authors' data analysis, 2017.

The sampled farmers are averagely aged (mean age is 47 years), which suggests that most of the farmers were in their productive age. The majority of the respondents did not complete primary education. Males constituted about 68 percent of the sample, with females constituting 32 percent. About 56 percent practiced Christianity as their religion, providing information on the representation of the two dominant religions in the country. Almost all (94 per cent) of the sampled farmers were married, and the average family size was 6 persons, which suggested financial responsibility for the household heads. An average farmer in the study sample cultivated 1.9 ha of land for rice production in the planting season preceding the survey year/period, while 37 percent lived in poor road network areas, an important infrastructural economic and sustainable development variable in our analysis.

The model results are presented in Table 3, respectively, for SG_1, SG_2, LG_1, and LG_2. The average values for each treatment were used in the analyses due to the high correlation between the panels within each treatment. The null hypotheses of the weak instruments were rejected, suggesting that the instrumental variables used were strong enough to obtain consistent estimates. The null hypotheses of the consistency of OLS were equally rejected in all of the risk models, implying that OLS may not yield consistent estimates. In addition, the Wald statistic, which was significantly different from zero for all the treatment models, attested to the overall goodness of fit of the models. The results corresponding to the 60 km are reported for SG_1, SG_2, LG_1, and LG_2, respectively, in line with [28,29], who reported a spatial dependence limit.

Table 3. The Effect of Spatial Dependence on Rice Farmers' Risk Preferences.

Variables	SG_1	SG_2	LG_1	LG_2
Spatial Dependence				
Spatial lag	0.0016 *** (0.0002)	0.0019 *** (0.0002)	0.0019 *** (0.0002)	0.0019 *** (0.0002)
Farm- and Farmer-Specific Factors				
Age	0.0032 *** (0.0009)	0.0021 *** (0.0008)	0.0022 ** (0.0010)	0.0019 ** (0.0008)
Education	0.0043 (0.0025)	0.0002 (0.0025)	0.0035 (0.0026)	0.0037 (0.0024)
Christian	0.0498 ** (0.0217)	0.0498 *** (0.0185)	−0.0263 (0.0207)	0.0217 (0.0191)
Household size	0.0021 (0.0031)	0.0027 (0.0034)	0.0016 (0.0034)	0.0004 (0.0037)
Farm size	−0.0163 ** (0.0075)	−0.0047 (0.0056)	−0.0155 * (0.0081)	−0.0074 (0.0069)
Male	0.0559 ** (0.0234)	0.0498 *** (0.0196)	0.0616 *** (0.0217)	0.0580 *** (0.0210)
Married	0.3103 *** (0.0611)	0.2214 *** (0.0556)	0.3089 *** (0.0587)	0.2137 *** (0.0509)
Location/Infrastructure				
Bad roads	0.1097 *** (0.0206)	0.0549 *** (0.0186)	0.1123 *** (0.0208)	0.0595 *** (0.0195)

Source: authors' data analysis, 2017. N = 328, ***, **, * respectively represent coefficients are significantly different from zero at 1%, 5% and 10%. Standard errors are in parentheses; diagnosis statistics: weak instruments: SG_1 = 30,562.80 ($p < 0.00$), SG_2 = 23,621.55 ($p < 0.00$), LG_1 = 25,382.5 ($p < 0.00$), LL = 18,951.71 ($p < 0.00$); OLS consistency: SG_1 = 29.15 ($p = 0.00$), SG_2 = 57.88 ($p = 0.00$), LG_1 = 71.4 ($p = 0.00$), LL = 36.59 ($p = 0.00$); Wald Tests: SG_1 = 787.9 ($p < 0.00$), SG_2 = 753.4 ($p < 0.00$), LG_1 = 738 ($p < 0.00$), LG_2 = 492.9 ($p < 0.00$).

Factors that significantly explain risk attitudes (SG_1) include age, religion, farm size, gender, marital status, bad roads, and spatial dependence, while age, religion, gender, marital status, bad roads, and spatial dependence significantly determined attitudes towards SG_2. Similarly, age, farm size, gender, marital status, bad roads, and spatial dependence were the determining factors for attitudes toward LG_1, while attitudes toward LG_2 were significantly explained by age, gender, marital status, bad roads, and spatial dependence. Note that the positive coefficients imply risk avoidance (risk aversion).

Willingness to take risks was spatially determined, as indicated by the significant coefficients of all the spatial lags in all of the risk treatments (Table 3). Similar studies observed the spatial parameter and reported that rho increases up to a particular distance, and later decreases [18,54]. We observed a similar pattern, with 60 km constituting the limit of spatial dependence. In short, statistical significance of rho suggests the existence of neighbourhood effects in risky decision making.

Accordingly, with respect to SG_1, a farmer's risk avoidance is positively associated with an increase in the distance weighted sum of all neighbors' SG_1. Taking the average neighbour's SG_1 for each farmer, we found that an increase in the average neighbour's SG_1 (i.e., risk avoidance, note that values are between 0 and 1) of 0.1 would mean an increase of 32.8 on the distance weighted sum of all neighbour's SG_1, $0.1 \times 328 = 32.8$. This means that an increase in the farmer's own SG_1 (risk avoidance) of 0.05 units ($32.8 \times 0.0016 = 0.05$).

The effect of a farmer's neighbours SG_2, LG_1, and LG_2 on their own SG_2, LG_1, and LG_2 was 0.06 for each, respectively. Older rice farmers avoided risk taking or were more risk averse relative to the younger farmers. The results also revealed that farmers practicing Christianity tended to avoid risk taking with respect to small stakes compared to Muslims and others. Small land holders were less willing to take risk relative to large-scale farmers, suggesting that increasing farm size might lead to taking risky decisions. For instance, an increase of 1 ha would lead to a decrease in SG_1 of 0.016 units. Male rice farmers were less likely to take risk compared to their female counterparts. Married farmers showed less willingness to take risks relative to singles. The results also showed that farmers living in the un-tarred bad road network areas were less willing to take risky decisions compared to those living in more accessible road areas. The directions of some of the variables and the results are therefore consistent with the expectations while others differ from previously expressed views on risk attitudes.

4. Discussion

The results show that unobserved spatial heterogeneity is associated with farmers' risk preferences. Although we cannot identify what exactly the heterogeneity is, by incorporating spatial dependency in the farmer's risk preference model, we can control for these spatially dependent effects (e.g., soil, topography, farmers emulating each other). It follows that given the observed socio-economic variables, farmers' risky behaviour is influenced by unobserved spatial attributes. Put differently, the closer the distance between the farmers, the more likely they would behave in a similar manner. This is plausible as it may reflect the geographical relationship as well as socio-economic conditions between and among individual rice farmers. For instance, farmers in Nigeria may exhibit similar behaviour, which may differ from their counterparts in Ghana, due largely to the different regional characteristics (e.g., soil, topography, climate, culture). In summary, our finding upholds the principle of proximity in similar patterns of attitudes and therefore agrees with Tobler [55], who posited that closer observations and individuals tend to behave in similar manners compared to distant ones; with important implications in sustainability and sustainable development.

Informal communication and interaction are common phenomena in both the urban and rural areas of most developing countries due largely to clustering. The revelation here shows that rice farmers are related in some ways climatically, geographically, economically, socially, culturally, and ecologically. This agrees with [5], and suggests that farmers living closely may behave similarly relative to distant ones. Evidence of spatial dependence may be reflected in decisions to adopt improved agricultural technology, as well as decisions relating to other investment opportunities. In fact, the adoption and diffusion of technological agricultural innovation may be accelerated or ride on the back of the information possessed by farmers' neighbours. Such geographical influences are often ignored by economic policy. In the study area, farmers share many personal and formal attributes/factors such as farm holdings, land use policy, educational institutions, and roads, as well as uncontrollable factors such as the weather and climate. It suggests that observed patterns of behaviours should not only be important for local interaction and interpersonal communication, but also instrumental in the decision-making processes with respect to local, national, and international agricultural policies. Furthermore, the degree of heterogeneity in risk attitudes is a complex process involving many uncontrolled variables. This makes incorporating spatial dependence in farmers' risk preferences important in controlling for these unobserved factors, which may vary from farmers' and farms' characteristics to institutional factors. Additionally, risk attitudes have found applications in different aspects of life such as health, finance, sports, and education. The use of education as an instrumental variable suggests that this variable is not only important in general economic policy, but also specifically key to farmers' spatial heterogeneous risky behavior. Such spatial attributes depict the universality of education irrespective of the geographical locations or place of residence of individuals. In short, there is evidence of spatial dependence in risky behaviour among the sampled farmers, an important novelty and revelation for policy in the field of agricultural and applied economics.

The finding supports previous findings, which reported a negative correlation between age and risk aversion [10,40]. It is, however, contrary to some findings that risk aversion decreases with age [9,15]. Older rice farmers may be less interested in taking up risky and productive investments due to their perceived old age. They may have a strong desire and expectation for enjoyment, being willing to enjoy the goodness of life since death is inevitable. On the contrary, the desire to invest in the youth for higher future outcomes and economic benefits may constitute a push factor for younger rice farmers who show more willingness to take risky decisions.

Although it may be difficult to infer how religious an individual is, the results indicate that Christians statistically and significantly behave differently by showing less willingness to taking risky decisions compared to others. This may partly and probably due to the small amount associated with the small stake lotteries. On the other hands, it may reflect attitudes toward certainty since a bird in hand worth more than hundred birds in the bush. Past studies have reported that religious farmers are risk averse [10,37]. Religion may drive farmers' beliefs as well as influencing their level of gambling and day-to-day activities, including investment decisions. Notwithstanding, politics may contribute to the preferences revealed by the subjects, and subsequently, farm decisions. In summary, the results confirm the heterogeneity of risk attitudes across religions. Farmers' risk preferences were correlated with farm size. More specifically, farmers who were risk averse were associated with small farms. This result is consistent with the expectation and previously reported finding [6]. There are two possible reasons for this finding. On one hand, small-scale farmers may require a significant amount of income to expand their scope of operation, which may make them reticent to taking risk. On the other hand, large farms may imply additional financial commitments, thus taking risks might be adoptable strategies for increasing farm income. If farm size is a proxy for wealth or income, it is safe to conclude that the result agrees with previous findings reporting the tendency of less risk aversion among wealthier farmers [5,6,8,10,37,40].

Male farmers were found to be more risk averse than female farmers. The result presents a contrary view to the previously reported findings that males are risk takers [49]. It is also opposed to the previous findings that female farmers are more averse to risk taking than their male counterparts [16,37,47]. More so, it disagrees with previously expressed views of financial risk behaviour that women are less financially tolerant and more financially risk averse compared to men [42–44]. It also disagrees with Harris, Jenkins and Glaser [45], who attributed gender differences in perceptions about outcomes and risk taking to low propensity in enjoyment among the women compared to men. Additionally, male rice farmers may perceive lotteries as liquidity capital compared to female farmers, who may attach more value to the monetary rewards offered by the lotteries. This proposition is based on the fact that, on average, male rice farmers cultivate more land for rice production compared to female farmers, indicating more income from farming. In addition, women tend to have higher expectations for social engagements and activities, which may drive their desire and willingness towards taking risk, irrespective of the size of the stake. Farmers' attitudes may also be viewed from the fact that males may have a strong attachment to the status quo or endowment effect; that is, not willing to lose the 'certain' yield from the traditional technology or be less willing to pay a price for the 'uncertain' yet higher yield from the improved technology.

Our finding indicate married rice farmers avoided risk taking compared to single rice farmers. As earlier noted, single individuals tend to view loss from a different perspective compared to the married individuals who may perceive a loss as a threat to livelihood due to additional family responsibility and financial commitments. Indeed, married farmers' avoidance of risk taking may be attributed to a fear of a loss of money. This is also in agreement with the popular saying that a bird in the hand is better than two in the bush, since married individuals have more pressing financial concerns and would probably do everything within their capacity to avoid losing money. Arguably, married farmers are expected to show more desire to take risks as an option for gaining more money to cater for their family's financial needs but our results show contrary which calls for policy concerns in relation to family size.

In both developed and developing countries, rural areas generally lack access to infrastructural facilities compared to urban areas. Bad road networks may limit movement and access to information and the market, thereby limiting the production and income potential of farmers residing in rural areas. It can therefore influence farmers' behaviour or attitudes during decision making processes. Additionally, country side including peri-urban areas are associated with low economic activities but dominated by agrarian economy. Furthermore, rural areas are often associated with poverty attributable to lack of access to social amenities and infrastructural facilities. Our result shows that poor road network, occasioned by poor infrastructure and low income influence low interest in risk taking. It is therefore aligns with past studies, which found poor farmers were more averse to risk taking [5,8,56]. Since roads are important infrastructure and economic development variables, it shows that this finding agrees with Harrison, Humphrey and Verschoor [9], who revealed that farmers living in low rainfall areas in Uganda showed a higher aversion to risk, on average, than farmers living in five other agro-climatic areas with relatively higher rainfall distribution.

The finding also aligns with those that attributed higher risk aversion to income variability [57,58]. Furthermore, farmers' risk aversion was reportedly negatively related to willingness to pay/adopt improved agricultural technology that may bring about sustainable intensification in East Africa [59]. This underscores the economic importance of risk aversion or risk avoidance in different aspects of economics including demand for improved farm practices, and in our case infrastructure facility. Infrastructure aids food supply and demand and thus constitutes push factor in the food supply chain. The revelation here may also be decoded as low tendency for risk taking in the rural areas is attributable to the less risky rural environment relative to the urban environment. In short, access to a good road network significantly explains farmers' risk aversion behaviour in the study area, the implications of which may be applied at national and regional levels. It therefore buttress the importance of road infrastructure not only in economic behavior but also in sustainability and sustainable development as it aids and accelerate economic growth through ease of movement of farm produce as well as other economic goods and services especially from the rural areas to the urban markets where farmers stand a better chance of earning higher profits

5. Conclusions

We provide insights into the role of unobserved spatial heterogeneity in explaining risk preferences among rice farmers in Nigeria by incorporating spatial dependency in a farmer's risk preference model.

We found that incorporating a spatial dependency term into farmers' risk preference models (i.e., SAR) can help to control for unobserved spatial heterogeneity in farmers' preferences. Although this type of heterogeneity is not observable and we cannot identify its source, controlling for it is important to avoid bias in the model coefficient estimates [30,31]. The non-observability of the spatial heterogeneity comes from the common lack of information on spatially dependent factors such as soil, topographic, climatic, and socio-economic conditions present in an area. Although there may be cases where some of this information may be available for the researcher (e.g., rainfall data), other types of information are rarely collected (e.g., whether farmers emulate each other or share information).

We found famers' risk preference heterogeneity due to their socio-demographic characteristics, such as age, gender, religion, and marital status; farm characteristics such as farm size; and local infrastructure (bad roads).

Our results may have important implications for policy design. Both observed and unobserved spatial heterogeneity may affect farmers' risk preferences and therefore farmers' decision-making processes especially relating to sustainable farm practices, sustainable intensification and subsequently sustainable development. Policies aiming to achieve sustainable development and food security usually involve some type of intervention (e.g., the promotion of a farmer's adoption of agricultural practices and technologies that contribute to achieving these objectives). Farmers' decisions to engage with these policy intervention programmes may depend on their risk preferences. Hence, in order to maximize the net benefits associated with these programmes, the design of farmers' engagement and behavioral factors seems crucial. Such design may require having different streams of action to account for observed and unobserved spatial heterogeneity in farmers' risk preferences. For instance, when promoting the adoption of new technologies in a region/country, there is a need to identify whether there are any socio-demographic, economic, or geographically determined conditions that may affect farmers' risk perceptions, which will eventually determine their decisions to adopt or not new technologies or improved farm practices that would enhance farm sustainability.

In the case that these exist, new technology adoption can be optimized by focusing the efforts into those characteristics and locations where farmers are less likely to be risk averse (i.e., more prone to engage with policy programmes). This means that specific interventions to more risk averse farmers may need to be designed to persuade farmers to change their perceptions and to adopt risky economic activities (e.g., adopting new technologies). However, the latter would only be economically viable if the expected benefits of the intervention are higher than the costs. Hence, both observed and unobserved spatial heterogeneity in risky decision making should be given special attention in the design and formulation of economic policies and programs that would improve the living conditions of rice farmers, especially in rural areas. Likewise, our results suggest that policies aiming at infrastructure improvement (e.g., road networks in the rural areas), which is associated with farmers being relatively less risk averse, may facilitate farmers engagement in policy programmes (e.g., adopting new technologies, sustainable agricultural practices). Good and accessible roads will not only increase farmers' level of awareness or information on improved agricultural technology, but also increase the chances of transporting and marketing farm produce at urban and international markets for farmers' economic benefits.

Author Contributions: Conceptualization, O.I.A.; methodology, O.I.A.; software, F.J.A.; validation, O.I.A., F.J.A. and N.G.; formal analysis, O.I.A. and F.J.A.; investigation, O.I.A.; resources, O.I.A.; data curation, O.I.A.; writing—original draft preparation, O.I.A.; writing—review and editing, O.I.A.; visualization, O.I.A.; supervision, F.J.A. and N.G.; project administration, O.I.A.; funding acquisition, O.I.A., F.J.A. and N.G. All authors have read and agreed to the published version of the manuscript.

Funding: This research was funded by Tertiary Education Trust Fund (TETFUND) of Nigeria while the APC was funded by Spanish Ministerio de Ciencia, Innovación y Universidades (RTI2018-096927-B-100) and Universitat Jaume I (UJI-B2018-76/77).

Institutional Review Board Statement: The study was conducted according to the guidelines of the Declaration of Helsinki, and approved by the Ethics Committee of the University of Reading (protocol code/ethical clearance application number P00140, approved on 12 February 2016).

Informed Consent Statement: Informed consent was obtained from all subjects involved in the study.

Data Availability Statement: The data used for the analyses would be made available on request.

Acknowledgments: The authors wish to acknowledge access to facility and support by the University of Reading, United Kingdom and Olabisi Onabanjo University, Nigeria, as well as the practical assistance provided by the research assistants during the field work. We particularly appreciate the funding support by TETFUND Nigeria and APC payment by the Spanish Ministerio de Ciencia, Innovación y Universidades (RTI2018-096927-B-100) and Universitat Jaume I (UJI-B2018-76/77). This article has undergone conferences including Agricultural Economics Society (AES 93rd Annual Conference, Dublin 2017), Center for Studies of African Economies (CSAE Annual Conference, University of Oxford 2019) and African Association of Agricultural Economists (AAAE) African Conference of Agricultural Economists (6th ACAE, Abuja 2019). The authors therefore expressed their profound appreciation to all participants in those conferences for their valuable comments and contributions.

Conflicts of Interest: The authors declare no conflict of interest. The funders had no role in the design of the study; in the collection, analyses, or interpretation of data; in the writing of the manuscript, or in the decision to publish the results.

References

1. World Bank. *Agriculture for Development: Overview, World Development Report 2008*; The World Bank: Washington, DC, USA, 2007. Available online: https://www.openknowledge.worldbank.org (accessed on 10 September 2015).
2. Tshoni, S. Analysis of Smallholders' farm diversity and risk attitudes in the Stellenbosch local municipal area. In *Family Farming Knowledge Platform*; Food and Agriculture Organization of the United Nations (FAO): Rome, Italy, 2015; pp. 1–88. Available online: http://scholar.sun.ac.za/bitstream/handle/10019.1/96677/tshoni_analysis_2015.pdfsequence=3&isAllowed=y (accessed on 10 January 2021).
3. Binswanger, H.P. Attitudes toward risk: Experimental measurement in rural India. *Am. J. Agric. Econ.* **1980**, *62*, 395–407. [CrossRef]
4. Binswanger, H.P. Attitudes toward risk: Theoretical implications of an experiment in rural India. *Econ. J.* **1981**, *91*, 867–890. [CrossRef]
5. Wik, M.; Aragie Kebede, T.; Bergland, O.; Holden, S.T. On the measurement of risk aversion from experimental data. *Appl. Econ.* **2004**, *36*, 2443–2451. [CrossRef]
6. Yesuf, M. Risk, Time and Land Management under Market Imperfections: Applications to Ethiopia. Ph.D. Thesis, Goteborg University, Gothenburg, Sweden, 2004.
7. Harrison, G.W.; Humphrey, S.J.; Verschoor, A. Choice under Uncertainty in Developing Countries, CEDEX Discussion Paper Series, No. 2005-18, 2005. Available online: http://www.nottingham.ac.uk/economics/cedex/papers (accessed on 12 November 2014).
8. Yesuf, M.; Bluffstone, R.A. Poverty, risk aversion, and path dependence in low-income countries: Experimental evidence from Ethiopia. *Am. J. Agric. Econ.* **2009**, *91*, 1022–1037. [CrossRef]
9. Harrison, G.W.; Humphrey, S.J.; Verschoor, A. Choice under uncertainty: Evidence from Ethiopia, India and Uganda. *Econ. J.* **2010**, *120*, 80–104. [CrossRef]
10. Tanaka, T.; Camerer, C.F.; Nguyen, Q. Risk and time preferences: Linking experimental and household survey data from Vietnam. *Am. Econ. Rev.* **2010**, *100*, 557–571. [CrossRef]
11. Brick, K.; Visser, M.; Burns, J. Risk aversion: Experimental evidence from South African fishing communities. *Am. J. Agric. Econ.* **2012**, *94*, 133–152. [CrossRef]
12. Reardon, T.; Taylor, J.E. Agroclimatic shock, income inequality, and poverty: Evidence from Burkina Feso. *World Dev.* **1996**, *24*, 901–914. [CrossRef]
13. Humphrey, S.J.; Verschoor, A. Decision-making under risk among small farmers in East Uganda. *J. Afr. Econ.* **2004**, *13*, 44–101. [CrossRef]
14. Nguyen, Q.; Leung, P. Do fishermen have different attitudes toward risk? An application of prospect theory to the study of Vietnamese fishermen. *J. Agric. Resour. Econ.* **2009**, *34*, 518–538.
15. Nguyen, Q. Does nurture matter: Theory and experimental investigation on the effect of working environment on risk and time preferences. *J. Risk Uncertain.* **2011**, *43*, 245–270. [CrossRef]
16. Tanaka, Y.; Munro, A. Regional variation in risk and time preferences: Evidence from a large-scale field experiment in rural Uganda. *J. Afr. Econ.* **2014**, *23*, 151–187. [CrossRef]
17. Nguyen, Q.; Leung, P. How nurture can shape preferences: An experimental study on risk preferences of Vietnamese fishers. *Environ. Dev. Econ.* **2010**, *15*, 609–631. [CrossRef]
18. Roe, B.; Irwin, E.G.; Sharp, J.S. Pigs in space: Modeling the spatial structure of hog production in traditional and nontraditional production regions. *Am. J. Agric. Econ.* **2002**, *84*, 259–278. [CrossRef]
19. Case, A. Neighborhood Influence and Technological Change. *Reg. Sci. Urban Econ.* **1992**, *22*, 491–508. [CrossRef]
20. Holloway, G.; Shankar, B.; Rahmanb, S. Bayesian spatial probit estimation: A primer and an application to HYV rice adoption. *Agric. Econ.* **2002**, *27*, 383–402. [CrossRef]
21. Krishnan, P.; Patnam, M. Neighbors and extension agents in Ethiopia: Who matters more for technology adoption? *Am. J. Agric. Econ.* **2014**, *96*, 308–327. [CrossRef]
22. Läpple, D.; Kelley, H. Spatial dependence in the adoption of organic drystock farming in Ireland. *Eur. Rev. Agric. Econ.* **2015**, *42*, 315–337. [CrossRef]
23. Tessema, Y.M.; Asafu-Adjaye, J.; Kassie, M.; Mallawaarachchi, T. Do neighbours matter in technology adoption? The case of conservation tillage in Northwest Ethiopia. *Afr. J. Agric. Resour. Econ.* **2016**, *11*, 211–225.
24. Holloway, G.; Lapar, M.; Lucila, A. How big is your neighbourhood? Spatial implications of market participation among Filipino smallholders. *J. Agric. Econ.* **2007**, *58*, 37–60. [CrossRef]
25. Benirschka, M.; Binkley, J.K. Land price volatility in a geographically dispersed market. *Am. J. Agric. Econ.* **1994**, *76*, 185–195. [CrossRef]

26. Bockstael, N.E. Modeling economics and ecology: The importance of a spatial perspective. *Am. J. Agric. Econ.* **1996**, *78*, 1168–1180. [CrossRef]
27. Weiss, M.D. Precision farming and spatial economic analysis: Research challenges and opportunities. *Am. J. Agric. Econ.* **1996**, *78*, 1275–1280. [CrossRef]
28. Areal, F.J.; Balcombe, K.; Tiffin, R. Integrating spatial dependence into stochastic frontier analysis. *Aust. J. Agric. Resour. Econ.* **2012**, *56*, 521–541. [CrossRef]
29. Kim, C.W.; Phipps, T.T.; Anselin, L. Measuring the benefits of air quality improvement: A spatial hedonic approach. *J. Environ. Econ. Mgt.* **2003**, *45*, 24–39. [CrossRef]
30. Anselin, L. Under the hood issues in the specification and interpretation of spatial regression models. *Agric. Econ.* **2002**, *27*, 247–267. [CrossRef]
31. LeSage, J.P.; Pace, R.K. *Introduction to Spatial Econometrics*; Balakrishnan, N., William, S.R., Eds.; CRC Press: London, UK, 2009; pp. 1–313.
32. Charness, G.; Gneezy, U.; Imas, A. Experimental methods: Eliciting risk preferences. *J. Econ. Behav. Org.* **2013**, *87*, 43–51. [CrossRef]
33. Bocqueho, G.; Jacquet, F.; Reynaud, A. Expected utility or prospect theory maximisers? Assessing farmers' risk behaviour from field-experiment data. *Eur. Rev. Agric. Econ.* **2014**, *41*, 135–172. [CrossRef]
34. Anselin, L. Lagrange multiplier test diagnostics for spatial dependence and spatial heterogeneity. *Geogr. Anal.* **1988**, *20*, 1–17. [CrossRef]
35. LeSage, J.P. An introduction to spatial econometrics. *Rev. d'économie Ind.* **2008**, *123*, 19–44. [CrossRef]
36. Fox, J.; Kleiber, C.; Zeileis, A. ivreg: Two-Stage Least-Squares Regression with Diagnostics. R Package Version 0.5-0. 2020. Available online: https://CRAN.R-project.org/package=ivreg (accessed on 5 April 2021).
37. Liu, E.M. Time to change what to sow: Risk preferences and technology adoption decisions of cotton farmers in China. *Rev. Econ. Stat.* **2013**, *95*, 1386–1403. [CrossRef]
38. Ihli, H.J.; Chiputwa, B.; Musshoff, O. *Do Changing Probabilities or Payoffs in Lottery-Choice Experiments Matter?* Evidence from Rural Uganda. Global Food Discussion Papers No. 24; Transformation of Global Agri-Food Systems: Trends, Driving Forces and Implications for Developing Countries; Georg-August-University of Göttingen: Göttingen, Germany, 2013.
39. Galarza, F. Choices under Risk in Rural Peru. Munich Personal RePEc Archive (MPRA); Working Paper No. 17708. October 2009. Available online: http://mpra.ub.uni-muenchen.de/17708/ (accessed on 25 May 2015).
40. Liebenehm, S.; Waibel, H. Simultaneous estimation of risk and time preferences among small-scale cattle farmers in West Africa. *Am. J. Agric. Econ.* **2014**, *96*, 1420–1438. [CrossRef]
41. Schubert, R. Analyzing and managing risks-on the importance of gender differences in risk attitudes. *Manag. Financ.* **2006**, *32*, 706–715. [CrossRef]
42. Charness, G.; Gneezy, U. Strong evidence for gender differences in risk taking. *J. Econ. Behav. Org.* **2012**, *83*, 50–58. [CrossRef]
43. Bannier, C.E.; Neubert, M. Gender differences in financial risk taking: The role of financial literacy and risk tolerance. *Econ Lett.* **2016**, *145*, 130–135. [CrossRef]
44. Fisher, P.J.; Yao, R. Gender differences in financial risk tolerance. *J. Econ. Psychol.* **2017**, *61*, 191–202. [CrossRef]
45. Harris, C.R.; Jenkins, M.; Glaser, D. Gender differences in risk assessment: Why do women take fewer risks than men? *Judgm. Decis. Mak.* **2006**, *1*, 48.
46. Stark, O.; Zawojska, E. Gender differentiation in risk-taking behavior: On the relative risk aversion of single men and single women. *Econ. Lett.* **2015**, *137*, 83–87. [CrossRef]
47. Ward, P.S.; Singh, V. Risk and Ambiguity Preferences and the Adoption of New Agricultural Technologies: Evidence from Field Experiment in Rural India. International Food Policy Research Institute (IFPRI) Discussion Paper 01324 (February 2014). Available online: https://ssrn.com/abstract=2392762 (accessed on 5 May 2021).
48. Sabater-Grande, G.; Georgantzís, N. Accounting for risk aversion in repeated prisoners' dilemma games: An experimental test. *J. Econ. Behav. Org.* **2002**, *48*, 37–50. [CrossRef]
49. García Gallego, A.; Georgantzís, N.; Jaramillo-Gutiérrez, A.; Parravano, M. The lottery-panel task for bi-dimensional parameter-free elicitation of risk attitudes. *Rev. Int. Sociol. Spec. Issue Behav. Exp. Econ.* **2012**, *70*, 53–72.
50. Attanasi, G.; Georgantzis, N.; Rotondi, V.; Vigani, D. Lottery- and survey-based risk attitudes linked through a multichoice elicitation task. *Theory Decis.* **2018**, *84*, 341–372. [CrossRef]
51. Barreda-Tarrazona, I.; Jaramillo-Gutierrez, A.; Navarro-Martinez, D.; Sabater-Grande, G. Risk attitude elicitation using a multi-lottery choice task: Real vs. hypothetical incentives. *Span. J. Financ. Account.* **2011**, *40*, 609–624. [CrossRef]
52. Barreda-Tarrazona, I.; Sabater-Grande, G.; Georgantzis, N. *Risk Elicitation through the S-GG Lottery Panel Task: Implementation note*; Working Paper 2020/23; LEE and Economics Department, Universitat Jaume I: Castellón, Spain, 2020.
53. Harrison, G.W.; Rutstrom, E.E. Risk aversion in the laboratory, Risk aversion in experiments. *Res. Exp. Econ.* **2008**, *12*, 41–196.
54. Bell, K.P.; Bockstael, N.E. Applying the generalized-moments estimation approach to spatial problems involving micro-level data. *Rev. Econ. Stat.* **2000**, *82*, 72–82. [CrossRef]
55. Tobler, W.R.A. Computer movie simulating urban growth in the Detroit region. *Econ. Geogr.* **1970**, *46*, 234–240. [CrossRef]
56. Lawrance, E.C. Poverty and the rate of time preference: Evidence from panel data. *J. Pol. Econ.* **1991**, *99*, 54–77. [CrossRef]
57. Guiso, L.; Paiella, M. Risk aversion, wealth and background risk. *J. Eur. Econ. Assoc.* **2008**, *6*, 1109–1150. [CrossRef]

58. Bezabih, M.; Sarr, M. Risk preferences and environmental uncertainty: Implications for crop diversification decisions in Ethiopia. *Environ. Resour. Econ.* **2012**, *53*, 483–505. [CrossRef]
59. Shee, A.; Azzarri, C.; Haile, B. Farmers' willingness to pay for improved agricultural technologies: Evidence from a field experiment in Tanzania. *Sustainability* **2020**, *12*, 216. [CrossRef]

sustainability

MDPI

Article

Law-Driven Innovation in Cereal Varieties: The Role of Plant Variety Protection and Seed Marketing Legislation in the European Union

Serena Mariani

Department of Law, University of Macerata, 62100 Macerata, Italy; s.mariani9@unimc.it

Abstract: The aim of this paper is to investigate the role of EU legislation in shaping innovation in cereal varieties. The research focuses on two fields of law and their relationship, i.e., intellectual property and agricultural law. More specifically, the normative legal investigation concerns the role played by Community plant variety protection and the EU legislation on the marketing of seed and plant propagating material in shaping innovation and stimulating plant breeding of new cereal varieties. The focus is on cereal varieties because innovation in this field has a great socio-economic impact, as well as strategic scientific and environmental implications. Breeding new cereal varieties is essential for the competitiveness of the seed and agricultural sector of the EU, and it can contribute to food security and the achievement of sustainable development goals. The study finds that it is necessary to simplify the existing legal framework by coordinating intellectual property and agricultural law, providing for legislative review and better coherence in order to effectively shape innovation and meet the changing demands of society and the sustainability challenges.

Keywords: agricultural innovation; sustainable agriculture; plant breeding; cereals; intellectual property; agricultural law; plant variety rights; seed marketing; European Union

Citation: Mariani, S. Law-Driven Innovation in Cereal Varieties: The Role of Plant Variety Protection and Seed Marketing Legislation in the European Union. *Sustainability* **2021**, *13*, 8049. https://doi.org/10.3390/su13148049

Academic Editor: Michael Blakeney

Received: 31 May 2021
Accepted: 15 July 2021
Published: 19 July 2021

Publisher's Note: MDPI stays neutral with regard to jurisdictional claims in published maps and institutional affiliations.

Copyright: © 2021 by the author. Licensee MDPI, Basel, Switzerland. This article is an open access article distributed under the terms and conditions of the Creative Commons Attribution (CC BY) license (https://creativecommons.org/licenses/by/4.0/).

1. Introduction

Nowadays, the bond between agriculture and innovation is more discussed than ever [1–3]. The agriculture of the third millennium is facing new and difficult challenges, and it needs a breakthrough — innovation represents a qualitative leap in this field. Overpopulation, scarcity of natural resources, climate change, and biodiversity conservation are some of the many factors to take into consideration during the debate on the role of innovation in agriculture. In this framework, plant breeders are required to use their Promethean "creative power" to develop new plant varieties, able to improve agricultural productivity and meet the new and challenging demands of society, especially the increasing demand for sustainability in agriculture [4,5].

The purpose of this study is to investigate the relationship in the European Union between two fields of law, i.e., intellectual property and agricultural law, in shaping innovation in cereal varieties. It must be underlined that in this paper the focus is not on the innovation concerning the techniques and methods used for breeding purposes. The investigation aims to deal with innovation concerning the characteristics of the end product of the breeding activity, i.e., the plant variety. Specifically, the investigation focuses on the relationship between Community plant variety protection (hereinafter referred to as "Community PVP", also as "CPVP") and the European Union legislation on the marketing of seed and plant propagating material (hereinafter referred to as "EU seed legislation") in fostering plant breeding on varieties of cereal species.

The focus is on cereal varieties because of their relevance from a social, environmental, and economic perspective. Firstly, cereals represent the basis of the human diet: they are "staple foods" and as such they are globally eaten on a regular basis [6]. Innovation in cereal breeding can play a key role since it is crucial to feed the world's increasing population with safe and sustainable food, while decreasing the environmental pressure of the agricultural production [7,8]. Furthermore, there is an alarming global trend towards genetic erosion of cereals, i.e., "the condition that results when a widely planted crop is uniformly susceptible to a pest, pathogen or environmental hazard as a result of its genetic constitution, thereby creating a potential for widespread crop losses" [9]. According to the FAO, at least 30 countries have reported genetic erosion of cereal crops in the last years, followed by vegetables, fruits, nuts, and legumes [9]. Furthermore, climate change and increasing temperatures will reduce global yields of some crops, including cereals [5,10]. In this context, innovation is fundamental to improving genetic diversity of plant varieties, increasing their resistance and reducing the use of plant protection products. Moreover, cereal breeding and production have a significant economic relevance, especially in the European Union. Indeed, the EU is a world leader in cereal production, especially wheat [6]. In addition, the European seed market is highly competitive, representing the third-largest seed market worldwide with more than 20% of the total global market for commercial seed [11]. Therefore, in the EU, innovation in cereal varieties is crucial for the competitiveness of the seed and agricultural sectors.

Innovation in cereal varieties is also fundamental to achieving sustainable food production systems. In this regard, the European Commission's Farm to Fork (F2F) strategy presented on 20 May 2020, cornerstone of the European Green Deal, points out that research and innovation are key drivers in accelerating the transition to sustainable, healthy, and inclusive food systems from primary production to consumption [12]. The F2F strategy will enable the transition to a healthier and more sustainable EU food system, ensuring food security, tackling climate change, protecting the environment, and preserving biodiversity. One of the most ambitious goals of the F2F strategy is to reduce the use and risk of pesticides by 50% and the use of fertilizers by 20%, by 2030. However, the strategy recognizes that climate change brings new threats to plant health. In this context, the sustainability challenge calls for innovation and seed security, ensuring access of farmers to quality seeds for plant varieties adapted to the pressures of climate change. Therefore, it is necessary to stimulate plant breeding of agricultural varieties, such as cereals, more resilient to climate change, high-yielding, and pest resistant, that can enhance agricultural productivity and food security, and safeguard the environment.

The study finds that plant breeding of certain agricultural species, such as cereals, is directly affected by the law. However, the achievement of the sustainable targets may be discouraged by the complexity and rigidity of the current (and outdated) legal framework, characterized by a lack of coordination between Community plant variety protection and the EU seed legislation.

2. Materials and Methods

This study originates from academic legal research based on normative legal investigation concerning the analysis of primary sources, jurisprudence, legal texts and doctrinal writings, in the form of a "library-based" research [13].

The focus is on the legislation of the European Union concerning, on the one hand, Community plant variety protection and, on the other, the EU legislation on marketing of seed and plant propagating material. The role of Community PVP in shaping innovation is examined vis-à-vis the EU seed legislation in light of the way the two areas of law impact the breeding of new cereal varieties. It follows that the investigation does not exclusively lie in the arena of intellectual property law but concerns its dialogue with agricultural law.

Intellectual property law and agricultural law have a complicated relationship and it has been pointed out that "despite its importance, there has been comparatively little written about intellectual property law and its intersection with agriculture" [14]. A potential pitfall is to consider plant variety protection as not connected to seed laws. On the contrary, the two areas of law share concepts and definitions and they also have mutual historical roots, since some of the first European laws on the topic both recognized a protection right for plant breeders and regulated the seed trade [15,16].

The primary sources of law employed in this study are the Council Regulation (EC) No 2100/94 of 27 July 1994 ("Basic Regulation") and the so-called EU seed legislation, which consists of 12 basic Directives covering the crops of major importance in the EU, including cereals. The Basic Regulation established the Community PVP and, unlike patents, it is considered a "technology-specific" intellectual property regime since it exclusively concerns "breeding achievements in the area of plant breeding" [17]. The seed legislation regulates the right to commercialize seeds and propagating material in the EU territory and is based on two pillars: the prior registration of the varieties in the Common Catalogues and the certification of the seed lots.

The two laws have different purposes: the rationale of Community PVP is to stimulate the breeding and development of new varieties, whereas the EU seed legislation has the purpose of ensuring the free marketing of quality seeds and propagating material throughout the EU territory.

The study focuses on the analysis of the protection and registration requirements and procedures in order to assess the points of contact and contrast between the two areas of law.

3. Results

3.1. Preliminary Remarks

It must be highlighted that a plant variety is defined according to Article 5 (2) of the Basic Regulation as "a plant grouping within a single botanical taxon of the lowest known rank, which grouping, irrespective of whether the conditions for the grant of a plant variety right are fully met, can be: 1. defined by the expression of the characteristics that results from a given genotype or combination of genotypes, 2. distinguished from any other plant grouping by the expression of at least one of the said characteristics, and 3. considered as a unit with regard to its suitability for being propagated unchanged".

Based on this definition, the notion of plant variety refers to a plant grouping defined by the "expression" of the genotype, i.e., the phenotype. It follows that two varieties with different genotypes, having similar phenotypical expression, are considered legally identical [1,18].

With regard to cereals, in 1966 the (then) European Economic Community Council adopted the Directive of 14 June 1966 on the marketing of cereal seed (66/402/EEC), last amended by Commission Implementing Directive (EU) 2020/177 of 11 February 2020. Pursuant to Article 1 thereof, this Directive shall apply to cereal seeds marketed within the EU. A legal definition of "cereals" in the seed sector is provided by Article 2 of Directive 66/402/EEC, which covers only the major cereal species. For the purposes of the Directive, cereals are defined as the plants of the following species intended for agricultural or horticultural production other than production for ornamental purposes: *Avena nuda* L.; *Avena sativa* L.; *Avena strigosa Schreb.*; *Hordeum vulgare* L. *Oryza sativa* L.; *Phalaris canariensis* L.; *Secale cereale* L.; *Sorghum bicolor* (L.) *Moench*; *Sorghum sudanense*; *xTriticosecale*; *Triticum aestivum* L.; *Triticum durum Desf.*; *Triticum spelta* L.; *Zea mays* L.

In the case of cereals, plant variety protection and seed marketing laws can significantly impact on each other's goals. The reason is that plant varieties are brought onto the market as reproductive material: in the case of cereals, as seeds. This means that the seed represents both a "commercial commodity" and a "technology carrier" [19], in which it is impossible to divide the tangible from the intangible subject matter.

This dual nature of the seed requires a dialogue between two fields of law. On the one hand, there is agricultural law, regulating seeds as agricultural products and establishing criteria for their marketing (i.e., the EU seed legislation). On the other hand, there is intellectual property law, regulating the new variety in its role as an innovative technology, creating a specific IP right for the variety resulting from the breeding activity (i.e., the Community plant variety protection).

In this context, a rule on the marketing requirements of the tangible subject matter can affect the exploitation of the exclusive rights granted to the IP holder on the intangible subject matter. The implications of this relationship are noteworthy. A lack of coordination could jeopardize the possibility for breeders to market their variety and, consequently, to secure a return on their investment, discouraging the expenditure of additional resources in breeding programs, thus discouraging innovation.

3.2. Community Plant Variety Protection

In the European Union, the plant variety protection system is governed by the Basic Regulation on Community plant variety rights, which also established the Community Plant Variety Office (CPVO) in Angers, France. Even though the (then) European Community became party to the 1991 Act of the Convention only in 2005, the Basic Regulation is consistent with the rules set out by the 1991 Upov Convention. Indeed, its Preamble states that "[...] this Regulation takes into account existing international conventions such as the International Convention for the Protection of New Varieties of Plants (Upov Convention)".

The plant variety protection system established by the Upov Convention and the Basic Regulation is a sui generis IP system, which has been defined a "legal hybrid" between patents and copyrights [20]. This is because it does not correspond to the other existing IP regimes, especially the dominant legal paradigms which protect "inventions" on the one hand, and "artistic works" on the other [21]. The reason lies in the very nature of the protected material, which is self-replicating, thus automatically able to reproduce itself, and cannot be reduced to a disclosure in writing. Therefore, an ad hoc proprietary regime specifically designed for plant varieties was required.

Community plant variety rights have uniform effect within the territory of the European Union and are the sole and exclusive form of Community industrial property rights for plant varieties, allowing breeders to protect their innovation throughout the European Union upon a single application.

Varieties of all botanical genera and species, including hybrids between genera or species, may form the object of Community PVRs. The result is that the protectable subject matter under the Community plant variety system is not limited to a certain number of botanical genera or species.

In order to be protectable under the Community PVP system, a plant variety has to be (a) distinct; (b) uniform; (c) stable; and (d) new, according to Article 6 of the Basic Regulation. Furthermore, the variety has to be designated by a denomination, in accordance with the provisions of Article 63.

The first three technical conditions form the so-called DUS requirements dealing with the phenotypical characteristics of the plant variety and they are examined through field trials. The new variety is not required to have any commercial or cultivation value [2].

Article 7 of the Basic Regulation concerns the distinctness requirement and specifies that distinctness has to be determined "by reference to the expression of the characteristics that results from a particular genotype or combination of genotypes". Therefore, the characteristics that distinguish a variety shall imply a certain degree of observability: they should be externally visible in the field during the technical examination. Distinctness has to be assessed in relation to other varieties whose existence is a matter of common knowledge on the date which a valid application for the grant of a PVR was received by the CPVO or by an entitled national agency.

By way of illustration, the registration into an official Catalogue or the granting of a PVR or even the application for a PVR, provided that the application has not been refused or withdrawn in the meantime, make the variety a matter of common knowledge.

The distinctness requirement has a remarkable meaning in the overall plant variety protection system since it represents the main pillar in providing the logical basis of IP protection for new plant varieties. It has been said that "when a candidate variety meets the distinctness requirement one could say that the variety has been added by the breeder to the plant kingdom or, to put it in patent terms, to the state of the art. This fact is the major justification for a possible protection of the variety with an industrial property right.

The simple copying or multiplication of a variety that has already existed as such, either in nature or in cultivated conditions, would not qualify for such a right" [16].

Article 8 of the Basic Regulation deals with the uniformity requirement. A candidate variety has to be "sufficiently uniform in the expression of those characteristics which are included in the examination for distinctness, as well as any others used for the variety description". Compliance with this requirement does not entail that the candidate variety need be entirely uniform. The use of the word sufficiently clarifies that a certain amount of variation is tolerated and expected, especially in the case of sexually reproduced plants. Indeed, the slight variability of the offspring depends particularly on the propagation process of the variety and is related to the very nature of plants as living organisms. Consequently, the level of uniformity and the determination of off-types differ from one variety to another, depending on the method of plant propagation [22].

Article 9 of the Basic Regulation lays down the stability condition, which concerns the capability of the plant to "remain unchanged after repeated propagation". It has been highlighted that it is particularly challenging to test this requirement. Several tests have to be performed and that is the reason why a specific technical examination on stability is not carried out by the entitled official bodies across EU [16]. Undeniably, it would be quite demanding to repeatedly propagate a variety for an unlimited number of life cycles in order to check whether and to what extent the crop changed its expressed characteristics after each propagation. For this reason, it has been argued that stability tends to be more "the expression of an expectation than the reflection of a fact" [16]. Nevertheless, stability has proved to be linked with uniformity, so it can be assessed in connection with it [22]. As a result, a candidate variety is deemed to be stable when it is proved to be uniform during field tests.

Some experts suggest that the requirements of "uniformity" and "stability" should be replaced by the criterion of "identifiability" [23–25]. This criterion has been introduced for improved traditional varieties by the Malaysian Plant Variety Protection Act, in place of "uniformity" and "stability". The purpose is to facilitate plant variety protection for varieties developed by Malaysian farmers, which are not required to pass the rigorous uniformity and stability test, provided that their new variety is identified by one special characteristic [26].

More specifically, the uniformity requirement has been the subject of severe criticism because it refers exclusively to the needs of conventional agriculture, encouraging dangerous homogeneity and discouraging the conservation and sustainable use of agricultural genetic diversity [23].

The DUS requirements are tested during a technical examination of the candidate variety, aimed at assessing whether these requirements are met. This technical examination is not conducted by the CPVO itself, but rather by the competent office or offices in at least one of the Member States "entrusted with responsibility for the technical examination of varieties of the species concerned" (so-called Examination Office), in accordance with Article 55 of the Basic Regulation. The relevant Examination Office is chosen on the basis of defined principles, e.g., the climatic and environmental growing conditions of the specific variety and the preference expressed by the breeder during the application process [16].

The CPVO Technical Unit provides the Examination Offices with several technical guidelines and protocols aimed at harmonizing the determination of whether a plant variety right should be granted or not. The Examination Offices must conduct the DUS test in accordance with those guidelines, following the instructions given by the Office, as provided for in Article 56 of the Basic Regulation.

Unlike the DUS requirement, the novelty condition cannot be tested through growing trials. This requirement deals with the limited "period of grace" accorded to plant breeders, establishing for how long a candidate variety can be marketed by or with the consent of the breeder before applying for protection. After that period, the variety can no longer be protected by Community plant variety rights. Article 10 of the Basic Regulation states the principle upon which, in order to be protected by the CPVR, variety constituents or harvested material of the candidate variety must not have been sold or otherwise disposed of to others for purposes of exploitation (rectius, commercial exploitation) by the breeder or with their consent, before the date of application for Community plant variety right. As required by Article 10, the acts jeopardizing the grant of CPVRs must not have taken place (a) earlier than one year before the date of application for Community plant variety right within the territory of the EU; (b) earlier than four years or, in the case of trees or of vines, earlier than six years before the said date, outside the territory of the EU.

As required by Article 6 of the Basic Regulation, in order to be protected, a variety must be also designated by a denomination in accordance with the provisions of Article 63 of the Basic Regulation. The variety denomination enables the identification of the variety and it is intended to be its generic designation [27].

When granting the Community PVR, the Office must examine the suitability of the variety denomination pursuant to Article 63, according to which there is an impediment to the designation of a variety denomination where: (a) its use in the territory of the EU is precluded by the prior right of a third party; (b) it may cause difficulties as regards recognition or reproduction; (c) it is identical or may be confused with a variety denomination under which another variety of the same or of a closely related species is entered in an official register of plant varieties or under which material of another variety has been marketed in a Member State or in a UPOV Member State, unless the other variety no longer remains in existence and its denomination has acquired no special significance; (d) it is identical or may be confused with other designations which are commonly used for the marketing of goods or which have to be kept free under other legislation; (e) it is liable to give offence in one of the Member States or is contrary to public policy; (f) it is liable to mislead or to cause confusion concerning the characteristics, the value or the identity of the variety, or the identity of the breeder or any other party to proceedings.

Furthermore, there is another impediment where, in the case of a variety which has already been entered in an official register of plant varieties or material thereof and has been marketed there for commercial purposes: (a) in one of the Member States; or (b) in a UPOV Member State; or (c) in another State for which it has been established in an EU act that varieties are evaluated there under rules which are equivalent to those laid out in the Directives on Common Catalogues; and the proposed variety denomination differs from that which has been registered or used there, unless the latter is the object of an impediment.

3.3. *EU Seed Legislation*

In order to be legally and freely marketed in the EU territory, certain plant species must meet the legal requirements set out by the EU seed legislation, concerning variety registration and seed certification.

The legal framework of the European Union on the marketing of seed and plant propagating material dates back to 1966 and has been gradually amended. The seed legislation has been used as a policy instrument by favoring the release of varieties that contributed to achieving the original EU policy goals, especially food security [28].

Nowadays, the EU seed legislation consists of 12 basic Directives, covering the crops of major importance in the EU (EU listed species): one horizontal Directive regulates the Common Catalogue of varieties of agricultural plant species (i.e., Council Directive 2002/53/EC of 13 June 2002), while the other 11 vertical marketing Directives establish rules for the marketing of specific crops of major importance in the EU territory. These crops are: fodder plant seeds, cereal seed, beet seed, vegetable seed, seed potatoes, seed oil and fiber plants, material for the propagation of the vine, propagating material of ornamental plants, vegetable material other than seed, fruit propagating material and fruit plants for fruit production, forest reproductive material.

The two main elements of the EU seed legislation are: 1. the registration of the varieties in the Common Catalogues; 2. the certification of the seed lots.

According to the first pillar of the seed legislation, plant varieties of certain species should be registered in the national variety list, which the EU Member States are obliged to establish, and then in the EU Common Catalogues. The registration on the national list leads to the inclusion in the Common Catalogues, which green-lights the commercialization of the variety in the territory of all the EU Member States and ensures its free movement within the European Union.

The first Directive concerning the registrations of plant varieties and the creation of a Common Catalogue was Council Directive 70/457/EEC of 29 September 1970 on the Common Catalogue of varieties of agricultural plant species. This Directive had the significant task of setting up uniform criteria and minimum requirements for the compilation of national catalogues of agricultural plant species. However, the Directive was frequently and substantially amended and, ultimately, it was replaced by Council Directive 2002/53/EC.

There are currently two Common Catalogues in the European Union: the Common Catalogue of varieties of agricultural plant species, regulated by Council Directive 2002/53/EC, and the Common Catalogue of varieties of vegetable plant species, established by Council Directive 2002/55/EC on the marketing of vegetable seed.

As a general rule, a variety that has not been admitted in the relevant Common Catalogue cannot be commercialized in the EU. This principle has been underlined also by the Court of Justice of the European Union in the Case C–59/11, Association Kokopelli v. Graines Baumaux SAS. In that case, the Court of Justice also highlighted that the primary objective of the rules on seed registration is to improve agricultural productivity in the European Union, which is part of the objectives of the common agricultural policy as provided for in Article 39 (1) (a) TFEU.

In order to be registered on the national list, plant varieties are required to undergo the distinctness, uniformity, and stability testing (DUS requirement), according to Article 5 of Council Directive 2002/53/EC, paragraphs 1 to 3.

The content of the DUS requirement enshrined in Council Directive 2002/53/EC is equivalent to the provisions laid down in the Basic Regulation: this equivalence also concerns the DUS testing process. Therefore, the DUS requirement has the same meaning under the EU seed legislation and the Community plant variety protection.

When the breeder of a new variety accepted in the Common Catalogue applies for Community PVP within the limited period of grace, the technical report produced by the national examination authorities and concerning the DUS assessment for the variety registration is generally taken over by the CPVO, if it has been performed in conformity with specific requirements [29].

In this context, national authorities in the European Union have agreed to use CPVO technical protocols for DUS testing not only for plant variety rights grant procedures but also for the procedures concerning variety registration. Usually, the take-over of technical reports by CPVO applies only to reports produced by entrusted examination offices within the EU for a variety already benefiting from national plant variety rights or entered for national listing in an EU Member State [30].

However, the exchange of technical reports among examination authorities has not yet been regulated at the EU level, even though the CPVO, the entrusted Examination Offices, and the European Seed Association (ESA, now called "Euroseeds") have been calling for the recognition of the "one key, several doors principle" for more than fifteen years. In 2005 the CPVO launched a Strategic Discussion on the future of DUS testing in the EU territory: participating Member States and stakeholders found the "one key, several doors principle" to be a fundamental goal. The purpose was to harmonize the DUS variety testing system throughout the European Union, in order to increase efficiency and to avoid duplicated costs for both breeders and national authorities. The principle entails that a plant variety whose DUS requirements have been officially tested according to well-defined quality requirements and results in a final DUS report should not be examined a second time for DUS in the EU territory.

Furthermore, Council Directive 2002/53/EC has established that specific denomination rules must be followed to identify the varieties: Article 9 (6) thereof requires that Article 63 of the Basic Regulation shall apply. Therefore, the same denomination rules apply to both Community plant variety protection and EU seed legislation.

In addition to the DUS requirement, varieties of agricultural species, including cereals, must be of satisfactory value for cultivation and use (VCU). This is necessary in order to be admitted to the national catalogues and, consequently, to the Common Catalogue of varieties of agricultural plant species, as provided for in Article 4 of Council Directive 2002/53/EC.

VCU is tested through field trials but, unlike the DUS examination whose focus is on morphological characters of the variety, in VCU testing the emphasis is on the assessment of the agronomic traits of the plant. These traits are related to crop production and performance. The purpose of this provision is to filter only those varieties having a specific economic value for farmers. It should ensure that only the finest varieties of agricultural species are registered, thus stimulating the breeding of improved crops. Therefore, only the varieties with a significant economic value are placed on the market, and this is deemed necessary to obtain a high-quality harvest to the greatest degree possible [16].

The reason is that farmers need to know well in advance how the variety is going to perform in the field. For this purpose, the marketing of economically valuable varieties is of utmost importance. Farmers must be able to get the most suitable varieties and the most reliable information on the agronomical value for cultivation, in order to optimize the growing practices and to achieve the best yield potential [31].

It must be noted that vegetable seeds are not subject to the VCU requirement. In this case, the Council Directive 2002/55/EC of 13 June 2002 on the marketing of vegetable seed is applied. According to Article 4 of said Directive, Member States shall ensure that a vegetable variety is accepted if it meets the DUS requirement. There is just one exception: in the case of species of industrial chicory, the varieties must be of satisfactory VCU.

The VCU requirement is not applied to vegetable varieties for different reasons. Firstly, because of "the large number of agronomic considerations and specific consumer preferences in these crops" [32]. It has also been underlined that the assessment of the VCU requirement would be complicated and costly because of the highly differentiated vegetable crop market [31].

With regard to the VCU requirement, the scope of the term "satisfactory" is determined by Article 5 (4) of Council Directive 2002/53/EC on the Common Catalogue of varieties of agricultural plant species. According to the article, a plant variety shall be regarded as of satisfactory VCU if, compared to other varieties accepted in the catalogue of the Member State in question, its qualities, taken as a whole, offer, at least as far as production in any given region is concerned, a "clear improvement" either (1) for cultivation or (2) as regards the uses which can be made of the crops or (3) the products derived therefrom.

In 2003, the Commission Directive 2003/90/EC was adopted, setting out implementing measures as regards the minimum characteristics to be covered during the examination and the minimum conditions for examining certain varieties of agricultural plant species. It also set out the conditions that the varieties have to comply with as regards the VCU requirement. These conditions, which are listed in Annex III of the Directive, are (1) yield; (2) resistance to harmful organisms; (3) behavior with respect to factors in the physical environment; (4) quality characteristics.

The broad meaning of these conditions required national authorities to specify the limits of those criteria: this led to dissimilarities in VCU assessment among Member States. Therefore, currently there is no uniform meaning for those criteria: the parameters for VCU assessment vary significantly between countries for the same variety [32].

The VCU examination involves replicated field trials and harvest tests that generally take a minimum of two years or a maximum of three years [32]. The legislation dealing with the VCU requirement seems not to consider the strong connection between the value for cultivation and use of a plant variety and the surrounding environment.

Seeds are living organisms in a living environment, where environmental factors have a substantial effect on the economically valuable characteristics of plant varieties, such as yield and quality features. More specifically, the outcomes of the VCU examination are affected by (1) the testing location and the specific climatic and soil conditions; (2) the agronomic practices; (3) the year, since the performance of the variety varies from one year to another. Therefore, the VCU of a particular variety can never be definitively established [31].

This is the reason why the role and effectiveness of the VCU testing have been questioned. It has been underlined that VCU testing does not necessarily predict the need of farmers for particular characteristics or the exact performance of the plant in a specific field: the results of the trials cannot represent a perfect assessment of the VCU of the variety. Therefore, it has been suggested that these trials should not be lengthy or complex in the pursuit of perfection [32].

Furthermore, the effect of the VCU assessment on the relevant crop sector, i.e., the agricultural sector, should be taken into consideration. The effectiveness of VCU may be questioned since "crop sectors where VCU is not compulsory and several third countries with no regulatory tests are considered as competitive as the regulated VCU crop sectors" [31]. Indeed, the productivity of the agricultural crop sector in the EU has increased not more or less than the productivity in other crop sectors where VCU is not compulsory (e.g., the vegetable sector) [31].

Another aspect related to the effectiveness of the VCU criteria concerns the burden for breeders to have their variety assessed for VCU, in particular the costs for such assessment. This could represent a limit for SMEs who want to register their varieties of agricultural crop species. Indeed, the evaluation of the value for cultivation and use does not come for free. In light of this, the International Federation of Organic Agriculture Movement (IFOAM) has suggested that VCU testing should no longer be compulsory but an optional requirement "for any species", not only for agricultural ones. According to this proposal, VCU should be used solely "as a marketing argument" by breeders and not as a compulsory requirement [33].

4. Discussion

The EU seed legislation establishes the requirements that certain plant species must meet in order to be freely marketed in the EU. In so doing, it affects the possibility for breeders to commercially exploit their IP right on the new plant variety. In fact, when a protected agricultural variety does not meet the registration requirement set out by Directive 2002/53/EC, the Community PVR holder is not able to perform the acts set out in Article 13 (2) of the Basic Regulation, e.g., offering for sale, selling or other marketing.

This restriction is provided for in Article 13 (8) of the Basic Regulation, stating that the exercise of the rights conferred by Community plant variety rights may not violate any provisions adopted on the grounds of public morality, public policy or public security, the protection of health and life of humans, animals, or plants, the protection of the environment, the protection of industrial or commercial property, or the safeguarding of competition, of trade or of agricultural production.

These interests could prevail over the breeder's right to perform the acts listed in Article 13 (2) and, thus, they could restrict the exercise of the Community plant variety right. In this case, the exercise of the Community plant variety right is restricted when the variety intended to be marketed in the EU territory does not comply with the seed legislation, which aims at safeguarding agricultural production.

As already stated, a cereal variety can be accepted in the Common Catalogue of varieties of agricultural plant species only if the variety is distinct, stable, and sufficiently uniform (DUS) and has a satisfactory value for cultivation and use (VCU), in addition to having a proper denomination.

It follows that cereal breeders that aim at protecting their new variety by Community PVP must take into consideration both the requirements for Community plant variety protection and those set out by the EU seed legislation to commercially exploit their innovation, as shown in Table 1.

Table 1. Requirements for seed marketing and plant variety protection of new cereal varieties in the EU.

Community Plant Variety Protection	EU Seed Legislation
Novelty	
Distinctness	*Distinctness*
Uniformity	*Uniformity*
Stability	*Stability*
Variety denomination	*Variety denomination*
-	*Value for cultivation and use*

More specifically, the breeding activities on cereals and other agricultural species in the EU are conditioned by the need for breeders to develop a variety that meets the VCU requirement, in addition to the DUS requirements.

However, the VCU is lightly defined at the EU level and the specific criteria adopted at a national level for the VCU assessment differ among Member States. The lack of harmonization can lead to distortions in the internal market and fail to guarantee the competition among EU breeders on a level playing field: differences include evaluation criteria, setup, and fee amount [28].

Furthermore, it should be highlighted that the VCU trial takes place in certain testing sites of each Member State. The living environment of each testing site can deeply affect the VCU assessment, so that the variety can perform differently in different environments. In fact, the VCU requirement does not have any pan-European value [31]. However, once the variety is registered in the Common Catalogue, its commercialization is not limited to a certain area but it can be marketed throughout the European Union, even in areas where the VCU has never been assessed. This leads to a paradox: the registration of the variety in the Common Catalogue determines the marketability of the variety throughout the EU territory, even though the VCU assessment, whose negative outcome can hinder that registration, is carried out in very specific testing sites.

Another critical connection occurs during the process for the grant of Community plant variety rights and concerns the technical examination for the DUS assessment, required both for plant variety protection and variety registration. It concerns, in particular, the lack of recognition of the "one key, several doors principle". Therefore, in the case where the variety has already been officially tested for DUS and this results in an official report, the same variety should not be examined a second time for DUS in the EU territory. The official technical report should be taken over by the EU Member State or the CPVO, regardless of its positive or negative outcome. Harmonizing the DUS testing system throughout the European Union will increase efficiency and avoid uncertainty and duplicated costs, both for breeders and national authorities. Indeed, one key (the official DUS report) would open several doors (variety registration and plant variety protection in the European Union).

Moreover, even though the VCU requirement and the uniformity criterion are considered as a useful tool for conventional agriculture, they are seen as an obstacle for the selection of varieties adapted to organic farming. On the one hand, VCU trials are not able to select low-input varieties used in organic farming [31]. On the other, the uniformity requirement discourages the marketing of heterogeneous groupings of plants, not taking into account the interests of local communities [24] and the needs of organic heterogeneous material, despite the F2F target to increase the share of organic agriculture to 25% by 2030 [12].

In light of the foregoing consideration, it is not surprising that in 2013 there was a proposal by the European Commission to replace the existing seed legislation, constituted by 12 Directives, by one single Regulation in order to update, simplify, and harmonize the current rules. The proposal had the purpose, inter alia, to extend the role of CPVO to plant variety registration in order to simplify the registration process and to allow the admission to the EU market either via registration on the national list or via direct application to the CPVO. This proposal represented a first step towards the coordination between EU seed legislation and Community plant variety protection [34].

However, the proposal was rejected in 2014 by the European Parliament. The Members of the European Parliament believed that the proposal gave too much power to the Commission and did not properly consider the different requirements of existing plant reproductive material, as highlighted in the report of the Committee on Agriculture and Rural Development and the opinion of the Committee on the Environment, Public Health and Food Safety (A7-0112/2014).

Nevertheless, a legislative revision could simplify and improve the legal framework on seed marketing, and it could also facilitate the relationship between the EU seed legislation and Community plant variety protection, thereby promoting innovation in the seed sector.

5. Conclusions

Innovation is the zeitgeist of contemporary society and, with regard to cereal varieties, it has a significant social, economic, and environmental weight. The EU breeding companies are actively engaged in cereal breeding activities and, in this context, the Community plant variety protection is supposed to stimulate the breeding and development of new varieties, and, ultimately, to foster innovation.

Although some legal practitioners initially thought that the Community plant variety protection was going to wither and die [35], Community PVP system is considered an incentive to develop new varieties, thus contributing to the achievement of the European Green Deal objectives [36]. However, the promotion of innovation in plant breeding may be discouraged by an outdated and uncoordinated legal framework.

As proposed by some authors [14], a better place for agriculture should be found in intellectual property law. Since agricultural law can affect the exercise of intellectual property rights on plant varieties, it is fundamental to update, simplify, and coordinate the existing laws.

A lack of coordination could lead to inadequate protection of the breeder's rights and, along with the obsolescence of certain legal provisions, could hinder innovation in the seed and agricultural sectors. In this context, only big firms have the capabilities required to overcome the obstacles of regulatory frameworks, whereas the developments of small companies or official institutions can be left out [1].

A peculiar aspect of the critical relationship between CPVP and EU seed legislation can be found in the requirements laid down for their specific purpose.

The two laws require that the variety must be tested for DUS, the content of which is equivalent. In light of this, the EU legislator should regulate the take-over of official DUS technical reports and recognize the "one key, several doors principle", in order to harmonize the DUS variety testing system throughout the European Union, thus increasing efficiency, avoiding uncertain outcomes and reducing costs.

However, the "one key, several doors principle" should not be perceived as a panacea that will alone guarantee the coordination between CPVP and EU seed legislation.

A legislative revision should concern also the VCU requirement and the procedure for its assessment. Even though the DUS report will be taken over, a protected cereal variety might not be allowed for marketing in the European Union if it does not meet the VCU requirement set out by the EU seed legislation, despite the fact that a Community plant variety right has been granted on it. A discussion should be initiated around the mandatory role of this requirement, considering its lack of compulsoriness for vegetable species. In addition, it is necessary to consider the dissimilarities in VCU assessment among Member States, which can affect competition on a level playing field in the internal market. It must be noticed that the EU Commission has recently published a study on the Union's options to update the existing legislation on the production and marketing of plant reproductive material [37].

Moreover, the current legal framework is considered outdated. It should take into consideration the current sustainability challenge and the need to implement the F2F strategy and achieve its ambitious goals, by promoting the breeding of new plant varieties capable of adapting to and mitigating the impact of climate change and used in low input production systems, such as organic production. Therefore, rules on a sustainable VCU should be introduced in the EU by adopting specific requirements for the marketing of plant varieties of agricultural species, in order to meet the F2F targets.

Since plant breeding is fundamental to achieving the goals set out by the F2F strategy, the revision of the legislation should consider the possibility of adding specific sustainability characteristics to the four criteria for VCU testing established in the current legislation. In particular, the target of breeding programs should not be focused solely on yield and productivity but on the climate resilience of the major crop species that play a key role in food security, such as cereals [38].

In addition, the protection and registration requirements must take into account not only the conventional agriculture system but also the specific needs of organic farming, promoting the breeding of varieties and plant reproductive material suitable for organic production according to the objectives of the Organic Regulation (EU) 2018/848, such as organic heterogeneous material.

In conclusion, the EU legislation is certainly a policy instrument and can shape innovation in the field of plant species, by favoring the protection of varieties with specific features and establishing the criteria varieties must meet in order to be marketed throughout the European Union territory. However, the primary objective of the current laws, adopted decades ago, is to improve agricultural productivity, focusing exclusively on the needs of conventional agriculture.

Nevertheless, current challenges differ from past ones. Nowadays, the legislation must drive innovation in order to accelerate the transition to sustainable, healthy, and inclusive food systems from primary production to consumption. Therefore, intellectual property and agricultural law must be coordinated, taking into consideration the changing demands of society and the current sustainability challenges highlighted by the F2F strategy.

In light of this, the strong relationship between the EU seed legislation and Community plant variety protection should be taken into account by the EU legislator, especially with regard to agricultural species. It is necessary to update, coordinate, and simplify the existing legal framework. Even though a legislative review could carry the risk of opening Pandora's box, in a fast-changing world, the law requires constant reconsideration in order to evaluate whether it is still able to achieve the current policy goals.

There is a need to stimulate innovation in cereal varieties to face the challenges related to food and agricultural production in the third millennium, as highlighted by the F2F strategy. In light of this, the European Union will only be really able to effectively foster innovation in plant breeding when the legislator recognizes the strong link between the fields of intellectual property and agricultural law, providing for legislative review and better coordination within the existing legal framework, especially in light of the current policy goals.

Funding: This research was partly funded by the University of Macerata, Agroservice Spa, and Regione Marche (EUREKA Ph.D. grant, D.R. n. 165—06/06/2016). The APC was funded by the University of Macerata, Department of Law.

Institutional Review Board Statement: Not applicable.

Informed Consent Statement: Not applicable.

Data Availability Statement: Not applicable.

Conflicts of Interest: The author declares no conflict of interest.

References

1. Rapela, M.A. *Fostering Innovation for Agriculture 4.0: A Comprehensive Plant Germplasm System*; Springer Nature: Cham, Switzerland, 2019. [CrossRef]
2. Blakeney, M.; Kadambot, H.M.S. (Eds.) *Local Knowledge, Intellectual Property and Agricultural Innovation*; Springer Nature: Singapore, 2020. [CrossRef]
3. Campos, H. (Ed.) *The Innovation Revolution in Agriculture*; Springer Nature: Cham, Switzerland, 2021. [CrossRef]
4. Fess, T.L.; Kotcon, J.B.; Benedito, V.A. Crop breeding for low input agriculture: A sustainable response to feed a growing world population. *Sustainability* **2011**, *3*, 1742–1772. [CrossRef]
5. Arora, N.K. Impact of climate change on agriculture production and its sustainable solutions. *Environ. Sustain.* **2019**, *2*, 95–96. [CrossRef]
6. FAO; IFAD; UNICEF; WFP; WHO. *The State of Food Security and Nutrition in the World 2019. Safeguarding Against Economic Slowdowns and Downturns*; FAO: Rome, Italy, 2019. Available online: http://www.fao.org/3/ca5162en/ca5162en.pdf (accessed on 20 May 2021).
7. FAO. *The Future of Food and Agriculture: Trends and Challenges*; FAO: Rome, Italy, 2017. Available online: http://www.fao.org/3/a-i6583e.pdf (accessed on 20 May 2021).
8. Cassman, K.G.; Dobermann, A.; Walters, D.T.; Yang, H. Meeting cereal demand while protecting natural resources and improving environmental quality. *Annu. Rev. Environ. Resour.* **2003**, *28*, 315–358. [CrossRef]
9. FAO. *The Second Report on the State of the World's Plant Genetic Resources for Food and Agriculture*; FAO: Rome, Italy, 2010. Available online: http://www.fao.org/documents/card/en/c/6ac34ffd-7a66-5d42-9573-3d09491ad39a/ (accessed on 20 May 2021).
10. Nawaz, M.A.; Chung, G. Genetic improvement of cereals and grain legumes. *Genes* **2020**, *11*, 1255. [CrossRef] [PubMed]
11. OECD. *Concentration in Seed Markets: Potential Effects and Policy Responses*; OECD Publishing: Paris, France, 2018. Available online: https://www.oecd.org/publications/concentration-in-seed-markets-9789264308367-en.htm (accessed on 20 May 2021).
12. European Commission. *A Communication from the Commission to the European Parliament, the Council, the European Economic and Social Committee and the Committee of the Regions—A Farm to Fork Strategy for a Fair, Healthy and Environmentally-Friendly Food System*; COM/2020/381 Final; European Commission: Brussels, Belgium, 2020. Available online: https://eur-lex.europa.eu/legal-content/EN/TXT/?uri=CELEX:52020DC0381 (accessed on 20 May 2021).
13. Mc Conville, M.; Hong Chui, W. (Eds.) *Research Methods for Law*; Edinburgh University Press: Edinburgh, UK, 2007.
14. Chapman, S.; Sherman, B. Finding a place for agriculture in intellectual property law. *Int. Rev. Intellect. Prop. Compet. Law* **2018**, *49*, 759–762. [CrossRef]
15. UPOV. *Seminar on the Nature and Rationale for the Protection of Plant Varieties under the UPOV Convention*; UPOV/PUB/697; UPOV: Geneva, Switzerland, 1990. Available online: https://www.upov.int/edocs/pubdocs/en/upov_pub_697.pdf (accessed on 20 May 2021).
16. Würtenberger, G.; Van Der Kooij, P.; Kiewiet, B.; Ekvad, M. *European Union Plant. Variety Protection*; Oxford University Press: Oxford, UK, 2015.
17. Straus, J. Plant Variety Protection. In *The Max Planck Encyclopedia of European Private Law*; Basedow, J., Hopt, K.J., Zimmermann, R., Eds.; Oxford University Press: Oxford, UK, 2012; Volume 2.
18. Blakeney, M. Agricultural Innovation: Patenting and Plant Variety Rights Protection. In *International Food Law and Policy*; Steier, G., Patel, K., Eds.; Springer International Publishing: Cham, Switzerland, 2016.
19. Louwaars, N.P. Seed policy, legislation and law: Widening a narrow focus. *J. New Seeds* **2002**, *4*, 1–14. [CrossRef]
20. Reichman, J.H. Legal hybrids between the patent and copyright paradigms. *Columbia Law Rev.* **1994**, *94*, 2432–2558. [CrossRef]
21. UPOV. *Actes des Conférences Internationales Pour la Protection des Obtentions Végétales 1957–1961*; UPOV/PUB/316; UPOV: Geneva, Switzerland, 1972. Available online: https://www.upov.int/edocs/pubdocs/fr/upov_pub_316.pdf (accessed on 20 May 2021).
22. UPOV. *General Introduction to the Examination of Distinctness, Uniformity and Stability and the Development of Harmonized Descriptions of New Varieties of Plants*; TG/1/3; UPOV: Geneva, Switzerland, 2002. Available online: https://www.upov.int/export/sites/upov/publications/en/tg_rom/pdf/tg_1_3.pdf (accessed on 20 May 2021).
23. Leskien, D.; Flitner, M. *Intellectual Property Rights and Plant Genetic Resources: Options for a Sui Generis System. Issues in Genetic Resources No. 6*; International Plant Genetic Resources Institute: Rome, Italy, 1997.
24. Seiler, A. Sui generis systems: Obligations and options for developing countries. *Biotechnol. Dev. Monit.* **1998**, *34*, 2–5.
25. Correa, C.M.; Shashikant, S.; Meienberg, F. *Plant Variety Protection in Developing Countries: A Tool for Designing a Sui Generis Plant Variety Protection System: An Alternative to UPOV 1991*; Berne Declaration, Development Fund, SEARICE, Third World Network; Association for Plant Breeding for the Benefit of Society (APBREBES) and its Member Organizations: Alfter, Germany, 2015.
26. Adhikari, K.; Jefferson, D.J. (Eds.) *Intellectual Property Law and Plant. Protection: Challenges and Developments in Asia*; Routledge: Abingdon, UK, 2020.
27. UPOV. *Explanatory Notes on Variety Denominations under the UPOV Convention*; UPOV/INF/12/4; UPOV: Geneva, Switzerland, 2012. Available online: https://www.upov.int/edocs/infdocs/en/upov_inf_12_4.pdf (accessed on 20 May 2021).
28. Osman, A.M.; Bonthuis, H.; Van den Brink, L.; Struik, P.C.; Almekinders, C.J.M.; Lammerts Van Bueren, E.T. Adapting value for cultivation and use testing to stimulate the release of improved varieties for the organic sector. The case of spring wheat in The Netherlands. *Org. Agric.* **2015**, *5*, 101–111. [CrossRef]

29. Kiewiet, B. *The Community Plant Variety Protection System*; Angers, France, 2009. Article Based on a Presentation Made on 29 May 2008 in Genoa, Italy. Available online: https://cpvo.europa.eu/sites/default/files/documents/articles/2009-07-10_Article_Italy.pdf (accessed on 20 May 2021).
30. CPVO. What is a 'Take-Over' of a DUS Report? Available online: https://cpvo.europa.eu/en/help-center/faq/what-take-over-dus-report (accessed on 25 May 2021).
31. Food Chain Evaluation Consortium. *Evaluation of the Community Acquis on the Marketing of Seed and Plant Propagating Material (S&PM). Assignment 5 of the Framework Contract for Evaluation and Evaluation Related Services—Lot 3: Food Chain*; Final Report; European Commission—DG SANCO: Brussels, Belgium, 2008. Available online: https://ec.europa.eu/food/system/files/2016-10/ppm_legis_review_s_pm_evaluation_finalreport.pdf (accessed on 25 May 2021).
32. Turner, M.; Bishaw, Z. *A Review of Variety Release Procedures and Related Issues with Recommendations for Good Practice*; ICARDA Working Papers: Beirut, Lebanon, 2016.
33. IFOAM. *Position Paper. Towards More Crop Diversity—Adapting Market Rules for Future Food Security, Biodiversity, and Food Culture*; IFOAM: Brussels, Belgium, 2013.
34. European Commission. *Proposal for a Regulation of the European Parliament and of the Council on the Production and Making Available on the Market of Plant Reproductive Material (Plant Reproductive Material Law)*; COM (2013) 262 Final; European Commission: Brussels, Belgium, 2013. Available online: https://eur-lex.europa.eu/legal-content/EN/TXT/?uri=CELEX%3A52013PC0262 (accessed on 20 May 2021).
35. Llewelyn, M.; Adcock, M. *European Plant. Intellectual Property*; Hart Publishing: Oxford, UK, 2006.
36. European Commission. *Communication from the Commission to the European Parliament, the Council, the European Economic and Social Committee and the Committee of the Regions. Making the Most of the EU's Innovative Potential. An Intellectual Property Action Plan to Support the EU's Recovery and Resilience*; COM(2020) 760 Final; European Commission: Brussels, Belgium, 2020. Available online: https://eur-lex.europa.eu/legal-content/EN/TXT/?uri=CELEX%3A52020DC0760 (accessed on 20 May 2021).
37. European Commission. *Commission Staff Working Document. Study on the Union's Options to Update the Existing Legislation on the Production and Marketing of Plant Reproductive Material*; SWD(2021) 90 Final; European Commission: Brussels, Belgium, 2021. Available online: https://ec.europa.eu/food/system/files/2021-04/prm_leg_future_prm-study_swd-2021-90.pdf (accessed on 25 May 2021).
38. Kahiluoto, H.; Kaseva, J.; Balek, J.; Olesen, J.E.; Ruiz-Ramos, M.; Gobin, A.; Kersebaum, K.C.; Takáč, J.; Ruget, F.; Ferrise, R.; et al. Decline in climate resilience of European wheat. *Proc. Natl. Acad. Sci. USA* **2019**, *116*, 123–128. [CrossRef] [PubMed]

sustainability

MDPI

Article

Contribution of Selected Factors on Farmers' Work Performance towards Fertilizer Application in Rice of Bangladesh

Thahamina Bagum [1,2], Md. Kamal Uddin [2,*], Salim Hassan [3], Nitty Hirawaty Kamarulzaman [4], Md. Zulfikar Rahman [5] and Ahmad Numery Ashfaqul Haque [2]

1. Department of Agricultural Extension & Information System, Sher-E-Bangla Agricultural University, Dhaka 1207, Bangladesh; tbegumr@gmail.com
2. Department of Land Resource Management, Universiti Putra Malaysia, Serdang 43400, Malaysia; numerybau@gmail.com
3. Department of Agriculture Technology, Universiti Putra Malaysia, Serdang 43400, Malaysia; salimhassan@upm.edu.my
4. Department of Agribusiness and Bioresource Economics, Universiti Putra Malaysia, Serdang 43400, Malaysia; nitty@upm.edu.my
5. Department of Agricultural Extension Education, Bangladesh Agricultural University, Mymensingh 2202, Bangladesh; zulfikarbau64@gmail.com

* Correspondence: mkuddin@upm.edu.my

Abstract: There is enormous possibility to increase rice yield in Bangladesh. Inefficient and often imbalanced fertilizer use impedes farmers from achieving expected yields. It is evident from past research that farmers have resorted to applying fertilizers at inappropriate rates that do not match well with the nutrient requirement of certain crops. Therefore, this study explores the contribution of selected factors that influence farmers' work performance and determine the highest contributing factors on farmers' work performance towards fertilizer application in rice. This research used a multistage simple random sampling method to select 355 farmers from twenty-one rice production areas of Bangladesh. Data, collected using a structured questionnaire, were subjected to multiple linear regression analysis to explore the contribution of selected factors and identify the highest contributing factors towards farmers' work performance. Results revealed that all the factors explained 56.1% of the variance in farmers' work performance. Motivation of farmers was found to be the highest contributing factor, followed by knowledge that influences farmers' work performance. The study concludes that farmers need to be equipped with essential knowledge and motivation crucial to strengthening their work performance as this will subsequently increase rice production.

Keywords: rice farmers; fertilizer; knowledge; attitude; ease of use; motivation; work performance

Citation: Bagum, T.; Uddin, M.K.; Hassan, S.; Kamarulzaman, N.H.; Rahman, M.Z.; Haque, A.N.A. Contribution of Selected Factors on Farmers' Work Performance towards Fertilizer Application in Rice of Bangladesh. *Sustainability* **2021**, *13*, 10795. https://doi.org/10.3390/su131910795

Academic Editor: Michael Blakeney

Received: 15 July 2021
Accepted: 18 September 2021
Published: 28 September 2021

Publisher's Note: MDPI stays neutral with regard to jurisdictional claims in published maps and institutional affiliations.

Copyright: © 2021 by the authors. Licensee MDPI, Basel, Switzerland. This article is an open access article distributed under the terms and conditions of the Creative Commons Attribution (CC BY) license (https://creativecommons.org/licenses/by/4.0/).

1. Introduction

Food security of Bangladesh largely depends on rice production. Rice plays the leading role in all crops by contributing 92% of the total food grain production [1]. Moreover, the country's agro-climatic conditions are perfect for cultivating rice all year [2]. However, the national average rice yield (2.60 t/ha) is much lower than the potential national yield (5.40 t/ha) and compared to other rice-growing countries [3]. A gap, ranging from 1 to 3 t/ha, still exists in between yields currently obtained by farmers and what could be achieved with improved management practices [4]. Apart from this, the population of Bangladesh is currently 162.7 million and projected to be 189.85 million by the year 2030, and thus it would require about 42.50 mt of rice [5]. In addition, the rate of population growth and the level of rice consumption are still relatively high. Simultaneously, rice-growing land needs to be share for cultivating new crops. Therefore, the current rice yield of 2.74 t/ha needs to be increased to 3.74 t/ha [6], to keep the production of rice in line with the growing population of the country.

Balanced fertilization is the key factor in enhancing the production of rice. Fertilizers have the quick capacity to restore depleted soil nutrients, improve soil fertility, and increase the yield of rice [7,8]. Moreover, with appropriate fertilizer management, farmers can potentially increase rice yields [9,10]. The results of long-term experiments also showed that fertilizers can increase output by 40–60% in grain production [11]. Therefore, timely application of fertilizer at the recommended rate is essential for improving rice yield.

However, the current level of applying fertilizer is significantly lower than the recommended levels for almost all food crops. Farmers have resorted to using imbalanced fertilization that does not match nutrient requirements and soil fertility rates [12]. About 51% of farmers applied fertilizer as recommended and the rest were less than the recommended rate [13]. As a result, a substantial gap still exists between the actual and recommended rate of all major fertilizers that are applied in rice cultivation at the farm level [2]. This gap between the recommended rate and the actual amount of fertilizer application is found much higher for TSP and MP than for urea. Farmers' application of fertilizer widely varies from the recommended doses as suggested by the Soil Resource and Development Institute (SRDI) and Department of Agriculture Extension (DAE) [2]. Farmers are habituated to apply imbalanced fertilizers that cannot meet the demand for soil nutrients and rice yield [12]. It implies that farmers' work performance to apply fertilizer is insufficient with the requirement of crops. Therefore, there is a gap between the actual rate and the recommended rate of fertilizer [2,14].

Individuals' performance towards efficient use of fertilizer can close the gap between actual and potential outputs [15]. Peoples' work performance is highly influenced by their ability to use knowledge and skills for a required task [16]. Therefore, farmers' work performance signifies their ability to carry out farm activities that lead to higher production [17]. Rice yields can be increased with the help of the timely applying of fertilizer. Hence, farmers need to apply the recommended dose of fertilizer at the proper time and using the appropriate method for sustainable production [18].

Besides, a wide range of factors such as farmers' demographical, psychological, and economic characteristics influence their performance behavior regarding agricultural practices [19]. Moreover, work performance is linked to the factors such as knowledge, skills, abilities, motivation, and personality traits [20,21]. In addition, identifying the highest contributing factors can facilitate the implication of the study findings in an efficient way. Previous researchers [15,17,22] also found the highest contributing factors on respondents' work performance and recommended guidelines for identifying areas that needed to be intensified for improved work performance of people concerned. Therefore, it is crucial to explore the factors that influence strengthening the work performance of farmers towards fertilizer application.

Several studies were conducted in Bangladesh on fertilizer issues [2,23]; however, research is very rare to study how various factors impact farmers' work performance towards fertilizer application in rice. Given this lack of research, this study, therefore, formulated the following objectives:

1. To explore the contribution of selected factors that influence farmers' work performance towards fertilizer application in rice;
2. To determine the highest contributing factors on farmers' work performance towards fertilizer application in rice.

The current study makes a unique contribution to the existing literature by providing a comprehensive analysis of the factors influencing the work performance of farmers towards fertilizer application in rice in Bangladesh.

The remaining sections of this paper are organized as follows: Section 2 reviews the existing literature on factors that influence farmers' work performance while Section 3 describes the methodology. Section 4 explains the results and discussion. Finally, the conclusions are presented in Section 5.

2. Review of Literature

Fertilizer is a kind of production input whose demand cannot be circumvented to obtain maximum yield and ensure sustainable crop production. Despite there were many empirical studies conducted nationally and internationally on the importance of using fertilizer for crop production, very few studies focused on farmers' work performance, especially on fertilizer application in rice.

Work performance is regarded as one of the most important dependent variables in several studies [24]. In general, work performance refers to an individual's quantifiable behaviors and outcomes that contribute to a specific goal [25]. Ufuophu-Biri and Iwu [26] described work performance as an instrument used to propel productivity while factors like motivation and skill serve as the processes of attaining productivity. According to Armstrong and Taylor [27], performance is the behavior that achieves results. Individual performance is a key variable in organizational behavior studies. In this study, the farmers' work performance is measured based on their objectives and the actual activities to achieve higher production.

Several interventions such as training, goal setting, feedback, incentives, and supervision may enhance work performance by improving individuals' knowledge, skill, and motivation [28]. Individuals' knowledge and attitude are treated as important determinants of performance [21,29]. In addition, adequate motivation is a crucial factor for higher performance [30]. Concerning the technological characteristics, the factor like perceived ease of use of technology is an important determinant for work performance [31]. Moreover, the quality of work performance is also related to specifications to fit when a person works. Knowledge, abilities, motivation, and other attributes are defined as specifications of work requirement [20].

2.1. Research Gap

Prior studies have identified a number of factors that might influence the performance of the concerned respondents. A brief list of empirical studies on respondents' performance has been mentioned in the following Table 1.

Table 1. List of empirical studies on respondents' performance.

Authors	Dependent Variable	Independent Variables	Major Findings
Demba [17]	Work Performance of Paddy Farmers	Decision making, investment, discipline, information seeker, risk taking, networking and problem solving	Discipline identified as the most contributing personality traits of farmers and regression model explained 58% of the total variance of paddy farmers' work performance.
Oluwatoyin [22]	Extension Agents' Work Performance	Technical skill, technology delivery skill, technology evaluation skill, leadership skill, decision making support skill and social skill	Technical skill, technology delivery skill, leadership skill, decision making support skill and social skill were found to be the significant predictors which jointly explained 59% of the total variance of extension agents' work performance.
Bagum [15]	Farmers' Performance	age, educational level, household size, farm size, annual income, extension media contact, training received, knowledge and attitudes	Farmers' age, household size, farm size, training received, extension media contact, knowledge and attitudes were identified as the most significant predictors that explained 45.3% of the variance of their fertilizer application performance
Nkari and Kibera [32]	Performance of commercial fruit and vegetable farmers	Farmer characteristics	Farmer characteristics accounted for 5.5% of the variation of the performance of commercial fruit and vegetable farmers.
Djomo et al. [33]	Performance of smallholder rice farmers	Farming experience, farm size, rice variety, extension visit, credit received, and rice output	Rice variety; extension visit; rice output; farming experience; farm size and credit received were found positive and significantly influenced the performance of smallholder rice farmers, explained 97.8% of the variation of their performance.
Maican et al. [34]	Farm economic performance	Farmers' motivation and job satisfaction	Farm's economic performance was influenced more by farmers' motivation than job satisfaction.

Table 1 summarizes personality traits, individual's skill, organizational commitment, job involvement, motivation, knowledge, job satisfaction, farm management practices, socio-economic characteristics of the respondents as the predictors of individuals' work performance. However, prior studies did not consider farmers' cognitive (i.e., knowledge) and affective (i.e., attitude and motivation) responses and technological characteristics (i.e., ease of use) in a single study rather examined their roles in separate studies, especially in the context of farmers. Therefore, the combined effect of those determinants was quite unknown. Nonetheless, farmers' knowledge, attitude, ease of use of technology, and motivation are crucial factors that need emphasis for the application agricultural technologies by the farming community. Farmers' knowledge and attitude are essential for assessing their technology using behavior [35]. Motivation plays a vital factor in the decision-making process at the farm level [36], while perceived attributes of certain practices such as being easy to apply had the greatest influence on their technology application decisions [37]. Moreover, in the context of Bangladesh, the influences of knowledge, attitude, ease of use of technology, and motivation of farmers on their work performance regarding fertilizer application have not been study before. Therefore, to fulfill this research gap, the researcher selected knowledge, attitude, ease of use of technology, and motivation as influential factors for determining farmers' work performance towards fertilizer application in rice.

2.2. *Development of Hypotheses*

The efficiency of knowledge in any process positively affects the individuals' performance [38]. Concerning farm management, agricultural knowledge is the main asset for farmers. With good knowledge, farmers would know what information they need, who to look for, and what to do in their tasks [39]. Empirical research executed by Chi et al. [40] on the effect of knowledge on performance showed that effective management of knowledge positively and significantly affected overall performance. Elsewhere, Masa'deh et al. [41] mentioned that knowledge is the enabler of firm performance. All these results extend the current understanding of knowledge as a predictor of performance. Based on the above discussion the following hypothesis has been developed:

Hypothesis 1 (H1). *Knowledge has a significant contribution to farmers' work performance towards fertilizer application in rice.*

Peoples' attitude towards certain objects or technology determines their behavioral intention and actual behavior performance [42]. Attitudes can be a strong predictor of behavior, i.e., performance [43]. Peoples' positive attitude influences their decision to adopt certain practices for their farming [44]. Moreover, the University of Minnesota Libraries Publishing [45] illustrates attitudes towards work as one of the major predictors of performance. The following hypothesis has been formed in light of the preceding discussion:

Hypothesis 2 (H2). *Attitude has a significant contribution to farmers' work performance towards fertilizer application in rice.*

Perceived ease of use (PEU) is one of the most important factors affecting technology use intention and a proxy to actual user behavior [46,47]. Farmers' decision to adopt technologies is dependent on how they perceive those technologies. Reimer et al. [48] informed that farmers' intention to perform certain practice in the United States Midwest region was found to be impacted by their subjective evaluation about the complexity of those practices. Bahramzadeh and Shokati [49] concluded that perceived ease of use is the most powerful factor in behavioral intention to perform a certain technology. Hence, it is clear that ease of use of technology is an important predictor of actual use of technology. Consistent with the previous discussion the following hypothesis has been formed:

Hypothesis 3 (H3). *Ease of use of technology has a significant contribution to farmers' work performance towards fertilizer application in rice.*

Motivation is a drive that stimulates people into action, direction to behavior and thus, their productivity [50]. An enormous impact of workers' motivation is visible on productivity and performance [51]. Previous studies provide evidence that motivation positively affects individual performance [52]. Hence, it is clear that workers' high motivation at work plays an important role in their satisfaction which ultimately reflects their higher work performance. In accordance with the preceding discussion, the following hypothesis has been developed:

Hypothesis 4 (H4). *Motivation has a significant contribution to farmers' work performance towards fertilizer application in rice.*

3. Methodology

The quantitative approaches seem to be the best when the focus of the study is to identify the factors that determine a certain behavior [53]. Hence, the researchers employed a quantitative approach to administering this study. A cross-sectional survey method was to collect data that helped the researchers to collects a larger set of data in relatively a shorter period of time [54,55]. In order to collect relevant data from a pre-determined sample, a structured questionnaire was carefully prepared, including open and closed form questions.

3.1. Location, Population, and Sample

A purposive and multistage random sampling technique was adopted to locate and select the respondents for the survey. This sampling technique was found to be successful in several cases [56,57]. This study chose a multistage sampling technique to minimize random errors and sample bias [57].

Gaibandha is considered one of the important districts in Bangladesh for rice production [58]. So, Gaibandha district was purposively selected as the study area. Then, considering the research cost, size of the area covered, time, human resources, accessibility, and availability of transportation, three Upazilas under the Gaibandha district, namely Gobindaganj, Palashbari, and Sadullapur, and seven villages from each Upazila were selected purposively. There were 3762 rice farmers identified in the 21 villages from three upazilas. Out of which, 355 farmers were randomly selected using Morgan recommendations [59]. A proportionate random sampling technique was used to determine the number of respondents from each selected village using the following formula:

$$n_1/N \times n_2 = s$$

where, N = selected total population of the study; n_1 = proportion population in respected village; n_2 = determined the total sample size of the study; and, s = sample size from respected village [60].

3.2. Measurement of Variables

In this study, all the quantitative data were coded into a numerical value. The suitable scoring procedures were applied to convert data to make them easier. Farmers' work performance towards fertilizer application in rice was the dependent variable while farmers' knowledge, attitude, ease of use of technology and motivation were the independent variables in the current study. All the independent variables were measured by using five-point Likert scale. Though the Likert scale is ordinal; however, Likert with five or more categories can often be used as continuous without any harm to the analysis that has been planned to apply [61,62]. In previous studies, scholars have treated their Likert scale questionnaire as interval scale [15,17,22]; hence researchers also follow the same in this study.

Fourteen statement items (I am able to apply recommended dose of all fertilizers to achieve targeted production, I am able to apply recommended dose of urea to increase plant growth and number of panicle in rice, I am able to apply recommended dose of Triple

Super Phosphate (TSP) fertilizer efficiently to improve yield of rice, I am able to apply recommended dose of Muriate of Potash (MP) efficiently to increase yield of rice, I am able to apply recommended dose of Gypsum (Sulphur fertilizer) efficiently to increase yield of rice, I am able to apply recommended dose of Zinc fertilizer ($ZnSO_4$) efficiently to increase yield of rice, I am able to apply urea in three equal splits to rice, I am able to apply first split of urea to rice as basal method, I am able to apply second and third split of urea to rice as topdressing method, I am able to apply Triple Super Phosphate (TSP), Muriate of Potash (MP), Gypsum (Sulphur fertilizer) and Zinc fertilizer ($ZnSO_4$) as basal method to rice, I am able to increase rice yield by improving the timing of fertilizer application, I am able to apply first split of urea final land preparation, I am able to apply second and third split of urea early tillering stage and just before panicle initiation stage of rice, I am able to apply Triple Super Phosphate (TSP), Muriate of Potash (MP) Gypsum (Sulphur fertilizer) and Zinc fertilizer ($ZnSO_4$) during final land preparation) were adapted from Demba [17] to measure the level of farmers' work performance. Farmers were asked to report their views with corresponding statements based on five-point Likert scale and specified five possible responses range from 1 to 5 (1 = Strongly Disagree, up to 5 = Strongly Agree) [15,63].

For the measurement of knowledge, respondents were requested to specify their opinion against 10 statement items (I know that first split of urea should be applied as basal dose and rest of two split as topdressing, I know that the recommended doses of fertilizer are important to optimal rice yield, I understand that using appropriate method of fertilization application is important to extend rice yield, I understand that rice plants require urea at the early and mid-tillering stage to maximize rice yield, I know that urea fertilizer increase plant growth of rice, I know that Triple Super Phosphate (TSP) fertilizer is useful for flowering and panicle initiation in rice, I know that Muriate of Potash (MP) fertilizer is responsible for grain size and weight of rice, I understand that urea should be applied in three equal splits in rice field for higher yield and I know that Triple Super Phosphate (TSP) and Muriate of Potash (MP) should be applied during final land preparation) that were adapted from Ntawuruhunga [64]. A five-point Likert scale was also used to specify respondents' responses range from 1 to 5 (1 = Very Low, to 5 = Very high).

Attitude also measured by adapting 11 statements items (I know the recommended dose of fertilizer and follow it in rice farming, I will get lower yield due to fail applying recommended rate of urea, Triple Super Phosphate (TSP) and Muriate of Potash (MP) in rice, I know the timing of fertilizer applications can increase the yield of rice and follow it in rice farming, I think fertilizer application method is important yield and follow appropriate method in rice, I know urea should apply in three equal splits to rice and follow it, I think timing of urea application is difficult for me to apply in rice, I think I can reduce fertilizer cost by improving the timing of urea application in rice field, I know urea application at once in rice field just before transplanting is easier and follow it in rice, I know organic manure (cow dung) is less important for higher yield and do not use it in rice, It is good to apply fertilizers based on own experienced rather than external advice and I know excessive use of fertilizer is bad for rice production and follow recommended doses of fertilizer) from Ghosh and Hasan [48]. Respondents were asked to respond based on five-point Likert scale ranging from 1 to 5, where 1 indicates strongly disagree and 5 indicates strongly agree.

In case of ease of use of technology, 10 statements items (Use of recommended dose of urea for rice, Use of recommended dose of Triple Super Phosphate (TSP) for rice, Use of recommended dose of Muriate of Potash (MP) for rice, Use of recommended dose of Gypsum (Sulphur fertilizer) for rice, Use of recommended dose of Zinc fertilizer ($ZnSO_4$) for Boro rice, Use of urea in three equal splits in rice field, Use of the first split of urea as basal after seedling establishment of rice, Use of the second split of urea at early tillering stage of rice, Use of the third split of urea at 5–7 days before panicle initiation of rice, and Use of all Triple Super Phosphate (TSP), Muriate of Potash (MoP), Gypsum (Sulphur fertilizer) and Zinc fertilizer as basal during final land preparation) were adapted from

Adrian [65]. Five-point Likert scale ranging from very difficult (1) to very easy (5) was used to measure farmers' ease of use of technology.

Finally, respondents were requested to specify their opinion against seven statements items (I apply fertilizer in my rice field as it gives higher yield, I use fertilizer in my rice field as it is easy to apply, Using fertilizer in my rice field gives me higher status in farming community, I apply fertilizer in my rice field because it's readily available, I use fertilizer in my rice field as I have received training on fertilizer application, I apply fertilizer in rice cultivation because my peers think I should use it, and I use fertilizer in my rice field as I have received sufficient extension support) in order to measure their motivation using a five-point scale ranging from "strongly disagree" (1) to "strongly agree" (5). Items of motivation were mostly adapted from Ryan et al. [66].

3.3. *Validity and Reliability Analysis*

In this study, the researcher adopted construct validity by measuring the content validation of the instruments. Content validity can be measured by seeking experts' opinions from the respected field of study to conform to the concept and measurements were clear and represented the concerned subject matter. In this procedure, experts' opinions were sought for all the items in the questionnaire and then validated by the supervisory committee. The questionnaire was finalized and sent to 35 non-sampled rice farmers who were randomly selected for pre-testing. Cronbach's Alpha test is used to measure the reliability of all the items under each construct in the questionnaire. Cronbach's Alpha of work performance, knowledge, attitude, ease of use of technology, and motivation was 0.862, 0.830, 0.770, 0.785, and 0.770, respectively. The value of Cronbach's alpha coefficient should be equal to or greater than 0.7 which means that the data is reliable and the internal consistency of the items in the scale is satisfactory [67]. Hence, the Cronbach's Alpha values of the items were found reliable.

3.4. *Data Collection and Statistical Analysis*

Data were collected by the first author of the paper in a face-to-face situation, given respondents' level of literacy and other factors like their preparedness for this type of study. Data were collected from March to May 2018. The collected data were coded, entered, and analyzed using SPSS v23 according to the objectives and hypothesis of the study. Multiple linear regression with 0.05 and 0.01 levels of probabilities were used to explore the contribution of the selected factors on farmers' work performance and determine the highest contributing factors on farmers' work performance towards fertilizer application in rice. In the current study, multiple regression works with the following formula:

$$Y = b_0 + b_1 (x_1) + b_2 (x_2) + \cdots\cdots + b_k (x_k) + \varepsilon_i \quad (1)$$

Here, Y is the dependent variable (farmers' work performance towards fertilizer application). $X_1, X_2 \ldots X_k$ indicates the independent variables (knowledge, attitude, ease of use of technology, and motivation of farmers); $b_1, b_2 \ldots b_k$ are the regression coefficients of independent variables and b_0 constant. Besides, ε_i indicates the error term.

4. Results and Discussion

This section is organized into two sub-sections. The first sub-section deals with the findings of the study and the second sub-section present the test of hypotheses. While the third and last sub-sections discusses the findings related to the contribution of independent variables (i.e., knowledge, attitude, ease of use of technology, and motivation of farmers) on the dependent variable (farmers' work performance towards fertilizer application in rice).

Multiple linear regression analysis was executed to explore the contribution of selected factors that influence farmers' work performance towards fertilizer application in rice and finds out the factor that has the highest contribution to farmers' work performance towards fertilizer application in rice. There were four independent variables that influence farmers' work performance, were selected as predictors of the mentioned dependent variable. These four independent variables—knowledge (X_1), attitude (X_2), ease of use of technology (X_3), and motivation (X_4) of farmers are expected to formulate a multiple linear regression model that could be explained the variation of work performance among farmers. Thus, the multiple linear regression equation of this study has been written as follows:

$$Y = b_0 + b_1 (X_1) + b_2 (X_2) + b_3 (X_3) + b_4 (X_4) + \varepsilon_i \quad (2)$$

where, Y = Work performance of farmers; b_0 = Constant; b_{1-4} = Regression coefficient; X_1 = Knowledge; X_2 = Attitude; X_3 = Ease of use of technology; X_4 = Motivation and ε_i = Error term.

4.1. Contribution of Selected Factors on Farmers' Work Performance

Table 2 represents the model summary of multiple linear regression analysis. It showed the first statistics R known as the multiple correlation coefficients between all predictor variables and farmers' work performance and obtained 0.749. The next statistic is R^2, the coefficient of determination that indicates the percentage of the total variance in a dependent variable explained by all the predictor variables. The value of R^2 is 0.561 indicated that all the independent variables were simultaneously explained 56.1% of the total variance of the dependent variable. The next statistic is Adjusted R^2, a modified version of R^2 that calculates R^2 using only those independent variables, which was significant for predicting the dependent variable. Here, the adjusted R^2 value (0.556) indicated that the significant predictor variables were simultaneously explained 55.6% of the total variance of farmers' work performance. In other words, the rest of 44.6% of the total variation of farmers' work performance has not been explained in the current study.

Table 2. Table of multiple linear regression model summary.

Multiple R	R^2	Adjusted R^2	Std. Error of the Estimate	F	df	*p*
0.749	0.561	0.556	0.49122	111.783	4	0.000 *

Significant: * $p < 0.05$.

In addition, the value of the F-test was 111.783 which is significant at $p < 0.05$. It implies that the multiple linear regression model has a significant influence over the dependent variable of the study. In other words, it could be said that the combination of independent variables as a predictor has a significant contribution to farmers' work performance. Thus, the regression model was good or fit to predict the contributions of independent variables.

4.2. Highest Contributing Factors on Farmers' Work Performance towards Fertilizer Application in Rice

Table 3 recognized the independent variables that have a significant value of $p < 0.05$. It implies those respected variables have a statistically significant and distinctive contribution to predict the dependent variable of the study. However, the variables, which do not have a significant value of $p < 0.05$, are not considered a significant predictor of the mentioned dependent variable [68].

Table 3 shows the unstandardized regression coefficient (b) and standardized regression coefficients (β) taken to examine the contributions of selected independent variables on farmers' work performance. The strength of the contribution of the respected independent variables was compared to each other based on their standardized coefficient (β). Standardize coefficient (β) was estimated in units of standard deviation and not in a unit of the respected independent variables. The standardized coefficient (β) was calculated by multiplying the unstandardized coefficient (b) with the standard deviation of the independent and dependent variables. Thus, standardized coefficient (β) becomes normalized as a unit-less coefficient, also known as z-score. According to Table 3, the motivation of farmers had the largest standardized coefficient (β) value of 0.478. It implies that the motivation of farmers showed the highest contribution to predict the work performance of farmers towards fertilizer application. The second highest β value was found for knowledge (0.265), followed by ease of use of technology (0.122), while attitude (0.073) had an insignificant contribution.

Table 3. Coefficients of multiple linear regression for farmers' work performance towards fertilizer application in rice.

(Y) Farmers' Work Performance					
Model	Unstandardized Coefficients		Standardized Coefficients		
Independent Variable	b	Std. Error	Beta (β)	t	p
(Constant)	−0.088	0.183		−0.481	0.631
(X_1) Knowledge	0.281	0.044	0.265	6.315	0.000 *
(X_2) Attitude	0.084	0.043	0.073	1.943	0.053
(X_3) Ease to use of technology	0.146	0.050	0.122	2.905	0.004 *
(X_4) Motivation	0.544	0.051	0.478	10.760	0.000 *

Significant: * $p < 0.05$.

The values of the unstandardized coefficients values for knowledge, attitude, ease of use of technology, and motivation of farmers were 0.281, 0.084, 0.146, and 0.544, respectively (Table 2). The unstandardized coefficients (b) value of the respected variables indicated the change amount in the dependent variable (Y) in accordance with the change of one unit of an independent variable (X). Thus, based on the estimated unstandardized coefficients (b), the multiple linear regression model has been obtained as follows:

$$Y = -0.088 + 0.281\ (X_1) + 0.084\ (X_2) + 0.146\ (X_3) + 0.544\ (X_4) + \varepsilon_i$$

Table 2 also revealed that knowledge ($t = 6.315$, $p = 0.000$), ease of use of technology ($t = 2.905$, $p = 0.004$) and motivation ($t = 10.760$, $p = 0.000$) significantly provide explanation of farmers' work performance towards fertilizer application in rice. In contrast, the contribution of attitude is insignificant to predict farmers' work performance as significant value (p) of attitude ($t = 1.943$, $p = 0.053$) is not <0.05.

4.3. Test of Hypotheses

Hypothesis 1 (H1). *Knowledge has a significant contribution to farmers' work performance towards fertilizer application in rice.*

According to the multiple linear regression analysis, the standardized coefficient (β) value for farmers' knowledge was 0.265 with a t value of 6.315 which was significant at $p < 0.05$. Therefore, the hypothesis (H1) of the study has been failed to reject and the null hypothesis (H0) has been rejected (Table 4).

Table 4. Summary of testifying the research hypotheses of the study.

Hypothesis	Value of Regression Coefficient (β)	p-Value	Result
Hypothesis 1	0.265	<0.05	Accepted
Hypothesis 2	0.073	>0.05	Rejected
Hypothesis 3	0.122	<0.05	Accepted
Hypothesis 4	0.478	<0.05	Accepted

Hypothesis 2 (H2). *Attitude has a significant contribution to farmers' work performance towards fertilizer application in rice.*

The value of standardized coefficient (β) for farmers' attitude towards fertilizer application was 0.073 with a t value of 1.943 which was significant at $p < 0.05$ ($p = 0.053$).

Therefore, the hypothesis (H2) of the study has been rejected (Table 4) and the null hypothesis (H0) has been failed to reject.

Hypothesis 3 (H3). *Ease of use of technology has a significant contribution to farmers' work performance towards fertilizer application on rice.*

The value of standardized coefficient (β) for ease of use of technology was 0.122 with a t value of 2.905 was significant at $p < 0.05$ ($p = 0.004$). Therefore, the hypothesis (H3) of the study has been failed to reject and the null hypothesis (H0) has been rejected (Table 4).

Hypothesis 4 (H4). *Motivation has a significant contribution to farmers' work performance towards fertilizer application in rice.*

The value of the β for farmers' motivation towards fertilizer application in rice was 0.478 with a t value of 10.760 was significant at $p < 0.05$ ($p = 0.000$). Therefore, the hypothesis (H4) of the study has been failed to reject and the null hypothesis (H0) has been rejected (Table 4).

4.4. Discussion

According to the regression model, R^2 (coefficient of determination) and adjusted R^2 are 0.561 and 0.556 respectively. Moreover, the F-value 111.783 was significant at $p < 0.05$. According to these findings, the regression model is a good fit. That means, the regression model's estimated result is satisfactory as 56.1% of the total variance of farmers' work performance has been explained by motivation, knowledge, ease of use of technology, and attitude of farmers simultaneously. Hence, it can be assumed that these independent variables have adequate power for the explanation. The adjusted R^2 (0.556) value also interpreted that only significant predictor variables have explained 55.6% of the total variance of farmers' work performance: motivation, knowledge, and ease of use of technology simultaneously. Therefore, it can be assumed that the regression model of the current study has explained a significant percentage of total variation that occurs in the work performance of the farmers towards fertilizer application in rice.

This finding is in line with Demba [17] executed a study on personality traits and work performance for paddy farmers and stated that coefficients for farmers' work performance model explained 59.5% of total variation on farmers' work performance in rice cultivation in the Gambia. Bagum et al. [15] also revealed that the regression model explained 49.2% of the total variance of farmers' performance regarding fertilizer application in Bangladesh. A similar trend is also found from the study conducted by Shah [63] and stated that the regression model explained 44.8% of the total variance of farmers' work performance in rice cultivation in Malaysia.

In this study, motivation was one of the significant predictors identified as the highest contributing factors to explain farmers' work performance towards fertilizer application. The value of β-coefficient for motivation suggests that with one standard deviation change in farmers' motivation, their work performance will be increased by 0.478 standard deviation. It indicates that the motivation of farmers mainly regulates their work performance towards fertilizer application. Motivation is the most important reason that influences farmers to practice a particular agricultural technology to achieve higher productivity [69]. Moreover, factors like education, experience, extension contract, and training help motivate farmers to improve work performance and increase output [70]. Since farmers are important for agricultural production, it is crucial to continually keep farmers' stimulation level up to perform the farming activity, especially fertilizer application.

Prior literature also mentioned that motivation is a significant predictor and the highest contributor to respondents' performance [71]. Ngima and Kyongo [72] also noticed a similar finding that motivation had a strong statistically significant influence on individuals' performance.

Knowledge the second-highest contributing factor was predicting farmers' work performance towards fertilizer application in rice. It indicates that the enhancement of knowledge can guide farmers to realize the appropriateness of using certain technology. Therefore, agriculture knowledge is vital for farmers to improve their performance to apply essential technologies and increase their productivity levels.

With references to knowledge, Bagum et al. [15] identified farmers' knowledge as an important predictor that significantly contributed to farmers' performance. Moreover, Campbell and Wiernik [28] argued that role-specific knowledge is one of the leading determinants of respondents' performance.

Farmers' attitude displayed an insignificant contribution to predicting farmers' work performance (p = 0.053) However, despite having found insignificant contribution, one should not ignore the importance of a favorable attitude in determining farmers' adoption decision of any farming practice [73]. Other studies provided evidence that attitude has a significant contribution to the performance of respondents [43,71]. However, sometimes, individuals' positive attitudes are not enough for performing a given behavior due to their different socio-economic circumstances. As per the researchers' observation, differences might be existed among the farmers according to knowledge, ability, attitudes, and these differences can influence their behavioral decision [74,75]. Such inconsistency might be prevailing among the respondent farmers in the study area. Thus, a farmer might possess a favorable attitude towards fertilizer application for higher yield yet not apply fertilizer at the recommended rate due to other factors like the high input cost or unavailability of fertilizers.

5. Conclusions

The overall findings of the multiple regression analysis explored the combination of significant predictors such as viz. knowledge, ease of use of technology, and motivation of farmers explained 55.6% of the total variance of farmers' work performance towards fertilizer application in rice. The rest of the variance of farmers' work performance may be explained by other factors that were not being considered in the current study. However, the estimated multiple linear regression model was good or fit to predict the contributions of independent variables. Therefore, the study concluded that the estimated regression model of farmers' work performance is suitable to predict the contributions of selected factors like knowledge, ease of use of technology and motivation of farmers in the current study.

Additionally, the motivation of farmers was recognized as the highest contributing factor followed by knowledge. Hence, it can be suggested that greater emphasis should be given to farmers' motivation and knowledge level to solve their problems and provide maximum effort for higher work performance.

Theoretically, this study will enhance the opportunity to execute new studies in the field of performance through providing critical literature support based on the significant contribution of knowledge, ease of use of technology, and motivation of farmers to their work performance. Besides, the current study's findings are significant to farmers as it focuses on the present level of farmers' work performance for establishing an effective working environment for them to ensure higher performance in applying agricultural practices and getting higher production of rice. Moreover, study findings will provide support as a basis of the national and local motivational campaign including training and technical support ought to be provided by the Department of Agricultural Extension (DAE) of Bangladesh, other GOs, and NGOs extension service providers to equip farmers with essential knowledge, high motivational level, and skill for strengthening their work performance.

Apart from this, the present study highlights only four variables: knowledge, attitude, ease of use of technology, and motivation of farmers that leads to better work performance of farmers towards fertilizer application in rice. Therefore, it is suggested that further research should be undertaken with other potential variables to explore the work performance of farmers. Moreover, other factors of rice cultivation such as irrigation, weed management, pest and disease management, and intercultural operations can be taken under consideration for future research on farmers' work performance.

Author Contributions: Conceptualization, T.B., M.K.U., S.H., N.H.K., M.Z.R. and A.N.A.H.; Methodology, T.B., M.K.U., S.H. and N.H.K.; Formal Analysis, T.B. and S.H.; Writing—Original Draft Preparation, T.B. and M.K.U.; Writing—Review and Editing, M.K.U., T.B., S.H., N.H.K., M.Z.R. and A.N.A.H.; Supervision, M.K.U., S.H., N.H.K. and M.Z.R. All authors have read and agreed to the published version of the manuscript.

Funding: This research paper was supported by Universiti Putra Malaysia and National Agricultural Technology Program (NATP): Phase-II Project (Vote No. 6282514), Bangladesh Agricultural Research Council.

Institutional Review Board Statement: Ethical review and approval were waived for this study due to collecting basic information from the respondents related to their farm practices. Disclose of these information were not harmful to any human. However, the authors were very careful to use these information just only for the study purpose.

Informed Consent Statement: Informed consent was obtained from all subjects involved in the study.

Data Availability Statement: Not applicable.

Acknowledgments: The authors would like to express their sincere gratitude to Organization for Women in Science for the Developing World (OWSD) and Swedish International Cooperation Agency (SIDA) for their support through the fellowship in this research project. The authors would also acknowledge the National Agricultural Technology Program (NATP): Phase-II Project, Bangladesh Agricultural Research Council for the financial support, and the Universiti Putra Malaysia, Malaysia for the research facilities.

Conflicts of Interest: The authors declare no conflict of interest.

References

1. Chowdhury, I.U.A.; Khan, M.A.E. The impact of climate change on rice yield in Bangladesh: A time series analysis. *Russ. J. Agric. Soc.-Econ. Sci.* **2015**, *40*, 12–28. Available online: https://ideas.repec.org/a/scn/031261/16082025.html (accessed on 30 June 2021).
2. Sultana, J.; Siddique, M.N.A.; Abdullah, M.R. Fertilizer recommendation for Agriculture: Practice, practicalities and adaptation in Bangladesh and Netherlands. *Int. J. Bus. Manag. Soc. Res.* **2015**, *1*, 21–40. [CrossRef]
3. Jahan, M.S.; Sarkar, M.; Barma, N.; Mondal, M.; Ferdousi, M. Grain yield, nutrient balance and economics of T. Aman rice cultivation as influenced by nutrients management. *Bangladesh J. Agric. Res.* **2015**, *40*, 17–34. [CrossRef]
4. Kabir, M.; Salam, M.; Chowdhury, A.; Rahman, N.; Iftekharuddaula, K.; Rahman, M.; Rashid, M.; Dipti, S.; Islam, A.; Latif, M.; et al. Rice Vision for Bangladesh: 2050 and Beyond. *Bangladesh Rice J.* **2015**, *19*, 1–18. [CrossRef]
5. Bangladesh Bureau of Statistics (BBS). *Yearbook of Agricultural Statistics of Bangladesh*; Bangladesh Bureau of Statistics, Ministry of Planning, Government of the People's Republic of Bangladesh: Dhaka, Bangladesh, 2016.
6. Bangladesh Rice Knowledge Bank. Bangladesh Rice Research Institute Home Page. Available online: http://www.brri.gov.bd/ (accessed on 24 January 2018).
7. Naher, U.A.; Othman, R.; Shamsuddin, Z.H.; Saud, H.M.; Ismail, M.R.; Rahim, K.A. Effect of root exuded specific sugars on biological nitrogen fixation and growth promotion in rice ('*Oryza sativa*'). *Aust. J. Crop Sci.* **2011**, *5*, 1210–1217. Available online: https://search.informit.org/doi/abs/10.3316/informit.746245793849135 (accessed on 21 January 2021).
8. Ghosh, M.; Swain, D.K.; Jha, M.K.; Tewari, V.K. Precision nitrogen management using chlorophyll meter for Improving Growth, Productivity and N Use Efficiency of Rice in Subtropical Climate. *J. Agric. Sci.* **2013**, *5*, 253–266. [CrossRef]
9. Semeon, G.; Garfield, M.; Meshesha, M.; Dessa, D. Agricultural knowledge Management systems in practice: The ability to support Wereda knowledge centers in Ethiopia. In Proceedings of the 19th Americas Conference on Information Systems, Chicago, IL, USA, 15–17 August 2013.
10. Hu, R.; Cao, J.; Huang, J.; Peng, S.; Huang, J.; Zhong, X.; Zou, Y.; Yang, J.; Buresh, R.J. Farmer participatory testing of standard and modified site-specific nitrogen management for irrigated rice in China. *Agric. Syst.* **2007**, *94*, 331–340. [CrossRef]
11. Aura, S. Determinants of the adoption of integrated soil fertility management technologies in Mbale division, Kenya. *Afr. J. Food Agric. Nutr. Dev.* **2016**, *16*, 10697–10710. [CrossRef]
12. Yousaf, M.; Li, J.; Lu, J.; Ren, T.; Cong, R.; Fahad, S.; Li, X. Effects of fertilization on crop production and nutrient-supplying capacity under rice-oilseed rape rotation system. *Sci. Rep.* **2017**, *7*, 1270. [CrossRef]
13. Hidayah, I. Farmers' behaviour in the implementation of component PTT (integrated plant and resource management) in irrigation paddy rice fields farming in Buru regency Maluku province Indonesia. *Int. J. Humanit. Soc. Sci.* **2013**, *3*, 129–138. Available online: http://www.ijhssnet.com/journal/index/1903 (accessed on 9 February 2021).

14. Basak, J.K.; Titumir, R.A.M.; Alam, K. Future Fertiliser Demand and Role of Organic Fertiliser for Sustainable Rice Production in Bangladesh. *Agric. For. Fish.* **2015**, *4*, 200–208. [CrossRef]
15. Bagum, T.; Uddin, K.; Hassan, S.; Kamarulzaman, N.H.; Rahman, Z. Determinant Factors of Farmers' Performance Regarding Fertilizer Application: An Overview from Bangladesh. *Asian J. Agric. Ext. Econ. Sociol.* **2019**, *34*, 1–12. [CrossRef]
16. Khan, I.; Nawaz, A. The leadership styles and the employees' performance: A review. *Gomal Univ. J. Res.* **2016**, *32*, 144–150. Available online: http://www.gomal.pk/GUJR/DECEMBER_%202016/_15_IRFAN_%20THE%20LEADERSHIP%20STYLES%20AND%20EMPLOYEES%20PERFORMANCE.pdf (accessed on 13 February 2021).
17. Demba, S. Personality Traits and Work Performance of Paddy Farmers in the Central River Region, Gambia. Master's Thesis, Universiti Putra Malaysia, Serdang, Malaysia, 2017.
18. Hameed, T.S.; Sawicka, B. Farmers' knowledge of fertilizers' practices on their farms. In Proceedings of the IX International Scientific Symposium "Farm Machinery and Processes Management in Sustainable Agriculture", Lublin, Poland, 22–23 November 2017; pp. 119–124. [CrossRef]
19. Zeweld, W.; Van Huylenbroeck, G.; Tesfay, G.; Speelman, S. Smallholder farmers' behavioural intentions towards sustainable agricultural practices. *J. Environ. Manag.* **2017**, *187*, 71–81. [CrossRef]
20. Guzman, J.J. Measuring Job Performance Associated to the Direct Effect of the Global Context. Ph.D. Thesis, University of Reigns, Ottawa, ON, Canada, 2008.
21. 21. Rani. A.B.A. Influence of Spiritual Intelligence, Age and Tenure on Work Performance of Nurses in Selected Hospitals, Peninsular Malaysia. Ph.D. Thesis, Universiti Putra Malaysia, Serdang, Malaysia, 2015.
22. Oluwatoyin, O. Perceptions of Cocoa Growers in Malaysia on Relationship between Transfer of Technology Skills, Human Resource Development Skills and Extension Agents' Work Performance. Ph.D. Thesis, Universiti Putra Malaysia, Serdang, Malaysia, 2019.
23. Rahman, K.M.A.; Zhang, D. Effects of Fertilizer Broadcasting on the Excessive Use of Inorganic Fertilizers and Environmental Sustainability. *Sustainability* **2018**, *10*, 759. [CrossRef]
24. Jankingthong, K.; Rurkkhum, S. Factors Affecting Job Performance: A Review of Literature. *Silpakorn Univ. J. Soc. Sci. Humanit. Arts.* **2012**, *12*, 115–128. Available online: https://so02.tci-thaijo.org/index.php/hasss/article/view/7140 (accessed on 29 December 2020).
25. Viswesvaran, C.; Ones, D.S. Perspectives on Models of Job Performance. *Int. J. Sel. Assess.* **2000**, *8*, 216–226. [CrossRef]
26. Ufuophu-Biri, E.; Iwu, C.G. Job Motivation, Job Performance and Gender Relations in the Broadcast Sector in Nigeria. *Mediterr. J. Soc. Sci.* **2014**, *5*, 191–198. [CrossRef]
27. Armstrong, M.; Taylor, S. *Armstrong's Handbook of Human Resource Management Practice*, 13th ed.; Kogan Page Publishers: London, UK, 2014; ISBN 9780749498283 (ebook).
28. Campbell, J.P.; Wiernik, B.M. The Modeling and Assessment of Work Performance. *Annu. Rev. Organ. Psychol. Organ. Behav.* **2015**, *2*, 47–74. [CrossRef]
29. Pringle, C.D. An Initial Test of a Theory of Individual Performance. *Psychol. Rep.* **1994**, *74*, 963–973. [CrossRef]
30. Choi, J.-H. The HR-performance link using two differently measured HR practices. *Asia Pac. J. Hum. Resour.* **2014**, *52*, 370–387. [CrossRef]
31. Kuo, Y.-F.; Yen, S.-N. Towards an understanding of the behavioral intention to use 3G mobile value-added services. *Comput. Hum. Behav.* **2009**, *25*, 103–110. [CrossRef]
32. Nkari, D.I.M.; Kibera, P.F.N. The influence of farmer characteristics on performance of commercial farmers in Kiambu County, Kenya. *Eur. J. Bus. Soc. Sci.* **2016**, *5*, 63–78. Available online: https://profiles.uonbi.ac.ke/fkibera/files/isaac_m._nkari._kibera_f.n._2016_the_influence_of_farmer_characteristics_on_performance_of_commercial_farmers_in_k (accessed on 23 December 2020).
33. Djomo, R.F.; Ndaghu, N.N.; Ukpe, U.H.; Innocent-Ene, E.O.; Onuigbo, I. Effect Farm Management Practices on Performance of Smallholder Rice Farmers in Cameroon: A Case of the West Region of Cameroon. *Agric. Res. Technol. Open Access J.* **2017**, *6*, 555686. [CrossRef]
34. Maican, S.; Muntean, A.; Paştiu, C.; Stępień, S.; Polcyn, J.; Dobra, I.; Dârja, M.; Moisă, C. Motivational Factors, Job Satisfaction, and Economic Performance in Romanian Small Farms. *Sustainability* **2021**, *13*, 5832. [CrossRef]
35. Meijer, S.S.; Catacutan, D.; Ajayi, O.C.; Sileshi, G.W.; Nieuwenhuis, M. The role of knowledge, attitudes and perceptions in the uptake of agricultural and agroforestry innovations among smallholder farmers in sub-Saharan Africa. *Int. J. Agric. Sustain.* **2015**, *13*, 40–54. [CrossRef]
36. Inwood, S.; Clark, J.K.; Bean, M. The Differing Values of Multigeneration and First-Generation Farmers: Their Influence on the Structure of Agriculture at the Rural-Urban Interface. *Rural Sociol.* **2013**, *78*, 346–370. [CrossRef]
37. Bwambale, N. Farmers' Knowledge, Perceptions, and Socioeconomic Factors Influencing Decision Making for Integrated Soil Fertility Management Practices in Masaka and Rakai Districts, Central Uganda. Ph.D. Thesis, Iowa State University, Ames, IA, USA, 2015. Available online: https://www.proquest.com/openview/055dd10a0a5b16abd3cea52e6a07b207/1?pq-origsite=gscholar&cbl=18750 (accessed on 20 March 2021).
38. Aksoy, Y.; Ayranci, E.; Gozukara, E. A Research on the Relationship between Knowledge Sharing and Employee Performance: The Moderating Role of Unethical Behaviors in Organizational Level. *Eur. Sci. J.* **2016**, *12*, 335–352. [CrossRef]
39. Lim, J.R.A.; Lee, A.S.H.; Lim, T.M. A study on knowledge quality and job performance of knowledge workers by analyzing content of social network sites using sentiment network analysis. *Inf. Manag. Bus. Rev.* **2013**, *5*, 525–530. Available online: http://eprints.sunway.edu.my/id/eprint/251 (accessed on 19 March 2021).

40. Chi, H.K.; Yeh, H.R.; Chen, Y.L. The mediating effect of knowledge management on customer orientation and job performance of salespeople. *J. Glob. Bus. Manag.* **2010**, *6*. Available online: http://www.jgbm.org/page/13%20Yu%20Ling%20Chen%20.pdf (accessed on 27 November 2020).
41. Masa'deh, R.; Gharaibeh, A.; Maqableh, M.; Karajeh, H. An empirical study of antecedents and outcomes of knowledge sharing capability in Jordanian telecommunication firms: A structural equation modeling approach. *Life Sci. J.* **2013**, *10*, 2284–2296. [CrossRef]
42. Ghosh, M.K.; Hasan, S.S. Farmers' attitude towards sustainable agricultural practices. *Bangladesh Res. Publ. J.* **2013**, *8*, 227–235. Available online: https://www.researchgate.net/profile/Mithun-Ghosh-10/publication/334603569_FARMERS\T1\textquoteright_ATTITUDE_TOWARDS_SUSTAINA-BLE_AGRICULTURAL_PRACTICES/links/5d35543d299bf1995b3f9e18/FARMERS-ATTITUDE-TOWARDS-SUSTAINABLE-AGRICULTURAL-PRACTICES.pdf (accessed on 15 January 2021).
43. Hettiararchchi, H.A.H.; Jayarathna, S.M.D.Y. The effect of Employee Work Related Attitudes on Employee Job Performance: A Study of Tertiary and Vocational Education Sector in Sri Lanka. *IOSR J. Bus. Manag.* **2014**, *16*, 74–83. [CrossRef]
44. Zhou, Q.; Hu, J.; Gao, S. Chemistry teachers' attitude towards ICT in Xi'an. *Procedia Soc. Behav. Sci.* **2010**, *2*, 4629–4637. [CrossRef]
45. University of Minnesota Libraries Publishing. *Principles of Management*, 2015th ed.; University of Minnesota: Minneapolis, MN, USA, 2015; ISBN 13: 9781946135186.
46. Ninsiima, D. Factors Affecting Adoption of an Information Communications Technology System for Agriculture in Uganda. Master's Thesis, Michigan State University, East Lansing, MI, USA, 2015. Available online: https://www.proquest.com/openview/42efda11d493bdb2c15cf5c6913c27a5/1?pq-origsite=gscholar&cbl=18750 (accessed on 17 October 2020).
47. Davis, F.D. Perceived Usefulness, Perceived Ease of Use, and User Acceptance of Information Technology. *MIS Q.* **1989**, *13*, 319–340. [CrossRef]
48. Reimer, A.P.; Weinkauf, D.K.; Prokopy, L.S. The influence of perceptions of practice characteristics: An examination of agricultural best management practice adoption in two Indiana watersheds. *J. Rural Stud.* **2012**, *28*, 118–128. [CrossRef]
49. Bahramzadeh, M.M.; Shokati, M.S. Cultural approach to IT: Compliance evaluation based on UTAUT. In Proceedings of the 8th International Conference on Management, Tehran, Iraq, 4 November 2010. Available online: https://www.civilica.com/paper-IRIMC08-IRIMC08%084.html (accessed on 30 September 2015).
50. Romando, R. Define Motivation. 2008. Available online: http://www.212articles.com/articles/38/1/Define-Motivation/Page1.htm (accessed on 8 March 2010).
51. Aziri, B. Job satisfaction: A literature review. *Manag. Res. Pract.* **2011**, *3*, 77–86. Available online: https://www.researchgate.net/publication/222103547 (accessed on 17 January 2021).
52. Sharma, N.; Sharma, A. Relationship between employee motivation and performance of the employees working in retail sector in Jaipur. *J. Manag. Eng. Inf. Technol.* **2017**, *4*, 10–17. [CrossRef]
53. Gowri, V. Sustainability Practices in Malaysian Grocery Retail Industry. Ph.D. Thesis, Universiti Putra Malaysia, Serdang, Malaysia, 2015.
54. Onwuegbuzie, A.J.; Collins, K.M. A Typology of Mixed Methods Sampling Designs in Social Science Research. *Qual. Rep.* **2007**, *12*, 281–316. [CrossRef]
55. Johnson, M. Social Science Research. *Soc. Stud. Sci.* **1989**, *19*, 759–762. [CrossRef]
56. Amegnaglo, C.J. Determinants of maize farmers' performance in Benin, West Africa. *Kasetsart J. Soc. Sci.* **2018**, *XXX*, 1–7. [CrossRef]
57. Omoro, P.A.; Shitandi, A.; Bitonga, R.O.; Basweti, E. Determination of farmer's choice of fertilizer application rate and its effect on the greenhouse technology performance in Gusii highlands, Kenya. *Int. J. Agric. Ext.* **2014**, *2*, 35–41. Available online: https://esciencepress.net/journals/index.php/IJAE/article/view/610/318 (accessed on 16 October 2020).
58. Zaman, M.A.U.; Pramanik, S.; Parvin, N.; Khatun, A. Crop Diversification in Rangpur Region. *Bangladesh Rice J.* **2018**, *21*, 255–271. [CrossRef]
59. Krejcie, R.V.; Morgan, D.W. Determining Sample Size for Research Activities. *Educ. Psychol. Meas.* **1970**, *30*, 607–610. [CrossRef]
60. Harun, Z.B. Knowledge, Attitude and Communication in Relation to the Goods and Services Tax Compliance among Wood Product Manufacturers in Peninsular Malaysia. Ph.D. Thesis, Univesriti Putra Malaysia, Serdang, Malaysia, 2016.
61. Johnson, D.R.; Creech, J.C. Ordinal Measures in Multiple Indicator Models: A Simulation Study of Categorization Error. *Am. Sociol. Rev.* **1983**, *48*, 398. [CrossRef]
62. Sullivan, G.M.; Artino, A. Analyzing and Interpreting Data from Likert-Type Scales. *J. Grad. Med. Educ.* **2013**, *5*, 541–542. [CrossRef] [PubMed]
63. Shah, S.B.S.K. Relationship between Personality Traits and Work Performance among Potential Paddy Farmers in Malaysian Granary Areas. Master's Thesis, Universiti Putra Malaysia, Serdang, Malaysia, 2016.
64. Ntawuruhunga, D.; Assefa, F.A.; Mambo, W.; Momanyi, H.M.; Omeli, M.R. Farmers' Knowledge, Attitude and Practice towards African Indigenous Vegetables in Kenya. 2016. Available online: http://repository.must.ac.ke/handle/123456789/1194 (accessed on 26 December 2019).
65. Adrian, A.M. Factors Influencing Adoption and Use of Precision Agriculture. Ph.D. Thesis, The Graduate Faculty of Auburn University, Auburn, AL, USA, 2006. Available online: https://www.proquest.com/openview/8ae0730e8876a63d87a6c158d81549fd/1?pq-origsite=gscholar&cbl=18750&diss=y (accessed on 26 December 2019).
66. Ryan, R.L.; Erickson, D.L.; De Young, R. Farmers' Motivations for Adopting Conservation Practices along Riparian Zones in a Mid-Western Agricultural Watershed. *J. Environ. Plan. Manag.* **2003**, *46*, 19–37. [CrossRef]

67. Hair, J.F.; Black, W.C.; Babin, B.J.; Anderson, R.E. *Multivariate Data Analysis: A Global Perspective*, 7th ed.; Prentice Hall: Upper Saddle River, NJ, USA, 2009.
68. Chiou, C.Y. Motivation and Self-Regulation as Predictors of E-Learning Experience among Undergraduates in a Malaysian Public University. Master's Thesis, Universiti Putra Malaysia, Serdang, Malaysia, 2015.
69. Benzing, C.; Chu, H.M. A comparison of the motivations of small business owners in Africa. *J. Small Bus. Enterp. Dev.* **2009**, *16*, 60–77. [CrossRef]
70. Mustapha, S.B.; Alkali, A.; Shehu, H.; Ibrahim, A.K. Motivation Strategies for Improved Performance of Agricultural Extension Workers in Nigeria. *Int. Acad. J. Org. Behav. Hum. Resou. Manag.* **2017**, *4*, 1–8. Available online: https://www.iaiest.com/abstract.php?id=6&archiveid=896 (accessed on 11 January 2021).
71. Khan, I.; Dongping, H.; Ghauri, T.A. Impact of attitude on employees' performance: A study of textile industry in Punjab, Pakistan. *World Appl. Sci. J.* **2014**, *30*, 191–197. [CrossRef]
72. Ngima, W.M.; Kyongo, J. Contribution of motivational management to employee performance. *Int. J. Humanit. Soc. Sci.* **2013**, *3*, 219–239. Available online: http://www.ijhssnet.com/journals/Vol_3_No_14_Special_Issue_July_2013/26.pdf (accessed on 22 December 2020).
73. Sadati, S.A.; Shaabanali Fami, H.; Asadi, A.; Sadati, S.A. Farmer's attitude on sustainable agriculture and its determinants: A case study in Behbahan County of Iran. *Res. J. Appl. Sci. Eng. Technol.* **2010**, *2*, 422–427. Available online: http://hdl.handle.net/10535/6663 (accessed on 17 January 2021).
74. Power, E.F.; Kelly, D.L.; Stout, J.C. Impacts of organic and conventional dairy farmer attitude, behaviour and knowledge on farm biodiversity in Ireland. *J. Nat. Conserv.* **2013**, *21*, 272–278. [CrossRef]
75. Austin, E.J.; Deary, I.J.; Willock, J. Personality and intelligence as predictors of economic behaviour in Scottish farmers. *Eur. J. Pers.* **2001**, *15*, S123–S137. [CrossRef]

MDPI
St. Alban-Anlage 66
4052 Basel
Switzerland
Tel. +41 61 683 77 34
Fax +41 61 302 89 18
www.mdpi.com

Sustainability Editorial Office
E-mail: sustainability@mdpi.com
www.mdpi.com/journal/sustainability

www.ingramcontent.com/pod-product-compliance
Lightning Source LLC
Chambersburg PA
CBHW040745250926
51850CB00007BB/1412